Handbook of
Foaming and Blowing Agents

2nd Edition

George Wypych

ChemTec Publishing

Toronto 2022

Published by ChemTec Publishing
38 Earswick Drive, Toronto, Ontario M1E 1C6, Canada

© ChemTec Publishing, 2017, 2022
ISBN 978-1-77467-000-2 (hard cover); ISBN 978-1-77467-001-9 (E-PUB)

Cover design: Anita Wypych

All rights reserved. No part of this publication may be reproduced, stored or transmitted in any form or by any means without written permission of copyright owner. No responsibility is assumed by the Author and the Publisher for any injury or/and damage to persons or properties as a matter of products liability, negligence, use, or operation of any methods, product ideas, or instructions published or suggested in this book.

Library and Archives Canada Cataloguing in Publication

Title: Handbook of foaming and blowing agents / George Wypych.
Other titles: Foaming and blowing agents
Names: Wypych, George, author.
Description: 2nd edition. | Includes bibliographical references and index.
Identifiers: Canadiana (print) 20210194480 | Canadiana (ebook) 20210194529
| ISBN 9781774670002 (hardcover) | ISBN 9781774670019 (PDF)
Subjects: LCSH: Plastic foams-Handbooks, manuals, etc.
| LCSH: Foamed materials-Handbooks, manuals, etc.
| LCSH: Foam-Handbooks, manuals, etc. | LCGFT: Handbooks and manuals.
Classification: LCC TP1183.F6 W964 2022 | DDC 668.4/93-dc23

Printed in Australia, United Kingdom and United States of America

Table of Contents

1	**INTRODUCTION**	1
2	**CHEMICAL ORIGINS OF BLOWING AGENTS**	3
2.1	Activators	5
2.2	Azodicarbonamide	6
2.3	Crosslinkers	7
2.4	Dinitroso pentamethylene tetramine compounds	8
2.5	Dispersions in polymers carriers	9
2.6	Foaming agent mixtures with other additive(s)	10
2.7	Gases	11
2.8	Hydrazides	12
2.9	Hydrocarbons	13
2.10	Hydrochlorcarbon	14
2.11	Hydrochlorofluorocarbon	15
2.12	Hydrofluorocarbons	16
2.13	Hydrofluoroolefins	17
2.14	Masterbatches	18
2.15	Microspheres	19
2.16	Mixtures of foaming agents	20
2.17	Nucleating agents	21
2.18	Proprietary	22
2.19	Salts of carbonic and polycarboxylic acids	23
2.20	Sodium bicarbonate	24
2.21	Sulfonylsemicarbazides	25
2.22	Tetrazoles	26
2.23	Water	27
3	**MECHANISMS OF ACTION OF BLOWING AGENTS**	29
3.1	Mechanisms of foaming by decomposing solids	29
3.2	Production of gaseous products by chemical reaction	34
3.3	Foaming by gases and evaporating liquids	35
4	**DISPERSION AND SOLUBILITY OF BLOWING AGENTS**	45
5	**PARAMETERS OF FOAM PRODUCTION**	51
5.1	Amount of blowing agent	51
5.2	Clamping pressure	54
5.3	Delay time	55
5.4	Desorption time	55
5.5	Die pressure	56
5.6	Die temperature	56

5.7	Gas content		57
5.8	Gas flow rate		58
5.9	Gas injection location		59
5.10	Gas sorption and desorption rates		59
5.11	Internal pressure after foaming		61
5.12	Mold pressure		62
5.13	Mold temperature		63
5.14	Operational window		64
5.15	Plastisol viscosity		64
5.16	Saturation pressure		64
5.17	Saturation temperature		66
5.18	Screw revolution speed		67
5.19	Surface tension		68
5.20	Time		69
5.21	Temperature		70
5.22	Void volume		70
6	**FOAM STABILIZATION**		**75**
7	**FOAMING EFFICIENCY MEASURES**		**83**
7.1	Cell size		83
7.2	Cell density		83
7.3	Cell wall thickness (average)		84
7.4	Foam density		84
7.5	Expansion ratio (by volume)		85
7.6	Open cell content		85
7.7	Void fraction		85
8	**MORPHOLOGY OF FOAMS**		**87**
8.1	Bimodal morphology		87
8.2	Cell density		94
8.3	Cell morphology		95
8.4	Cell size		98
8.5	Cell wall thickness		100
8.6	Closed cell		102
8.7	Core & skin thickness		104
8.8	Morphological features of foams		106
8.9	Open cell		108
9	**FOAMING IN DIFFERENT PROCESSING METHODS**		**111**
9.1	Blown film extrusion		111
9.2	Calendering		111
9.3	Clay exfoliation		111
9.4	Compression molding		112
9.5	Depressurization		114

9.6	Extrusion	115
9.7	Free foaming	119
9.8	Injection molding	120
9.9	Microwave heating	124
9.10	Rotational molding	125
9.11	Solid-state foaming	126
9.12	Supercritical fluid-laden pellet injection molding foaming technology	128
9.13	Thermoforming	128
9.14	UV laser	129
9.15	Vacuum drying	129
9.16	Wire coating	130
10	**SELECTION OF FOAMING AND BLOWING AGENTS FOR DIFFERENT POLYMERS**	**133**
10.1	Acrylonitrile-butadiene-styrene	133
10.2	Acrylonitrile-butadiene-acrylate	134
10.3	Bismaleimide resin	135
10.4	Bromobutyl rubber	136
10.5	Cellulose acetate	137
10.6	Chitosan	138
10.7	Cyanoacrylate	139
10.8	Epoxy	140
10.9	Ethylene chlorotrifluoroethylene	142
10.10	Ethylene-propylene diene rubber	143
10.11	Ethylene-vinyl acetate	145
10.12	Fluorinated ethylene propylene	147
10.13	Hydroxypropyl methylcellulose	148
10.14	Melamine resin	149
10.15	Phenol formaldehyde	150
10.16	Poly(3-hydroxybutyrate-co-3-hydroxyvalerate)	152
10.17	Poly(butylene succinate)	153
10.18	Poly(ε-caprolactone)	155
10.19	Polyacrylonitrile	156
10.20	Polyamide	157
10.21	Polycarbonate	158
10.22	Polycarbonate/ABS	161
10.23	Polychloroprene	162
10.24	Polydimethylsiloxane	163
10.25	Polyetherketone	166
10.26	Polyetherimide	167
10.27	Polyethersulfone	169
10.28	PES/PEN blends	170
10.29	Polyethylene	171
10.30	Poly(ethylene-co-octene)	174
10.31	Poly(ethylene terephthalate)	175

10.32	Polyimide	177
10.33	Poly(lactic acid)	179
10.34	Polymethylmethacrylate	182
10.35	Polyoxymethylene	184
10.36	Polypropylene	185
10.37	Polystyrene	189
10.38	Polyurethane	192
10.39	Polyvinylalcohol	197
10.40	Polyvinylchloride	199
10.41	Poly(vinyl chloride-co-vinyl acetate)	203
10.42	Polyvinylidenefluoride	204
10.43	Natural rubber	207
10.44	Starch	208
11	**ADDITIVES**	**209**
11.1	Activators, accelerators, and kickers	209
11.2	Catalysts	210
11.3	Crosslinking agents	211
11.4	Curing agents	212
11.5	Diluents	214
11.6	Exfoliated additives	215
11.7	Fibers	216
11.8	Fillers	217
11.9	Fire retardants	221
11.10	Foam stabilizers	222
11.11	Nucleating agents	222
11.12	Plasticizers	225
11.13	Polymeric modifiers	226
11.14	Retarders	227
11.15	Surfactants	227
12	**EFFECT OF FOAMING ON PHYSICAL-MECHANICAL PROPERTIES OF FOAMS**	**231**
12.1	Compression set, strength, and modulus	231
12.2	Crystallinity	235
12.3	Deformation recovery	237
12.4	Density	237
12.5	Elastic modulus	238
12.6	Elongation	239
12.7	EMI shielding	240
12.8	Expansion ratio	242
12.9	Flexural modulus	243
12.10	Glass transition temperature	243
12.11	Impact strength	245
12.12	Relative permittivity	246

12.13	Resistivity	247
12.14	Rheology	248
12.15	Shape memory	249
12.16	Shear modulus	250
12.17	Shrinkage	251
12.18	Sound absorption	252
12.19	Surface roughness	254
12.20	Surface tension	254
12.21	Tear strength	255
12.22	Tensile modulus	255
12.23	Tensile strength	256
12.24	Thermal conductivity	256
13	**ANALYTICAL TECHNIQUES USEFUL IN FOAMING**	**261**
13.1	Cell density	261
13.2	Cell size	262
13.3	Density	263
13.4	Differential scanning calorimetry	263
13.5	Fourier transform infrared	263
13.6	Optical expandometry	264
13.7	Polarizing optical microscope	264
13.8	Scanning electron microscopy	265
13.9	Transmission electron microscopy	266
13.10	Confocal laser scanning microscopy	266
13.11	X-ray analysis	267
14	**HEALTH AND SAFETY AND ENVIRONMENTAL IMPACT OF FOAMING PROCESSES**	**269**
	INDEX	**273**

1

INTRODUCTION

Material foaming is one of the important processes, which generally lead to lighter and more cost-effective materials, but frequently help to develop unique products for unique applications.

This field, similar to many other technical fields, makes use of terms, which are sometimes not very well defined and, therefore, misused. Before addressing the subject matter in a comprehensive way, it will be useful to review and define some of the most frequently used terms. They include:

- A foaming agent is a detergent that reduces surface tension, useful in converting a liquid to foam. Foaming agent (or surfactant) increases the colloidal stability of foam, produced in its presence by reducing bubbles' coalescence. In another definition, a foaming agent is a material that facilitates the formation of foam such as a surfactant or a blowing agent. This differs from the previous definition because it includes two different groups of materials: surfactants and blowing agents, considering that they both produce foams. The last definition seems reasonable, especially that the terms of foaming and blowing agents are frequently used interchangeably.
- As defined in encyclopedias and dictionaries, the blowing agent is a substance that is capable of producing a cellular structure via a foaming process in a variety of materials that undergo hardening or phase transition, such as polymers, plastics, and metals. Chemical and physical blowing agents are distinguished. Chemical blowing agents are mixed into the plastics at lower temperatures. Then, above a certain temperature specific to each blowing agent, gaseous reaction products evolve, which help to produce a foamed structure. Physical blowing agents are metered into the plastics, which are the most frequently in the form of a melt, and they form bubbles by various means discussed in detail in Chapter 10. The blowing agents are further divided into endothermic and exothermic foaming agents. The endothermic chemical foaming agents are chemical compounds that take heat away from the chemical reaction. This produces foams with a much smaller cell structure, resulting in improved appearance and better physical properties. The exothermic chemical foaming agents generate heat during the decomposition process. They liberate more gas per gram of foaming agent than endothermic agents and produce higher gas pressures.
- Accelerators are additives increasing the rate of decomposition of a blowing agent or reducing its decomposition temperature. In PVC, they are usually a part of stabilizing systems, and they are called kickers. The term accelerator is also used in curing systems, for example, in polyurethanes and rubbers.

- Activators are used for the same purpose as accelerators to reduce the decomposition temperature of the blowing agent. Zinc oxide is the most common activator.
- The catalyst is used in polyurethane formulation to regulate the rate of production of gaseous products of the reaction between isocyanates and water (blowing agent). Catalysts are also used in polyurethanes to regulate the rate of cure due to the reaction between isocyanate and polyol. Both reaction rates are synchronized to produce foams of expected quality.
- Nucleating agents are used to control foam density and provide uniform cell size. Talc is the most common example. Nucleating agents also have other functions useful in foam production as they increase the crystallization rate and reduce spherulite size.

Foaming technology makes use of all three states of matter. Carbon dioxide is the best example, considering its use in the form of gas, supercritical liquid, and solid (dry ice). All these forms are used on an industrial scale. Water is also a popular blowing agent because of its low cost and affinity to many materials in which it can form hydrogen bonding, which helps in its uniform distribution in the matrix material.

The liquid form of the blowing agent is very useful because it simplifies homogenization. Solid materials, especially azodicarbonamide, are also very popular in many applications.

Foaming processes can be controlled by many parameters, including type and amount of foaming agent, additives, saturation pressure, desorption time, die pressure, die temperature, feed ratio, gas contents, its flow rate and injection location, internal pressure after foaming, mold pressure, mold temperature, viscosity of composition under processing conditions, surface tension, time-temperature regime, and many other.

The selection of formulation depends on the mechanisms of action of blowing agents and foaming mechanisms, as well as dispersion and solubility of foaming agents and foam stabilization requirements.

Foaming processes are most frequently used in extrusion and injection molding. Still, many other processing methods are also used, such as compression molding, depressurization, free foaming, microwave expansion, hot press, rotational molding, or the simple addition of pre-expanded spheres, or spheres which can be expanded during normal production conditions.

At least forty polymers and their blends are involved in applications of foaming technologies, but the field is much broader, including such diverse materials as concrete, many food products, metal products, insulating materials, sponges, etc. The foaming processes usually decrease mechanical properties and increase compression set in materials, but shape memory foams are also produced.

The above and many other topics are discussed in the fourteen chapters of this book, which includes theoretical and practical information. Data on the raw materials used in their production are included in the **Databook of Blowing and Auxiliary Agents**, which is a very useful companion to the present book.

2

CHEMICAL ORIGINS OF BLOWING AGENTS

The basic properties of blowing agents and auxiliary agents include general information, physical and chemical properties, health effects, ecological impact, and use. The detailed information on more than 360 blowing and auxiliary agents can be found in a separate book entitled **Databook of Blowing and Auxiliary Agents**.[1]

This book provides tables characterizing the average properties of additives included in 23 major groups of products used in these applications. It is important to consider how additives were grouped in this section. Blowing agents are very frequently mixtures of several components, including in addition to the blowing agent(s) also some auxiliary agents required to reach performance criteria selected by the product designer. Up to 15 components are known to be present in the commercial additives. In some cases, products contain only one component or just a few. The tables below are arranged in the alphabetical order groups of blowing and auxiliary agents to assist the reader in an easier finding of the information. Blowing agents are grouped according to their major chemical components. If the composition is not known, products are qualified as "Proprietary" or "Masterbatch," and these groups are more diverse and have a broad range of properties. Also, they are, by far, the most abundant groups of all blowing agents. The auxiliary agents included in the sample are not so numerous (e.g., activator, crosslinker, kicker, etc.), and the types of data differ from blowing agents to stress the general properties of such additives.

The data in this chapter are included for comparative purposes to assist in the general selection of the products, but for particular purposes of the application of these additives, **Databook of Blowing and Auxiliary Agents** should be consulted, or information can be obtained by searching open literature. The following tables are included below:

- activators
- azodicarbonamides
- crosslinkers
- dinitroso pentamethylene tetramines
- dispersions in polymer carriers
- foaming agent mixtures with other additive(s)
- gases
- hydrazides
- hydrocarbons
- hydrochlorofluorocarbon
- hydrofluorocarbons
- hydrofluoroolefins
- masterbatches

- microspheres
- mixture of foaming agents
- nucleating agents
- proprietary
- salts of carbonic and polycarbonic acids
- sodium bicarbonate
- sulfonylsemicarbazides
- tetrazoles
- water

The sequence and groups are the same as used in the **Databook of Blowing and Auxiliary Agents**.

2.1 ACTIVATORS

GENERAL	
Chemical name(s): Ba/Zn carboxylate, carbamide (urea), K/Zn carboxylate, urea, zinc 2-ethylhexanoate, zinc benzenesulfinate, zinc carbonate, zinc ditolyl sulfinate, zinc oxide, zinc stearate	**RTECS number**: YR6250000
Composition: up to 98 wt%	
PHYSICAL CHEMICAL PROPERTIES		
Melting, °C: 40-265	**Decomposition**, °C: 133-142	**pH**: 7-10
State: solid or dispersion	**Color**: colorless, white, yellow	**Odor**: odorless to slight odor
Density at 25°C, kg/m^3: 965-1460		**Particle size**, μm: 3-20
HEALTH & SAFETY	
Flash temp., °C: 127	**Oral**, rat LD50, mg/kg: >4200 to >5000
UN risk: R36/37/38	**UN safety**: S26,S36/37/39
ECOLOGICAL IMPACT	
Atmospheric life, years: 14	**Partition**, log K_{ow}: -2.97 to -2.26
Global warming potential: 1320	**Ozone depletion potential**: 0
Algae, LC50, mg/l: 10000/72H	
USE
Recommended for polymers: EPDM, NBR, PE, PVC, PVC/NBR, rubber, SBR
Recommended for products: automobiles, carpet backing, electric and electronic products, flooring, housing products, leather cloth, vibration-proof material, cushioning material, sealing material, heat insulating material, noise insulation material, closure sealing gaskets for food containers, paper and paperboard, wallcoverings
Processing methods: extrusion, blow molding, calendering, spread coating, dip coating, rotational molding
Concentrations used, phr: 1-4.5

2.2 AZODICARBONAMIDE

$$H_2N-\underset{O}{\underset{\|}{C}}-N=N-\underset{O}{\underset{\|}{C}}-NH_2$$

GENERAL		
Chemical name(s): azodicarbonamide		**Character**: exothermic
Composition: 10-99.1 wt%; azodicarbonamide pasted with mineral oil; coated with dispersing agent; in PS, NBR, silicone, LDPE, EVA or EPDM carrier; in combination with other blowing agent (e.g., DNPT, OBSH, or TSH)		
PHYSICAL CHEMICAL PROPERTIES		
Decomposition, °C: 110-230		
State: paste, solid	**Color**: white to yellow to orange	
Density at 25°C, kg/m³: 1030-1780		**Max. gas yield**, °C: 210
Total gas evolution, ml/g: 44-320 (basic grades 230-320)		**Max. gas evolution**, °C: 199-208
Main gaseous product: N_2, NH_3, CO, CO_2, H_2O		**pH**: 6.5-7.5
Refractive index at 20°C: 1.762		**Particle size**, μm: 2.6-17
Water solubility at 20°C, g/l: 0.2		
Vapor pressure at 20°C, kPa: 0.9456		
HEALTH & SAFETY		
HMIS, F/H/R: 2/1/0	**NFPA**, F/H/R: 1/1/0	**Carcinogenicity**: no
Oral, rat LD50, mg/kg: >6400		
UN risk: R42,R44	**UN safety**: S2,S22,S24,S37	**UN/NA class**: 1325/3242
ECOLOGICAL IMPACT		
Global warming potential: 4000		
USE		
Recommended for polymers: ABS, acrylics, CR, IIR, LDPE, LLDPE, MDPE, PA-6, PE, PP, PS, PVC rigid, PVC-plastisols, PVC/NBR, PVDC, EVA, EPDM, NR, SBR, NBR, silicone rubber, TPE, TPV, crosslinked PE		
Recommended for products: apparel, automotive applications, carpet, expanded sheet, flooring sealants, foam coated wire, foamed film, food trays, gaskets, leather, lunch boxes, pipe, rod, shoe midsole, tool handles, upholstery, vinyl wall covering, wire insulation		
Processing methods: rotational molding, PE: injection molding, as well pressure, extrusion, powder & atmospheric; EPDM & EVA extrusion molding; PP pressure molding, calender molding, fluidized bed coating		
Concentrations used, wt%: 0.5-3 and up		

2.3 CROSSLINKERS

GENERAL		
Chemical name(s): azobisisobutyronitrile, dicumyl peroxide; 3,3,5,7,7-pentamethyl-1,2,4-trioxepane		
Composition: ~40% dicumyl peroxide; ~95% 3,3,5,7,7-pentamethyl-1,2,4-trioxepane		
PHYSICAL CHEMICAL PROPERTIES		
State: solid or liquid	Color: white to off-white	Odor: faint
Density at 25°C, kg/m^3: 950-1770		Start of decomposition, °C: 30
HEALTH & SAFETY		
Flash temp., °C: 41.5		UN/NA class: 3077
ECOLOGICAL IMPACT		
Partition, log K$_{ow}$: 2.4		
USE		
Recommended for polymers: ABS, LLDPE, HDPE, PA, PBT, PC, PPE, PMMA, PP, PVC, natural and synthetic rubbers, silicone, unsaturated polyester		
Recommended for products: gaskets, insulating sheets, rollers, profiles, wire & cable		

2.4 DINITROSO PENTAMETHYLENE TETRAMINE COMPOUNDS

GENERAL		
Chemical name(s): N,N´-dinitrosopentamethylentetramine		**Character**: exothermic
Composition: dinitroso pentamethylene tetramine and surface treated urea activator		
PHYSICAL CHEMICAL PROPERTIES		
State: solid	**Color**: white to light yellow	**Decomposition**, °C: 120-210
Density at 25°C, kg/m^3: 1450-1540		
Total gas evolution, ml/g: 111-285		
Main gaseous product: N_2, NO_x, CO, HCHO		
pH: 7-9	**Water solubility** at 20°C, g/l: slightly soluble	
HEALTH & SAFETY		
Oral, rat LD50, mg/kg: 240-940		**UN/NA class**: 3224
USE		
Recommended for polymers: CR, EPDM, epoxy, EVA, NR, SBR, NBR, polyester, PB, PE, PVC, silicone		
Recommended for products: pressurized foam, cellular rubber, soles for beach sandals, sponges		
Processing methods: extrusion, molding		

2.5 DISPERSIONS IN POLYMERS CARRIERS

GENERAL	
Chemical name(s): acid and base components, azodicarbonamide, backing soda, 4,4'-oxybis(benzenesulfonyl hydrazide), sodium salts of carbonic and polycarboxylic acids	**Character**: endothermic, exothermic, endo/exo
Carrier: ABS, EP rubber, EPDM, EVA, LDPE, PS, SBR	
PHYSICAL CHEMICAL PROPERTIES	
Decomposition, °C: 135-250	**Color**: white, off-white to light orange
State: solid/pellets	**Total gas evolution**, ml/g: 10-165
Main gaseous product: CO, CO_2, N_2, NH_3	**pH**: 7
HEALTH & SAFETY	
HMIS, F/H/R: 0/3/0	**NFPA**, F/H/R: 0/3/0-2	**Carcinogenicity**: no
Flash point, °C: 210 to non-flammable		
USE		
Recommended for polymers: ABS, EMA, EPDM, EVA, HIPS, IIR, NBR, NR, PBT, PC, PE, PET, PP, PPO, PPS, PS, PVC, PVDV, SAN, SBR, silicone, TPE, TPO, TPU		
Recommended for products: boards, decorative ribbons, film, food trays, gaskets, lunch boxes, profiles, sheet, tapes, wine corks		
Processing methods: blow molding, direct gassing, extrusion, gas-assist molding, gas counter-pressure foam molding, injection molding, nucleation, rotational molding, slush molding, structural foam molding		
Processing temperature	°C	120-315
Concentrations used, wt%: 0.15-4		

2.6 FOAMING AGENT MIXTURES WITH OTHER ADDITIVE(S)

GENERAL		
Chemical name(s): azodicarbonamide, p-toluenesulfonyl hydrazide		**Character**: endo/exo
Additives: activator, dispersing agent, paraffinic mineral oil, self-nucleating agent		
PHYSICAL CHEMICAL PROPERTIES		
State: solid/pellets	**Color**: yellow	**Decomposition**, °C: 105->200
Density at 25°C, kg/m^3: 1250	**Total gas evolution**, ml/g: 84-170	
Main gaseous product: CO_2, N_2, NH_3		
USE		
Recommended for polymers: EPDM, NBR, NR, polyester, SBR, thermoplastics (except PC)		
Processing methods: extrusion, injection molding		
Concentrations used, wt%: 1-15		

2.7 GASES

GENERAL		
Chemical name(s): carbon dioxide (gas, liquid, solid), nitrogen		
RTECS number: FF6400000, QW9700000		
PHYSICAL CHEMICAL PROPERTIES		
State: gas, liquid solid	**Boiling**, °C: -210.01 to -78.5	**Odor**: odorless
Color: colorless to white (solid)		**pH**: 3.2-3.7
Density at 25°C, kg/m^3: 1.1848-1.8714 (gas); 1256.74 liquid at -20°C; 1562 (solid)		
Dissociation constant (pK$_a$): 6.35 and 10.33		**Relative permittivity**: 1-1.6
Refractive index at 20°C: 1	**Water solubility** at 20°C, g/l: 0.02-2	
Surface tension, mN/m: 8.85-16.2		
Specific heat, kJ/kg K: 0.709-1.4		
Thermal conductivity, W/m K: 0.0146-0.0238		
Vapor pressure at 20°C, kPa: 5729.1		
Viscosity at 0°C, mPa s: 1.37E-04; 1.66E-04		
Critical point, kPa/°C: 7377/30.98; 3396/-146.96		
HEALTH & SAFETY		
Autoignition, °C: not applicable		**Flash point**, °C: not applicable
HMIS, F/H/R: 0/0-3/0-1	**NFPA**, F/H/R: 0/0-3/0-1	**Carcinogenicity**: no
OSHA, PEL, ppm: 5000	**Inhalation**, rat LC50, mg/kg: 90000/5M	
UN/NA class: 1013, 2187, 1845; 1066, 1977		
ECOLOGICAL IMPACT		
Ozone depletion potential: 1		**Partition**, log K$_{ow}$: 0.83; 0.67

2.8 HYDRAZIDES

GENERAL		
Chemical name(s): p,p´-oxybisbenzenesulfonylhydrazide; p-toluenesulfonylhydrazide		**Character**: exothermic
RTECS number: DB7321000; MW0210000		
Composition: hydrazide pasted with mineral oil (up to 25%); up to 98 wt% hydrazide; 30-75 wt% in carrier resin (e.g., EPDM or EPR)		
PHYSICAL CHEMICAL PROPERTIES		
Melting, °C: 60-103	**Boiling**, °C: 125	**Decomposition**, °C: 105-164
State: solid	**Color**: colorless to white	**Odor**: geranium-like
Density at 25°C, kg/m³: 1250-1550		**pH**: 6-8
Total gas evolution, ml/g: 120-150		
Main gaseous product: N_2, CO_2, H_2O		
Water solubility at 20°C, g/l: insoluble		
Max. gas evolution, °C: 150-160		
HEALTH & SAFETY		
HMIS, F/H/R: 2-3/1-2/0-2	**NFPA**, F/H/R: 2-3/1-2/2	**Carcinogenicity**: no
Oral, rat LD50, mg/kg: 283->2300		**UN/NA class**: 3226
USE		
Recommended for polymers: CR, EPDM, EVA, IIR, NR, NBR, PE, PVC, SBR, silicone rubber		
Recommended for products: sponge rubber cosmetic puffs, electrostatographic toners, buoyant seats, insulation and packing materials, adhesives, sealants, and leather substitutes, high-performance rubber sheet, NBR-PVC pipe insulation, wire insulation, neoprene-blended EPDM for automotive, press-cured NBR-PVC for flotation and athletic padding, SBR rubber rug underlay, casting for PVC carpet underlay, EVA shoe mid-sole and latex foaming, coaxial cables, cellular rubber articles in which the bright color of the final product is important		
Processing methods: rotational molding, pressure molding, extrusion, injection molding		
Concentrations used, wt%: 0.5-5		

2.9 HYDROCARBONS

GENERAL	
Chemical name(s): cyclopentane, dimethoxymethane, isobutane, isopentane, methyl formate, neopentane	Composition: 98-100%
RTECS number: EK4430000, GY2390000, LQ8925000, PA8750000	
PHYSICAL CHEMICAL PROPERTIES	
Melting, °C: -160 to -16.6 Boiling, °C: -12-49	State: gas, liquid
Refractive index at 20°C: 1.343-1.4065	
Odor: etheral, sweet, chloroform-like, gasoline-like, odorless	
Density at 25°C, kg/m^3: 590-980	Color: colorless
Water solubility at 20°C, g/l: 15.6-300	
Thermal conductivity, W/m K: 0.02-10.7	
Vapor pressure at 20°C, kPa: 45-146	
HEALTH & SAFETY	
Autoignition, °C: 361-450 Flash temp., °C: -51 to -18	OSHA, PEL, ppm: 150-600
HMIS, F/H/R: 3-4/1-3/0 NFPA, F/H/R: 3/1-2/0-2	Carcinogenicity: no
UN risk: R11, R12, R36/38, R51/53, R65, R66, R67	
UN/NA class: 1146, 1234, 1265, 1969, 2044	
ECOLOGICAL IMPACT	
Atmospheric life, days: 3 Kyoto compliant: yes	Montreal compliant: yes
Global warming potential: 0-11	Ozone depletion potential: 0
USE	
Recommended for polymers: PS, PU	
Recommended for products: appliances (freezers, refrigerators), bedding, construction panels, coolers, doors, furniture, pillows, propellants, seating, spray foam, XPS board	

2.10 HYDROCHLORCARBON

GENERAL		
Chemical name(s): *trans*-1,2-dichloroethylene		
Composition: 97.0% *trans*-1, 2-dichloroethylene, 0.3% *cis*-1, 2-dichloroethylene		
PHYSICAL CHEMICAL PROPERTIES		
State: liquid	Color: bolorless	Boiling point, °C: 48
Density at 25°C, kg/m^3: 1252-1260		Melting point, °C: -60
Refractive index at 20°C: 1.447		
Vapor pressure at 20°C, kPa: 24-36.3		
HEALTH & SAFETY		
UN risk: R11,R20,R52/53,R39/23/24/25, R23/24/25		Autoignition, °C: 460
UN/NA class: 1160	UN safety: S7,S16,S29,S61,S45,S36/37	Flash point, °C: 2.2
ECOLOGICAL IMPACT		
Global warming potential: <11		
Ozone depletion potential: 0		
USE		
Recommended for polymers: PU, phenolic foam system, sprayed polyurethane foam (SPF), polyisocyanurate boardstock foam		
Recommended for products: as an addition to HFC and HFO foam formulations; building and construction foam, appliance foam		

2.11 HYDROCHLOROFLUOROCARBON

$$Cl-\underset{Cl}{\overset{F}{C}}-CH_3$$

GENERAL		
Chemical name: 1-chloro-1,1-difluoroethane, 1,1-dichloro-1-fluoroethane		
PHYSICAL CHEMICAL PROPERTIES		
Melting, °C: -103.5	Boiling, °C: -40.8 to 76	State: gas, liquid
Density at 25°C, kg/m³: 1020-1430		Color: colorless
Water solubility at 20°C, g/l: 2.6-4		Odor: ether-light
Viscosity at 25°C, mPa: 0.43-0.61		Relative permittivity: 8.16
Thermal conductivity, W/m K: 0.019		
Surface tension, mN/m: 19.3	Vapor pressure at 20°C, kPa: 76.3-333	
HEALTH & SAFETY		
Autoignition, °C: 375-532	Inhalation, rat LC50, mg/kg: 300/4H; 62,000/4H ppm	
Dermal, rabbit LD50, mg/kg: >2000		Carcinogenicity: no
Oral, rat LD50, mg/kg: 5000	UN risk: R36/37/38,R52/53,R59	
UN safety: S26,S36/37/39,S59,S61		OSHA, PEL, ppm: 500
ECOLOGICAL IMPACT		
Atmospheric life, years: 9.2-17.9	Ozone depletion potential: 0.055-0.11	
Global warming potential: 630-2310		Partition, log K_{ow}: 1.85
Algae, LC50: 44/72H	*Bluegill sunfish*, LC50, mg/l: 126	
Daphnia magna, LC50, mg/l: 31.2		
USE		
Recommended for polymers: phenolic, PS, PU		
Recommended for products: insulating foams, insulated metal panels and doors, refrigeration warehouses, roofing, supermarket display cases, storage rooms, small appliances		

2.12 HYDROFLUOROCARBONS

GENERAL	
Chemical name(s): *cis*-1,1,1,4,4,4-hexafluoro-2-butene, 1,1,1,3,3-pentafluoropropane, 1,1,1,3,3-pentafluorobutane, 1,1 difluoroethane, 1,1,1,2-tetrafluoroethane	RTECS number: KV9400000
PHYSICAL CHEMICAL PROPERTIES	
Density at 25°C, kg/m^3: 0.9-5.84 (gas), 1020-1430 (liquid)	Color: colorless
Odor: odorless, ether-like	Refractive index at 20°C: 1.2825-1.447
Surface tension, mN/m: 12.1-27.5	Melting, °C: -138 to -27
Vapor pressure at 20°C, kPa: 1.23-107.84	Boiling, °C: -26.4 to 76
Water solubility at 20°C, g/l: 0.2-7.18	State: gas, liquid
Thermal conductivity, W/m K: 0.019-0.0824	
Viscosity at 20°C, mPa: 0.202-0.01-0.489	
HEALTH & SAFETY		
Autoignition, °C: 380 to >743	Flash point, °C: -50 to 28	OSHA, PEL, ppm: 200-1000
HMIS, F/H/R: 0-3/2/0	NFPA, F/H/R: 1-3/0-3/0-1	Oral, rat LD50, mg/kg: >5000
Inhalation, rat LC50, ppm: >102900/4H to 437000/4H		
UN risk: R11,R18,R20,R36/37/38,R52/53		
UN safety: S7,S9,S16,S23,S24/25,S26,S29,S33,S36/37/39,S45,S61		
UN/NA class: 1018, 1030, 1150, 1993, 3159, 3163		
ECOLOGICAL IMPACT	
Atmospheric life, days: 18-16112	Partition, log K$_{ow}$: 1.06-2.3
Global warming potential: 2-2310	Kyoto compliant: no/yes
Ozone depletion potential: 0-0.076	Montreal compliant: yes
Bluegill sunfish, LC50, mg/l: 777	
Daphnia magna, LC50, mg/l: 2.55-980	
Fathead minnow, LC50, mg/l: 2.75-95.7	
Rainbow trout, LC50, mg/l: 38-460	
USE	
Recommended for polymers: PE, PP, XPS, rigid PU, isocyanurate, phenolic	
Recommended for products: spray foam insulation, wall sheathing, roofing, appliance foams, pipe insulation, insulation board, food packaging, integral skins, graphics arts, industrial, transportation, shoe soles, decorative display items and other specialty parts, building and construction	

2.13 HYDROFLUOROOLEFINS

GENERAL		
Chemical name(s): *cis*-1,1,1,4,4,4-hexafluoro-2-butene, *trans*-1-chloro-3,3,3-trifluoropropene		
PHYSICAL CHEMICAL PROPERTIES		
State: gas, liquid	Color: colorless	Odor: odorless
Density at 25°C, kg/m^3: 1092-1413		Boiling point, °C: -30-33
Water solubility at 20°C, g/l: 0.78-3.8		Melting point, °C: -150 to -90
Refractive index at 20°C: 1.2835		Viscosity, mPas: 0.15-0.49
Vapor pressure at 20°C, kPa: 80-683		
HEALTH & SAFETY		
HMIS, F/H/R: 2-3/0-3/0	NFPA, F/H/R: 2-3/0-3/0	Carcinogenicity: no
UN risk: R11,R18,R36/37/38	UN safety: S9,S16,S23,S24/25,S26,S33, S36/37/39,S45	UN/NA class: 1993, 3163
ECOLOGICAL IMPACT		
Global warming potential: 1-7		
Ozone depletion potential: 0-0.0005		
USE		
Recommended for polymers: HDPE, PBT, phenolic, PP, PS, PTFE, PU, PVC		
Recommended for products: appliance foams, construction panels and doors, insulating foams, integral skin, panels, pipe insulation, pour in place, residential and commercial appliances, spray foam, trailers, wall sheathing		

2.14 MASTERBATCHES

GENERAL	
Chemical name(s): azodicarbonamide and non-azodicarbonamide	**Character**: endothermic, exothermic
Active component, %: 10-50	
PHYSICAL CHEMICAL PROPERTIES	
Total gas evolution, ml/g: 16-110	**State**: solid:
Main gaseous product: CO_2	
USE	
Recommended for polymers: ABS, HIPS, PA11, PA12, PBT, PE, PET, POM, PP, PS, PVC, SAN, TPE, TPO, TPU	
Recommended for products: thick-walled olefin handles	
Processing temperature	°C	175-295
Concentrations used, wt%: 0.1-3	

2.15 MICROSPHERES

GENERAL		
Chemical name(s): expandable microspheres, expanded microspheres		
Composition: 6.0-11.0% isobutane; >80.0% copolymer; 64.0-66.0% unexpanded microspheres in copolymer of ethylene vinylacetate; 15.0-20.0% isopentane; >75.0% copolymer; solids content 13-99 wt%		
PHYSICAL CHEMICAL PROPERTIES		
Decomposition, °C: >180	**Density** at 25°C, kg/m^3: 1000-1300	
State: solid, slurry	**Color**: white, off white	**Odor**: odorless
Start of decomposition, °C: 85-148		**Max. gas yield**, °C: 115-200
Foam density, kg/m^3: 6.5-46	**pH**: 3-4	
Water solubility at 20°C, g/l: insoluble		
HEALTH & SAFETY		
Autoignition, °C: 170	**Oral**, rat LD50, mg/kg: >1900	
UN risk: R12,R51/53,R65,R66, R67		**UN/NA class**: 1325, 2211
USE		
Recommended for polymers: acrylics, polyesters, PUR, PVC plastisols, PVDC, polysulfides, silicones		
Recommended for products: automotive, textile, paint ink, wallpaper, shoes sole injection, paints, reflective coatings, leather finishing, concrete, cable, sealants		
Processing methods: extrusion, molding		

2.16 MIXTURES OF FOAMING AGENTS

GENERAL	
Chemical name(s): azodicarbonamide, calcium salts of polycarboxylic acids, carbonic and phosphoric acids, N, N'-dinitrosopentamethylenetetramine, 4,4'-oxybis(benzenesulfonyl hydrazide, p-toluenesulfohydrazide	**Character**: exothermic

PHYSICAL CHEMICAL PROPERTIES		
Decomposition, °C: 125-260	**Total gas evolution**, ml/g: 26-180	
State: solid	**Color**: white to light orange	**Odor**: faint
Main gaseous product: CO, CO_2, N_2, NH_3		**Particle size**, μm: 30

HEALTH & SAFETY		
HMIS, F/H/R: 3/0/0	**NFPA**, F/H/R: 3/0/2	**Carcinogenicity**: no
Oral, rat LD50, mg/kg: 5000	**Flash point**, °C: 210	

USE
Recommended for polymers: acrylics, EVA, PE, PP, PS, PVC plastisols, TPE
Recommended for products: pressurized foam, rubber for table tennis, shoes, wall calendars, wallpaper
Processing methods: extrusion, injection molding, pressure foaming

2.17 NUCLEATING AGENTS

GENERAL		
Chemical name(s): nucleating agent		
PHYSICAL CHEMICAL PROPERTIES		
Decomposition, °C: 135-220		
State: solid	**Color**: off white	**Odor**: odorless
Max. gas yield, °C: 216	**Total gas evolution**, ml/g: 27-46	
pH: 7	**Water solubility** at 20°C, g/l: negligible	
HEALTH & SAFETY		
Flash temp., °C: non-flammable		**Carcinogenicity**: no
Oral, rat LD50, mg/kg: 5000	**UN/NA class**: not regulated	
USE		
Recommended for polymers: PE, PET, PP, PS, PU		
Recommended for products: coaxial and CAT V cable, monofilament, film, sheet, board		
Processing methods: extrusion		
Concentrations used, wt%: 0.5-2		

2.18 PROPRIETARY

GENERAL		
Chemical name(s): formulations known by manufacturers		**Character**: exothermic, or endothermic, or mixed
Composition: some indicate that they contain nucleating agents or carrier resins			
PHYSICAL CHEMICAL PROPERTIES		
Decomposition, °C: 125-250	**Total gas evolution**, ml/g: 80-170	
State: solid	**Color**: white, gray, yellow	**Foam density**, kg/m^3: 35-40
Max. gas evolution, °C: 160-304			
Main gaseous product: CO_2, N_2			
USE		
Recommended for polymers: ABS, EPDM, EVA, HDPE, HIPS, LDPE, LLDPE, MDPE, PA, PBT, PC, PC/ABS, PET, PP, PPO, PS, PVC, SEBS, TPE, TPU			
Recommended for products: celuka sheets, sheets, pipe and other profile, drip hose, ribbons, wire & cable, coaxial cable, vinyl wall covering, PVC 3-layer pipe, food containers, food tray, lunch boxes, synthetic cork			
Processing methods: direct gas extrusion, extrusion, gas counter pressure molding, injection molding, rotational molding, thermoforming			
Concentrations used, wt%: 0.2-6			

2.19 SALTS OF CARBONIC AND POLYCARBOXYLIC ACIDS

GENERAL	
Chemical name(s): magnesium, sodium, or zinc salts of carbonic and polycarboxylic acids	**Character**: endothermic
Composition: some are dispersed in polymeric carrier; some contain addition of azodicarbonamide	
PHYSICAL CHEMICAL PROPERTIES	
Max. gas evolution, °C: 176-224	**Decomposition**, °C: 120-177
State: solid	**Color**: white, off white, gray, light orange	**Odor**: odorless to faint
Total gas evolution, ml/g: 24-145	
Main gaseous products: N_2, CO, CO_2, (NH_3)	
pH: 7	**Water solubility** at 20°C, g/l: negligible
HEALTH & SAFETY	
Flash temp., °C: 210 to non flammable	**OSHA**, PEL, ppm: 15
HMIS, F/H/R: 3/0/0	**NFPA**, F/H/R: 3/0/2	**Carcinogenicity**: no
Oral, rat LD50, mg/kg: 5000	**UN/NA class**: 3242	
USE	
Recommended for polymers: ABS, EVA, HIPS, PBT, PC, PE, PET, POM, PP, PPE/PPO, PPS, PS, PVC, SAN, TPE, TPO, TPU	
Recommended for products: food packaging, labels, fine sheet products, monofilament, film, board	
Processing methods: extrusion, injection molding	
Concentrations used, wt%: 0.5-4	

2.20 SODIUM BICARBONATE

GENERAL		
Chemical name(s): sodium hydrogen carbonate	**Character**: endothermic	**RTECS number**: VZ0950000
Composition: monosodium citrate, sodium hydrogen carbonate, partially hydrogenated oil, tricalcium phosphate, synthetic amorphous silica, polyethylene carrier, and ethylene homopolymer carrier; 20% sodium hydrogen carbonate in olefin carrier; 99% sodium hydrogen carbonate		
PHYSICAL CHEMICAL PROPERTIES		
Boiling, °C: 851		**Decomposition**, °C: >50-260
State: solid	**Color**: white, off white	**Odor**: odorless
Density at 25°C, kg/m^3: 2160-2200		**pH**: 6-8
Total gas evolution, ml/g: 40-170		
Max. gas evolution, °C: 150-183		
Main gaseous product: CO_2, H_2O		
Refractive index at 20°C: 1.3344		
Water solubility at 20°C, g/l: 90		
Vapor pressure at 25°C, kPa: 3.44E-06		
HEALTH & SAFETY		
Flash temp., °C: non-flammable		**OSHA**, PEL, ppm: 5
HMIS, F/H/R: 0/0/1	**NFPA**, F/H/R: 0/0/1	**Carcinogenicity**: no
Oral, rat LD50, mg/kg: 4220-8290		
Inhalation, rat LC50, mg/kg: 4.75		
UN risk: R34	**UN safety**: S24/25	**UN/NA class**: not regulated
ECOLOGICAL IMPACT		
Daphnia magna, LC50, mg/l: 2350		
Algae, LC50, mg/l: 650/102H	*Rainbow trout*, LC50, mg/l: 8250-9000	
USE		
Recommended for polymers: ABS, EVA, PE, PET, PP, PS, PVC, TPE, TPO		
Recommended for products: celuka sheets, profiles, block foam, sheet (packaging, building materials), insulation coating material, various types of cushioning material		
Processing methods: extrusion, injection molding, rotational molding		
Concentrations used, wt%: 0.5-2		

2.21 SULFONYLSEMICARBAZIDES

GENERAL	
Chemical name(s): p-toluenesulfonylsemicarbazide	Character: exothermic
Composition: >95.0% p-toluenesulfonyl semicarbazide	
PHYSICAL CHEMICAL PROPERTIES	
Density at 25°C, kg/m^3: 1381	Decomposition, °C: 215-235
State: solid	Color: white	pH: 6.5-7.5
Total gas evolution, ml/g: 120-140	
Main gaseous product: N_2, CO_2, CO	
Refractive index at 20°C: 1.588	
HEALTH & SAFETY	
Oral, rat LD50, mg/kg: 980	Carcinogenicity: no
UN risk: R22,R36/37/38	UN safety: S26,S36/37
USE	
Recommended for polymers: ABS, HDPE, PA, PS, PVC, modified polyphenylene oxide, polysulfone	
Recommended for products: ABS extruded tubing, profiles and sheet, rigid PVC profile	
Processing methods: injection molding, extrusion	

2.22 TETRAZOLES

GENERAL		
Chemical name(s): 5-phenyltetrazole		
Composition: >99.0% 5-phenyltetrazole		
PHYSICAL CHEMICAL PROPERTIES		
State: solid	**Color**: white	**Decomposition**, °C: 215-230
Density at 25°C, kg/m³: 1420	**Total gas evolution**, ml/g: 190-210	
Main gaseous product: N_2, CO_2		
HEALTH & SAFETY		
Oral, rat LD50, mg/kg: 1900	**Flash temp.**, °C: 161.3	**Carcinogenicity**: no
UN risk: R11,R22,R44	**UN safety**: S16,S22,S26,S37/39	
UN/NA class: 1325		
USE		
Recommended for polymers: ABS, PA, PBT, PC, PPE, PPS		
Processing methods: extrusion, injection molding, rotational, slush molding		

2.23 WATER

GENERAL		
Chemical name(s): water	**Character**: endothermic	**RTECS number**: ZC0110000
PHYSICAL CHEMICAL PROPERTIES		
Melting, °C: 0	**Boiling**, °C: 100	**pH**: 7, typical 6.5-8.5
State: liquid	**Color**: colorless	**Odor**: odorless
Dissociation constant (pK$_a$): 13.995		**Surface tension**, mN/m: 71.97
Density at -4°C, kg/m^3: 999.972; 917 (ice)		**Main gaseous product**: H$_2$O
Refractive index at 20°C: 1.3325		**Specific heat**, kJ/kg K: 1.996
Thermal conductivity, W/m K: 0.016 (steam)		
Vapor pressure at 25°C, kPa: 3.169		**Viscosity** at 25°C, mPa: 0.89
Relative permittivity: 80.1	**Volume resistivity**, Ω-cm: 1.82E+07	
HEALTH & SAFETY		
Flash temp., °C: not applicable	**Oral**, rat LD50, mg/kg: 90000	
HMIS, F/H/R: 0/0/0	**NFPA**, F/H/R: 0/0/0	**Carcinogenicity**: no

More information on nucleating agents and data for specific nucleating agents can be found in **Databook of Nucleating Agents**,[3] **Handbook of Nucleating Agents**,[4] and **Databook of Blowing and Auxiliary Agents**.[1]

REFERENCES

1   Wypych, G., **Databook of Blowing and Auxiliary Agents**, 2nd Edition, *ChemTec Publishing*, Toronto, 2022.
2   Wypych, G., **PVC Degradation and Stabilization**, 4th Edition, *ChemTec Publishing*, Toronto, 2020.
3   Wypych, A; Wypych, G, **Databook of Nucleating Agents**, 2nd Edition, *ChemTec Publishing*, Toronto, 2021.
4   Wypych, G., **Handbook of Nucleating Agents**, 2nd Edition, *ChemTec Publishing*, Toronto, 2021.

3

MECHANISMS OF ACTION OF BLOWING AGENTS

In an attempt to characterize the mechanisms of foaming in polymer systems, we will concentrate on the following processes:
- foaming with the use of solid blowing agents which are decomposed to the gaseous products by application of heat
- production of gaseous products by chemical reaction
- foaming by gases and evaporating liquids.

3.1 MECHANISMS OF FOAMING BY DECOMPOSING SOLIDS

Azodicarbonamide is the most common, classical example of a solid blowing agent that decomposes under the influence of heat with the production of gases that either escape from a system or form bubbles.

Chemical decomposition mechanisms of azodicarbonamide are fairly well known for more than three decades. The heating of azodicarbonamide produces a mixture of gases and a solid residue.[1] The solid residue (urazole, NHCONHCONH, and hydrazodicarbinamide, $NH_2CONHNHCONH_2$) serves as a nucleating agent of bubble formation, and gaseous mixture (cyanic acid, HNCO, ammonia, NH_3, carbon monoxide, CO, and nitrogen, N_2) feeds bubble growth. The main gaseous products include nitrogen and carbon monoxide.[2]

The following reactions characterize the typical decomposition process:[2]

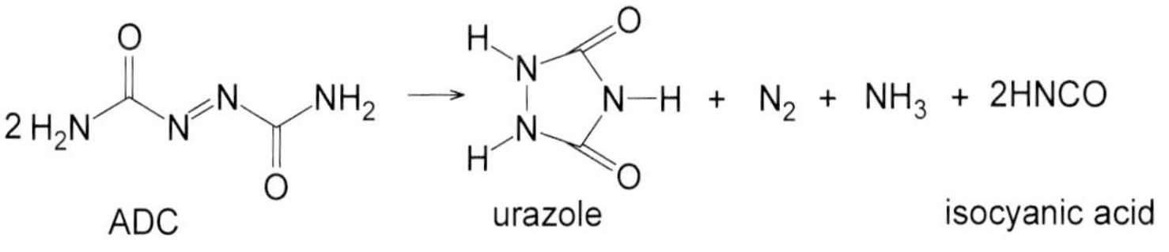

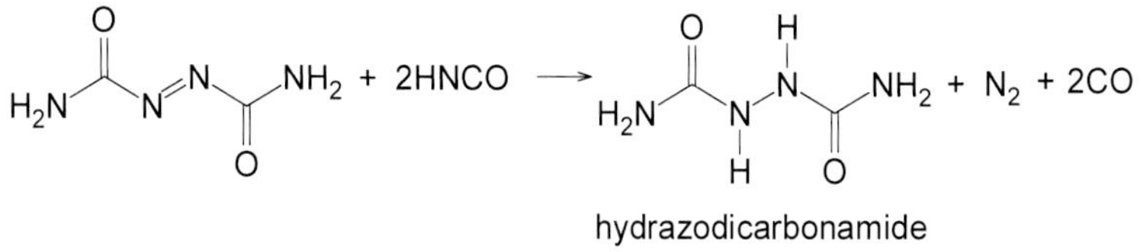

$$2\ H_2N\text{-CO-N=N-CO-NH}_2\ (\text{ADC}) \rightarrow \text{urazole} + N_2 + NH_3 + 2\ HNCO\ (\text{isocyanic acid})$$

$$H_2N\text{-CO-N=N-CO-NH}_2 + 2\ HNCO \rightarrow \text{hydrazodicarbonamide} + N_2 + 2\ CO$$

Depending on the conditions of the decomposition reaction, the composition of degradation products may vary.[3,4] ADC does not completely decompose. 35% of gas, 40% of solid residue, and 25% of sublimate are formed.[4] The gas consists of 65% nitrogen, 32%

carbon monoxide, and 3% other gases, including ammonia and carbon dioxide, with ammonia forming at high temperatures (220°C and higher).[4]

The reaction is exothermic, and decomposition begins at 169°C. The decomposition reaction is autocatalytic since the reaction products further react with unreacted ADC, increasing the gas formation rate.[4] The reaction rate is described by the following equation:[3]

$$\frac{d\alpha}{dt} = (k_1 + k_2\alpha^n)(1-\alpha)^m \quad [3.1]$$

where:
- α extent of reaction
- n,m   exponents independent of temperature
- n+m   reaction order
- k_1, k_2 rate constants which are functions of temperature and can be obtained from Arrhenius relationship

Because the reaction of decomposition of azodicarbonamide is exothermic, it is more difficult to control and may produce the nonuniform bubbles because some locations may be overheated due to the production of heat from the exothermic reaction.[5] To reduce the influence of this effect, injection molded parts are extensively cooled immediately after the reaction is initiated.[5]

It is also important to mention in the discussion of mechanism that the bubbles are initially filled with a gas composition, which was produced as a result of decomposition, but eventually, these gases are replaced by the atmosphere of surrounding gas (usually air).[5]

Cell density depends on the amount of blowing agent, time, and temperature, but also on gas retention in the polymer matrix. The balance of gas pressure in the bubble and surface tension and viscosity of formulation plays an essential role, but also some additives actively influence the process. For example, an increase in the content of stearic acid in wood-flour in the EVA matrix prevented gas from escaping.[6] With increasing content of stearic acid, not only more gas remained in the EVA matrix but also the amount of gas available for cell nucleation and growth increases.[6] As a result, average cell size and cell density increased (Figure 3.1).[6] Activators are used to regulate the temperature of decomposition of blowing agent, and catalysts, crosslinkers, and curatives affect rheological properties of the matrix. Both types of additives affect the outcome of blowing processes. All additives and their effects are discussed in Chapter 11.

The formation and uniformity of cells and their density depend on the amount of blowing agent and the frequency of bubble nucleation. For example, an increase of fiber content improved the heterogeneous cell nucleation and thus cell morphology (higher cell density, smaller cell sizes) (Figure 3.2).[7]

Coalescence and coarsening influence cell density and cell size.[8] Coalescence is caused by rupturing of cell walls during foaming and must be suppressed in order to achieve a uniform and controlled cell growth.[6] Cell coarsening leads to a reduction in cell density due to the diffusion of gas from small cells to the larger ones.[8] For example, excessive foaming time can promote this effect.[8]

The knowledge of the gelation and fusion processes and the thermal decomposition of the blowing agent in plastisol gives a better understanding of the complex dynamic

3.1 Mechanisms of foaming by decomposing solids

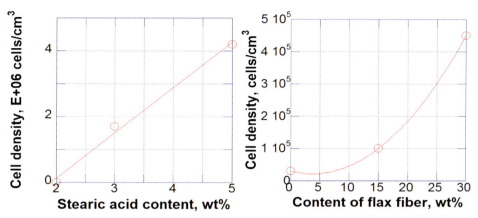

Figure 3.1. Effect of stearic acid on cell density of EVA/wood flour foams. [Data from Kim, J-H; Kim, G-H, *J. Appl. Polym. Sci.*, **131**, 40894, 2014.]

Figure 3.2. Cell density of HDPE foam containing different amounts of flax fibers. [Data from Tissandier, C; Gonzalez-Nunez, R; Rodrigue, D, *J. Cellular Plast.*, **50**, 5, 449-73, 2014.]

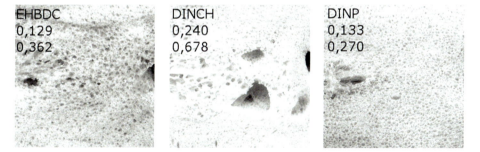

Figure 3.3. Photographs of foams obtained with EHBDC, DINCH, and DINP plasticizers. Number on photographs are average cell size (upper row) and standard deviation (lower row). Both values in mm². [Adapted, by permission, from Zoller, A; Marcilla, A, *J. Appl. Polym. Sci.*, **128**, 354-62, 2013.]

behavior of foaming systems.[9] Diisononyl cyclohexane-1,2-dicarboxylate, DINCH, is less compatible with polymer compared with the other two studied plasticizers (bis(2-ethylhexyl)-1,4-benzenedicarboxylate, EHBDC, and diisononyl phthalate, DINP). Because of this, its plastificates cannot withstand pressure evolved by the released gases during the foaming process yielding foams of poorer quality (Figure 3.3).[10]

N,N'-dinitroso pentamethylene tetramine is one popular chemical blowing agent frequently used in food-contact applications. Exposed to heat N, N'-dinitroso pentamethylene tetramine produces numerous gaseous products.

$$\text{N,N'-dinitroso pentamethylene tetramine} \longrightarrow N_2 + H_2O + CHO + NH_3$$

The rate of gas volume produced depends on temperature (Figure 3.4) but also on the amount of blowing agent (Figure 3.5). Initially, the blowing ratio increases with the

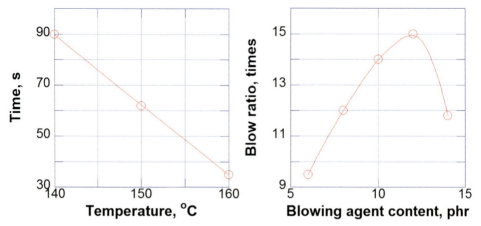

Figure 3.4. Time required to produce constant amount of gas volume (35 ml/0.2 g of blowing agent) vs. temperature of decomposition. [Data from Guan, L T; Du, F G; Wang, G Z; Chen, Y K; Xiao, M; Wang, S J; Meng, Y Z, J. Polym. Res., 14, 245-51, 2007.]

Figure 3.5. Blowing ratio vs. the amount of blowing agent. [Data from Guan, L T; Du, F G; Wang, G Z; Chen, Y K; Xiao, M; Wang, S J; Meng, Y Z, J. Polym. Res., 14, 245-51, 2007.]

amount of blowing agent increased, but at a certain point, the amount of blowing agent (and subsequently amount of the produced gas) is too large that it causes bubble ruptures and gas escape.[11]

In some cases, a blowing agent not only produces gaseous substances to foam polymer but also participates in the cure mechanism. Such is a case of p,p'-oxybis benzene sulfonyl hydrazide in EPDM system:[12]

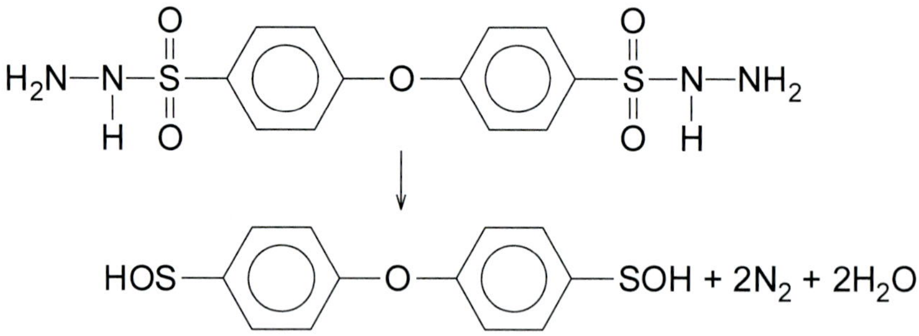

In the first step, p,p'-oxybis benzene sulfonyl hydrazide is decomposed to a sulfur-containing compound, nitrogen, and water vapor.[12] After that, the sulfur-containing compound reacts with a crosslinking agent during the curing reaction.[12]

The blowing agent induces several differences in the vulcanization reaction: it decreases reaction temperature while increasing reaction heat.[12] It eliminates the exothermic peak before vulcanization and decreases the fully cured resin's glass transition temperature.[12] The presence of the blowing agent reduces the operational window.[12]

p-Toluenesulfonyl semicarbazide decomposes to 55% nitrogen, 37% carbon dioxide, 3% ammonia, and 2% carbon oxide, with solid residues (3%) such as ditolyl disulfide and

3.1 Mechanisms of foaming by decomposing solids

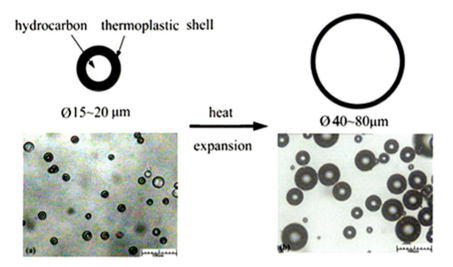

Figure 3.6. Expansion of microspheres. [Adapted, by permission, from Hou, Z; Xia, Y; Qu, W; Kan, C, *Int. J. Polym. Mater. Polym. Biomater.*, **64**, 427-31 2015.]

ammonium p-toluenesulfonate at ~220°C.[13] The decomposition temperature can be lowered to 190°C using activators such as lead stearate or urea.[13]

Sodium hydrogen carbonate is an endothermic blowing agent. Its decomposition stops when the temperature falls below its decomposition temperature.[5] Decomposition temperature can be influenced by the addition of an activator (e.g., acetic acid).[14]

On thermal exposure, sodium bicarbonate decomposes:

$$2NaHCO_3 \rightarrow Na_2CO_3 + H_2O + CO_2$$

This reaction begins at 50°C, and at lower temperatures, it is slow (the fast process of decomposition occurs at 200°C). Reaction with acetic acid is instantaneous:

$$NaHCO_3 + CH_3COOH \rightarrow CH_3COONa + H_2O + CO_2$$

Ammonium chloride can be used to promote and accelerate the blowing process of sodium bicarbonate according to the following equations:[15]

$$NaHCO_3 + NH_4Cl \rightarrow NaCl + NH_3 + CO_2$$

$$Na_2CO_3 + 2NH_4Cl \rightarrow 2NaCl + 2NH_3 + CO_2 + H_2O$$

Polymeric materials may become porous because of the addition of microspheres. Microspheres may contain a blowing agent that expands on heating and produces small inclusions according to the mechanism given in Figure 3.6.[16]

In the case presented, the thermoplastic expandable microspheres have a core-shell structure produced by suspension polymerization of vinylidene chloride, acrylonitrile, and methylmethacrylate as monomers and isobutane as a blowing agent.[16] The maximum expansion volume was 25 times of the original volume at about 111-120°C. The blowing agent content in microspheres was ~21.5 wt%.[16] This is possibly the simplest and easy to conduct a mechanism of blowing.[16] Although, it may also be affected by the density of the matrix, which may counteract pressure within microspheres and retard foaming.[16] Also, the use of bulkier blowing agents having a lower diffusion rate through the polymer shell improve expansion rate.[17]

Glass microspheres performed very well in shape memory foams as compared with blown foams and expanded microspheres.[18] This may be due to the fact that glass microspheres may also have reactive surface groups, which can reinforce the matrix.[18]

3.2 PRODUCTION OF GASEOUS PRODUCTS BY CHEMICAL REACTION

In the previous section, the reaction between sodium bicarbonate and acetic acid was one example of a chemical reaction that is further discussed in this section. Baking powder (different than baking soda, which contains only sodium hydrogen carbonate) is one popular example of chemical composition producing gaseous products by chemical reaction. In addition to sodium hydrogen carbonate, it contains an acidic reagent, such as calcium acid phosphate, sodium aluminum phosphate, or potassium bitartrate, which undergo chemical reaction producing carbon dioxide as one of the products of the reaction.

The most widely known polymeric foam in which gaseous product is formed by a chemical reaction is obtained from two-component polyurethanes, which contain at least polyol, isocyanate, catalyst, and water. In these foams, isocyanate performs two roles:
- chain extender and crosslinker
- carbon dioxide generating additive.

$$ONCRNCO + 2H_2O \longrightarrow H_2NRNH_2 + 2CO_2$$

Both functions combined together permit expansion of liquid composition by 10-30 times of the original liquid volume with simultaneous curing due to the chain extension and crosslinking (formed amine groups contribute to fast curing considering that isocyanate is usually partially reacted to form prepolymer). Crosslinking depends on the chemical structure of both polyols (or prepolymers) and isocyanates (two functional components will increase the molecular weight of the polymer chain; three or more functional components will, in addition to the chain extension, contribute to the formation of crosslinks). The reaction of the isocyanate group with the hydroxyl group of water and/or polyol is relatively slow, and that is why it requires specially formulated catalysts, which are selected in such a way that they balance both reactions of foaming and curing, as their rates have to complement each other to produce required expansion rate and required rigidity of polyurethane foam. Rigidity (or elasticity) of polyurethane foams varies in a very broad range, and it can be regulated by the chemical structure and molecular weight of the polyol and crosslinking (the density of hardblocks). The expansion rate of foam is regulated by elements of a catalyst system, which promotes the reaction of isocyanate groups with water. The uniformity of cellular structure is tailored with the help of surfactants (usually silicone products).

The method of foaming by reaction with water is considered relatively environmentally friendly because it produces only carbon dioxide (although carbon dioxide contributes to global warming), but it is quite expensive because isocyanates (and catalysts), which are expensive, are used for the production of carbon dioxide (or gaseous foaming product). For this reason, most polyurethane foams are now produced either by a combination of water and liquid blowing agents or (more likely) by liquid blowing agents. Typical blowing agents for the production of PU foam are propane, isobutane, and dimethyl ether. These are, for example, used in very popular "spray foams." It is quite obvious that the process is very dangerous because blowing agents are highly flammable materials. They also have a higher global warming potential than carbon dioxide.

This brief introduction to polyurethane foams is not expanded any further since it is a topic for a separate monographic source only partially connected with the subject of this book.

3.3 FOAMING BY GASES AND EVAPORATING LIQUIDS

Carbon dioxide and four groups of liquids (hydrocarbons, hydrofluorocarbons, methyl formate, and water) are discussed in this section as they are a few remaining blowing agents from scores used a few decades ago. The selection of these is based on two environmental factors: global warming potential and ozone depletion potential. All these groups of products are accepted according to the Montreal and Kyoto agreements. These materials do not cause ozone depletion and have a limited effect on global warming. Carbon dioxide is a reference material of global warming having global warming potential equal unity, and only hydrocarbons have higher global warming potential (mostly 3). Carbon dioxide used in these processes (unlike in chemical blowing and reactive blowing, discussed in the previous sections) does not increase global warming because it is taken from the existing resources of carbon dioxide and returned back (unlike formation in the processes discussed in the previous two sections).

We will begin this discussion with processes conducted with the use of carbon dioxide. The carbon dioxide can be used in one of the three forms: gas, liquid, or solid. In an extrusion foaming process using gaseous carbon dioxide, the polymer is first completely melted to gain good liquidity.[19] Figure 3.7 shows a schematic diagram of an extruder adapted to the production of foam.[19] The required amount of carbon dioxide is injected

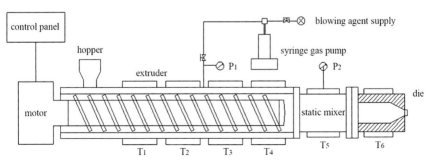

Figure 3.7. Schematic diagram of extruder adapted to foaming with gaseous carbon dioxide. [Adapted, by permission, from Fan, C; Wan, C; Gao, F; Huang, C; Xi, Z; Hu, Z; Zhao, L; Liu, T, *J. Cellular Plast.*, **53**, 3, 277-98, 2016.]

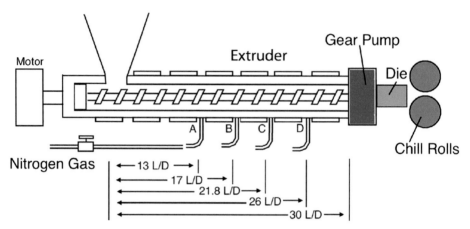

Figure 3.8. Extrusion foaming process utilizing four different blowing agent injection locations. [Adapted, by permission, from Ghandi, A; Bhatnagar, N, *Polym. Plast. Technol. Eng.*, **54**, 1812-8, 2015.]

into the mix with relatively high pressure.[19] A strong shear mixing is then applied by a well-designed screw.[19] This mixing should cause big gas bubbles to break up into a number of tiny bubbles.[19] Subsequently, the homogeneous blend of polymer melt and gas should be achieved through diffusion and dissolution of gas.[19] At the die exit, the high pressure drop induces thermodynamic instability, which causes bubbles to nucleate and grow.[19] On cooling, the bubbles gradually form a stable structure when the pressure inside the bubble equilibrates to the atmospheric pressure.[19] The rheological properties play in this process very important roles, including:[19]
- the shear viscosity of the polymer melt determines the foaming extrusion process inside the extruder
- the shear and extension flow determine the melt strength, help to maintain high pressure, and avoid premature nucleation in the die
- depending on the rheological properties of melt, the pores will either grow or burst
- elasticity contributes to the pore deformation during growth (too strong elasticity limits the pore growth)
- the extensional behavior of the polymer melt, specifically the strain hardening contributes to the stability of pore structure; the strain hardening behavior is the key factor to achieve good foaming material.

Based on the above points, the polymers suitable for extrusion foaming acquire moderate viscosity, favorable elasticity, and the presence of strain hardening behavior in a reasonable range (i.e., high melt strength).[19]

The presence of carbon dioxide helps to increase crystallinity. as it was demonstrated in the case of poly(lactic acid).[20] Crystallinity increase was observed during carbon dioxide saturation, foaming, and annealing processes, and the maximum crystallinity of 38.2% was achieved in the PLA foams.[20]

In the study on the effect of blowing agent residence time on morphological transformations in extrusion foaming, the extruder was equipped with four blowing agent injection location (Figure 3.8).[21] The injection location providing higher gas residence time

3.3 Foaming by gases and evaporating liquids

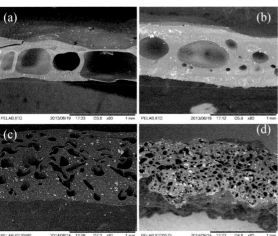

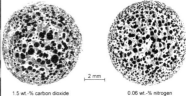

Figure 3.10. EPDM profile cross-sections with carbon dioxide and nitrogen as blowing agents. [Adapted, by permission, from Michaeli, W; Westermann, K; Sitz, S, *J. Cellular Plast.*, **47**, 5, 483-495, 2011.]

Figure 3.9. Influence of blowing agent injection location on foam morphological attributes; die temperature of 150°C, screw rotational speed of 30 rpm with 0.1 wt% N_2 injected at (a) 26 L/D, (b) 21.8 L/D, (c) 17 L/D, and (d) 13 L/D. [Adapted, by permission, from Ghandi, A; Bhatnagar, N, *Polym. Plast. Technol. Eng.*, **54**, 1812-8, 2015.]

resulted in foams with smaller cell size, higher expansion ratio, and enhanced cell density.[21] Figure 3.9 explicitly demonstrates this influence.[21] The longer residence times of blowing agent enhanced interaction between gas molecules and the polymer matrix.[21] Higher concentrations of blowing agent and lower residence times resulted in foams with big pores. The longer residence times resulted in the fine-cell microstructure, higher expansion ratio, and higher cell density.[21] Higher shear resulted in the intense cell nucleation producing foams with smaller cells, higher cell density, and large expansion ratio.[21]

Figure 3.10 shows that foam structures of the nitrogen-produced extrudates are considerably more homogeneous as compared to the application of carbon dioxide.[22] Nitrogen-blown foams have 21 cells/mm^3 compared with 13 cells/mm^3 for carbon dioxide.[22] It is explained by a higher diffusivity of carbon dioxide in EPDM rubber.[22] The mass transport of nitrogen through the matrix, therefore, causes that the desorbing gas diffuses faster into existing cells.[22] The diffusion characteristics of the blowing agent affect the nucleation and the bubble growth.[22]

Foam structure was regulated by the use of alcohol as a co-blowing agent for supercritical CO_2 foaming polystyrene.[23] Solubility parameter of blowing agent and the diffusion coefficient of CO_2 increased with the growth of chain length of alcohol.[23] Decanol decelerated the desorption rate of the blowing agents to maintain a higher content of blowing agents in PS, which resulted in high cell density.[23] Butanol improved the volume expansion ratio of the foams due to plasticization effect on PS.[23] By combining the advantages of these two alcohols as the co-blowing agents, an optimized foam structure was obtained with a smaller cell size (5.82 μm) and larger volume expansion ratio than that obtained when foaming with pure CO_2.[23]

The porosity, pore size distribution, and density of the poly(ether imide) foams were affected by pressure drop, pressure loss rate, and temperature at the die.[24] Significant increases in porosity and pore sizes and decreases in density were observed when pressure

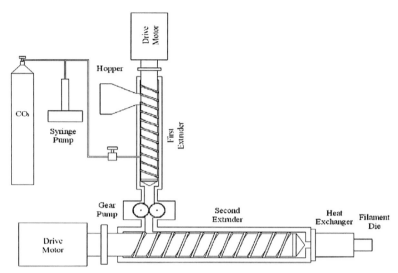

Figure 3.11. Tandem extrusion system. [Adapted, by permission, from Wang, K; Wu, F; Zhai, W; Zheng, W, J. Appl. Polym. Sci., 129, 2253-60, 2013.]

imposed on CO_2 became greater than the critical pressure values for CO_2 (temperature was always greater than the critical temperature of CO_2 in the extruder and the die).[24]

In another study, nitrogen lead to smaller cell sizes as compared to supercritical carbon dioxide.[25] But, if the nucleating agent and the amount of filler were increased, the quality of the foam sheets was worse with nitrogen.[25] This was because of the higher degassing pressure of nitrogen compared to carbon dioxide.[25]

Ultra-light, heat-resistant, flexible, and thermally insulating graphene-fluororubber foam was prepared using supercritical N_2 as a blowing agent.[26] Fluororubber foam had closed-cell structure as general rubber foam, and density of 9 mg/cm^3.[26] Thermal conductivity in the range of 0.020-0.032 W/mK meets requirements for thermal insulation applications.[26]

Compared with the use of gases (carbon dioxide and nitrogen), the supercritical carbon dioxide is by far more popular as a blowing agent. The batch foaming process is common because it can produce foams with small size and high cell density.[27] Two types of batch foaming processes are used: temperature-induced and pressure-induced phase separation.[27] In the case of the temperature-induced process, the material is heated just above the melting point after saturation, the near-surface region melts quickly and stays molten, whereas the interior remains intact, thus leading to nonuniform cells in the foams.[27] In pressure-induced phase separation, the material in the high-pressure vessel is at the polymer-softening temperature range, which is narrow for many polymers, especially polypropylene.[27] In the modified process, the polymer was saturated with supercritical carbon dioxide at high temperature to disrupt crystalline regions.[27] Then the temperature was decreased to the foaming temperature, and the system was depressurized to obtain PP foam.[27]

The continuous extrusion process using supercritical CO_2 as the blowing agent was used for foaming of linear polypropylene (Figure 3.11).[28] The first extruder is for plasti-

3.3 Foaming by gases and evaporating liquids

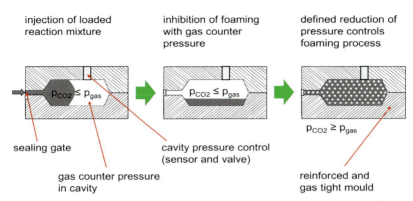

Figure 3.12. Principle of using gas counter pressure for molded PU foaming with high amounts of CO_2 as a physical blowing agent. [Adapted, by permission, from Hopmann, C; Latz, S, *Polymer*, **56**, 29-36, 2015.]

cizing the polymer resin and the second extruder to provide mixing and cooling in order to completely dissolve carbon dioxide in the polymer melt.[28] In this process, polytetrafluoroethylene was blended with polypropylene to improve polypropylene foaming behavior.[28] The presence of PTFE improved cell morphology of PP foams and broadened the foaming window of PP.[28]

Gas saturation and desorption studies revealed that the diffusion of CO_2 in PPS/PES blends was hindered by the crystal barriers in the semi-crystalline PPS component.[29] Higher saturation pressure caused the production of smaller cells and increased cell density in the microcellular PPS/PES blends due to the higher gas concentration.[29] This study showed that the crystalline structure could be used to regulate the morphology of a foam.[29]

The modified spherical ordered mesoporous silica particles were dispersed in the polystyrene matrix.[30] The porous structure and high specific area of the spherical ordered mesoporous silica particles made them good reservoirs for supercritical carbon dioxide.[30] The heterogeneous nucleation effect of spherical ordered mesoporous silica particles was more significant when the foaming temperature and saturation pressure were low.[30]

Gas counter pressure was used to improve the flexibility of foams with high amounts of carbon dioxide.[31] A gas counter-pressure inside the mold cavity enables precise control of the physical foaming process (see Figure 3.12).[31] This foaming technology offers a new degree of freedom in adjusting the hardness of lightweight, flexible foams using an environmentally friendly process.[31]

The bubble size and the bubble density can be effectively affected by the concentration of a dissolved CO_2 and backpressure in foaming injection molding.[32] A lower nucleation energy barrier, faster supersaturation, and high nucleation rate result in smaller bubble size and higher bubble density when CO_2 concentration is increased.[32] The small cell size with high bubble density can be obtained by pressure drop rate controlled by the setting of backpressure.[32] A smaller bubble size and a higher bubble density can be achieved with low mold temperature and lower injection speed.[32] Nano-$CaCO_3$ provides good nucleating properties, reducing cell size, and increasing cell density.[32]

Supercritical fluid-laden pellet injection molding foaming technology, SIFT lowers equipment costs without sacrificing the production rate.[33] SIFT is a good candidate for the mass production of foamed injection molded parts.[33] Both N_2 and CO_2 can be used in this

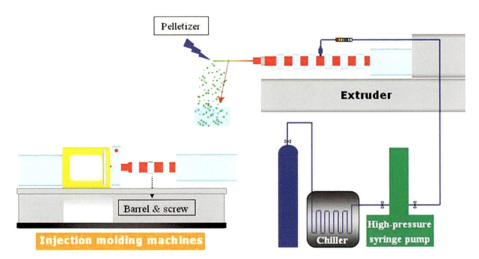

Figure 3.13. The SIFT extrusion/injection molding process. [Adapted, by permission, from Sun, X; Turng, L-S, *Polym. Eng. Sci.*, **84**, 899-913, 2014.]

process as the physical blowing agents.[33] The gas-laden pellets are produced in a single-screw extruder (Figure 3.13).[33] High precision pump maintains the precise injection of liquid gas.[33] This process can be performed with a co-blowing agent.[33] Using this method, the foam cell morphology for low-density polyethylene, polypropylene, and high-impact polystyrene is significantly improved.[33] The cell morphology and glass fiber orientation in foamed parts were influenced by cooling and shear effects.[34]

A computer simulation for rigid polyurethane foaming reactions was compared with experimental data for six blowing agents, including methyl formate and C5-C6 hydrocarbons.[35] Evaporation of the blowing agent was modeled as an overall mass transfer coefficient multiplied by the difference in activity of the blowing agent in the gas foam cells influenced by resin walls of cells.[35]

Co-blowing agents n-pentane and/or cyclopentane were used in CO_2 batch foaming.[36] The agents increased foam porosity and induced a bimodal cell size distribution.[36] It was possible to foam at near room temperature, and porosities of almost 80% were reached while maintaining cell diameter <10 μm.[36]

Hydrocarbons are injected into the extruder downstream, where they are incorporated into the melt.[37] In a single screw extruder, a two-stage screw is used, and the hydrocarbon liquid is injected through the vacuum vent.[37] In a twin-screw extruder, the liquid is injected downstream after a good melt seal on the screw typically formed by a rearward conveying element.[37] Good ventilation is required to remove hydrocarbon gas that escapes from the foamed structure as it expands.[37] In addition to ventilation at the end of the extruder, foamed material continues to give off hydrocarbons for a few days after the extrusion is completed.[37] Hydrocarbon gases are flammable, and safety precautions are of paramount importance.[37] The use of hydrocarbons injected into the extruder provides parts having a lower density.[37]

The solubilities of hydrofluorocarbons, butane, and isobutane in molten low-density polyethylene were measured using a volumetric method at pressures of up to 3.4 MPa and

3.3 Foaming by gases and evaporating liquids

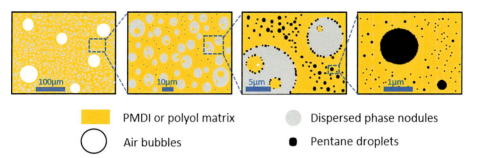

Figure 3.14. Morphology of bubble formation during foaming with hydrocarbons. [Adapted by permission, from Reignier, J; Alcouffe, P; Méchin, F; Fenouillot, F, *J. Colloid Interface Sci.*, **552**, 153-65, 2019.]

at a temperature range of 383.15-473.15K.[38] Their solubilities increased with decreasing temperature and increasing pressure.[38] Liquid blowing agents became excellent foaming agents when they had good solubility in the polymer matrix and high diffusion coefficient.[39]

The morphological study of the growing polyurethane rigid foams during the very first seconds of the process was performed by cryogenic scanning electron microscopy (cryo-SEM) performed at −150°C.[40] Sub-micron size physical blowing agent droplets (isopentane), μm size dispersed phase nodules, and large air bubbles dispersed in a continuous matrix were present in the initial mixture (Figure 3.14).[40]

The isopentane liquid droplets (undissolved part of the physical blowing agent) did not vaporize to create their own bubbles.[40] Isopentane molecules initially dissolved in the continuous phase diffused into the pre-existing air bubbles with no energy barrier to overcome (non-classical nucleation) whereas the isopentane droplets simply acted like reservoirs.[40]

Figure 3.15 shows a line for the production of extruded large-cell polystyrene foam in a continuous process using zero-ozone depletion potential hydrofluorocarbon.[41] Production of large cells requires fewer nucleation sites.[41] With fewer nucleation sites, slower overall expansion occurs.[41] This causes that there is[41]

- more dissolved gas available for cell expansion due to their fewer numbers
- longer diffusion distances between the nucleated cells

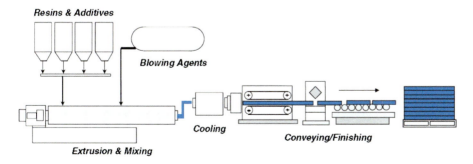

Figure 3.15. Continuous extrusion foaming for Styrofoam using hydrofluorocarbon as blowing agent. [Adapted, by permission, from Fox, R; Frankowski, D; Alcott, J; Beaudoin, D; Hood, L, *J. Cellular Plast.*, **49**, 4, 335-49, 2013.]

- extended plasticization/T_g depression because of the longer residence of the solubilized blowing agent within the polymer.

The majority of soluble blowing agents possess physical characteristics for the large-cell generation, but excessive solubility at processing temperatures can subsequently result in degraded mechanical foam behavior.[41] The 1,1,1,2-tetrafluoroethane, R-134a, was found to be the most suitable for this application.[41]

The 1,3,3,3-tetrafluoropropene, HFO 1234ze, was developed as a "fourth generation" hydrofluorocarbon to replace R-134a as a blowing agent for foam and aerosol applications because it is a low global warming blowing agent. Talc as nucleating agent together with HFO 1234ze was used in foam extrusion of thermoplastic cellulose acetate resulting in homogeneous fine foam morphologies with closed cells.[42] Depending on the blowing agent content and talc content, average cell size ranges from 1 to 0.12 mm, and foam density ranges between 100 and 400 kg/m^3.[42]

Water would be an ideal choice from an environmental perspective, but its solubility in polystyrene is low, and the amount, which can be employed prior to the onset of pinholes and non-uniform or open cells, is minimal.[41]

Extrusion foaming using water as a blowing agent offers a green process.[43] But because of low water solubility in hydrophobic polymers such as polypropylene, it is difficult to produce low-density foams without the addition of an effective water carrier.[43] Thermoplastic starch was added as water carrier to polypropylene to produce open cell foam.[43] The foaming process can be divided into three stages: nucleation, cell growth, and solidification.[43] With starch, the polypropylene forms immiscible polymer blends. The blend morphology before foaming influences the foaming process.[43] The typical sea-island morphology of the blend initially formed in extruder changes into short starch fibers under the shear and elongation flow during the extrusion process.[43] Nucleation starts at the interface between starch and PP or occasionally within the starch domains.[43] Then, cell growth proceeds rapidly when the blowing agent (water vapor) diffuses into the nucleus.[43] The process ends with the shrinkage of extrudate and solidification of the PP matrix when the melt temperature decreases below the crystallization point of PP.[43] Figure 3.16 shows the schematic diagram of the stages.

Morphology

Nucleation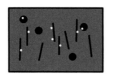

Cell growth

Cell coalescence/ stabilization

Figure 3.16. Schematic diagram of expansion mechanism of propylene matrix containing thermoplastic starch as a water carrier. [Adapted, by permission, from Xu, M Z; Bian, J J; Han, C Y; Dong, L S, *Macromol. Mater. Eng.*, **301**, 149-59, 2016.]

Porous bionanocomposite based on halloysite nanotubes as nanofiller and plasticized starch as the polymeric matrix was prepared by melt-extrusion.[44] Foaming was performed by water as a natural blowing agent, and the increase of the die temperature.[44] A high water content helps to reduce the foaming temperature, and halloysite acts as a nucleating agent and barrier agent, increasing the number of small cells.[44]

The vapor-foamed microcellular injection molding process was used to fabricate polycarbonate parts using water as the blowing agent.[45] Active carbon, a powder obtained from charcoal, was used as a nucleating and reinforcing agent.[45] Obtained foam had a small cell size of about 50 μm.[45] The molecular weight and polydispersity of polycarbonate showed minor material degradation, likely due to the short contact time between PC and water in the machine barrel.[45]

The use of glycerol as hydroxyl component of polyurethane synthesis and cellulose micro/nanocrystals as reinforcement increased water absorption of the polyurethane system, which made it possible to use water as a blowing agent.[46]

In one technological process, poly(vinyl alcohol) acts as a foaming matrix and macromolecular carbon source, melamine phosphate acts as a gas source, acid source, and heterogeneous nucleating agent, and water acts as a plasticizer, blowing agent, and "synergistic flame retardant".[47]

The relationship between phase transition and foaming behavior of starch-based material was established by investigating the effects of water content on melting temperature, crystallinity, foaming process, cell structure, as well as mechanical properties.[48] There was a critical point of water content (between 16-18%) at which expansion changed significantly since the cell structure changed from open to closed.[48] The lower the water content, the higher the T_m of starch-based materials, causing lower melting strength and open cell structure.[48] Conversely, higher water content and lower T_m resulted in higher melting strength and closed-cell structure.[48] The closed-cell structure prevented moisture evaporation during foaming and lowered T_m, which caused foam shrinking when the material was cooled down.[48] In the open-cell foam, the water evaporated during the foaming process, resulting in higher T_m and stable rigid foam structure, with a unit density of about 26 kg/cm^3.[48] Water formed hydrogen bonds with hydroxyl chains of starch.[48] The bound water was a good plasticizer but an inefficient blowing agent for preparing starch-based foams.[48]

The isocyanate-functionalized silica nanoparticles were chemically incorporated into the polyurethane during the synthesis of flexible PU foam from polypropylene glycol and toluene diisocyanate by a one-component method using water as the blowing agent.[49] Shape memory foam can be produced by this combination of materials.[49]

Water with dissolved salt was fed through the hopper of an injection molding machine at a preset rate and mixed with polycarbonate in the machine barrel.[50] The salt crystals of 10-20 μm recrystallized during molding acted as nucleating agents in the PC foamed parts.[50]

Water bound to carrying substances, such as silica or carbon black, was used as a blowing agent.[51] A stable, reproducible foaming process was established with this application.[51]

REFERENCES

1   Bhatti, A S; Dollimore, D; Goddard, R J; O'Donnell, G, **Thermochim. Acta**, 76, 273-86, 1984.
2   Michałowski, S; Prociak, A; Zajchowski, S; Tomaszewska, J; Mirowski, J, *Polym. Testing*, **64**, 229-234, 2017.
3   Robledo-Ortiz, J R; Zepeda, C; Gomez, C; Rodrigue, D; Gonzalez-Nunez, R, *Polym. Testing*, **27**, 730-5,

2008.
4   Krutko, I; Danylo, I; Kaulin, V, *Voprosy khimii i khimicheskoi tekhnologii*, **2019**, 1, 26-34, 2019.
5   Garbacz, T; Palutkiewicz, P, *Cellular Polym.*, **34**, 4, 189-214, 2015.
6   Kim, J-H; Kim, G-H, *J. Appl. Polym. Sci.*, **131**, 40894, 2014.
7   Tissandier, C; Gonzalez-Nunez, R; Rodrigue, D, *J. Cellular Plast.*, **50**, 5, 449-73, 2014.
8   Saiz-Arroyo, C; Rodriguez-Perez, M A; Tirado, J; Lopez-Gil, A; de Saja, J A, *Polym. Int.*, **62**, 1324-33, 2013.
9   Zoller, A; Marcilla, A, *J. Appl. Polym. Sci.*, **121**, 1495-1505, 2011.
10 Zoller, A; Marcilla, A, *J. Appl. Polym. Sci.*, **128**, 354-62, 2013.
11 Guan, L T; Du, F G; Wang, G Z; Chen, Y K; Xiao, M; Wang, S J; Meng, Y Z, *J. Polym. Res.*, **14**, 245-51, 2007.
12 Restrepo-Zapata, N C; Osswald, T A; Hernandez-Ortiz, J P, *Polym. Eng. Sci.*, **55**, 2073-88, 2015.
13 Coste, G; Negrell, C, Caillol, S, *Eur. Polym. J.*, **140**, 110029, 2020.
14 Wan Hamad, W N F; Teh, P L; Yoeh, C K, *Polym. Plat Technol. Eng.*, **52**, 754-60, 2013.
15 Fauzi, M S; Lan, D N U; Osman, H; Ghani, S A, *Sains Malaysiana*, **44**, 6, 869-74, 2015.
16 Hou, Z; Xia, Y; Qu, W; Kan, C, *Int. J. Polym. Mater. Polym. Biomater.*, **64**, 427-31 2015.
17 Jonsson, M; Nordin, O; Larson Kron, A; Malmstrom, E, *J. Appl. Polym. Sci.*, **117**, 384-92, 2010.
18 Ellson, G; Di Prima, M; Ware, T; Tang, X; Voit, W, *Smart Mater. Struct.*, **24**, 055001, 2015.
19 Fan, C; Wan, C; Gao, F; Huang, C; Xi, Z; Hu, Z; Zhao, L; Liu, T, *J. Cellular Plast.*, **53**, 3, 277-98, 2016.
20 Ji, G; Wang, J; Zhai, W; Lin, D; Zheng, W, *J. Cellular Plast.*, **49**, 2, 101-17, 2013.
21 Ghandi, A; Bhatnagar, N, *Polym. Plast. Technol. Eng.*, **54**, 1812-8, 2015.
22 Michaeli, W; Westermann, K; Sitz, S, *J. Cellular Plast.*, **47**, 5, 483-495, 2011.
23 Qiang, W; Zhao, L; Liu, T; Gao, X; Hu, D, *J. Supercritical Fluids*, **158**, 104718, 2020.
24 Aktas, S; Gevgilili, H; Kucuk, I; Sunol, A; Kalyon, D M, *Polym. Eng. Sci.*, **54**, 2064-74, 2014.
25 Geissler, B; Feuchter, M; Laske, S; Fasching, M; Holzer, C; Langecker, G R, *J. Cellular Plast.*, **52**, 1, 15-35, 2016.
26 Zhang, Z X; Wang, Y M; Phule, A D, *Colloids Surf. A: Physicochem. Eng. Aspects*, **604**, 125310, 2020.
27 Ding, J; Ma, W; Song, F; Zhong, Q, *J. Appl. Polym. Sci.*, **130**, 2877-85, 2013.
28 Wang, K; Wu, F; Zhai, W; Zheng, W, *J. Appl. Polym. Sci.*, **129**, 2253-60, 2013.
29 Ma, Z; Zhang, G; Shi, X; Yang, Q; Li, J; Liu, Y; Fan, X, *J. Appl. Polym. Sci.*, **132**, 42634, 2015.
30 Yang, J; Huang, L; Zhang, Y; Chen, F; Zhong, M, *J. Appl. Polym. Sci.*, **130**, 4308-17, 2013.
31 Hopmann, C; Latz, S, *Polymer*, **56**, 29-36, 2015.
32 Xi, Z; Chen, J; Liu, T; Zhao, L; Turng, L-S, *Chinese J. Chem. Eng.*, **24**, 180-9, 2016.
33 Sun, X; Turng, L-S, *Polym. Eng. Sci.*, **84**, 899-913, 2014.
34 Xi, Z; Sha, X; Liu, T; Zhao, L, *J. Cellular Plast.*, 50, 5, 489-505, 2014.
35 Al-Moameri, H; Zhao, Y; Ghoreishi, R; Suppes, G J, *J. Appl. Polym. Sci.*, **132**, 42454, 2015.
36 Nistor, A, Sovova, H; Kosek, J, *J. Supercritical Fluids*, **130**, 30-9, 2017.
37 Wagner, J R; Mount, E M; Giles, H F, **Foam Extrusion. Extrusion**. 2nd Ed., *WilliamAndrew*, 2014, pp. 603-7.
38 Wang, M; Sato, Y; Iketani, T; Takashima, S; Masuoka, H; Watanabe, T; Fukasawa, Y, *Fluid Phase Equilibria*, **232**, 1-2, 1-8, 2005.
39 Vo, C V; Fox, R T, *J. Cellular Plast.*, **49**, 5, 423-38, 2013.
40 Reignier, J; Alcouffe, P; Méchin, F; Fenouillot, F, *J. Colloid Interface Sci.*, **552**, 153-65, 2019.
41 Fox, R; Frankowski, D; Alcott, J; Beaudoin, F; Hood, L, *J. Cellular Plast.*, **49**, 4, 335-49, 2013.
42 Zepnik, S; Hendriks, S; Radusch, H-J, *J. Mater. Res.*, **28**, 17, 2013.
43 Xu, M Z; Bian, J J; Han, C Y; Dong, L S, *Macromol. Mater. Eng.*, **301**, 149-59, 2016.
44 Schmitt, H; Creton, N; Prashantha, K; Soulestin, J; Lacrampe, M-F; Krawczak, P, *J. Appl. Polym. Sci.*, **132**, 41341, 2015.
45 Peng, J; Sun, X; Mi, H; Jing, X; Peng, X-F; Turng, L-S, *Polym. Eng. Sci.*, **55**, 1634-42, 2015.
46 Mosiewicki, M A; Rojek, P; Michalowski, S; Aranguren, M I; Prociak, A, *J. Appl. Polym. Sci.*, **132**, 41602, 2015.
47 Guo, D; Bai, S; Wang, Q, *J. Cellular Plast.*, **51**, 2, 145-63, 2015.
48 Meng, L; Liu, H; Lin, X, *Ind. Crops Products*, **134**, 43-9, 2019.
49 Kang, S M; Kim, M J; Kwon, S M; Park, H; Jeong, H M; Kim, B K, *J. Mater. Res.*, **27**, 22, 2837-43, 2012.
50 Peng, J; Turng, L-S; Peng, X-F, *Polym. Eng. Sci.*, **52**, 1464-73, 2012.
51 Hopmann, C; Lemke, F; Binh, Q N, *J. Appl. Polym. Sci.*, **133**, 43613, 2016.

4

DISPERSION AND SOLUBILITY OF BLOWING AGENTS

Solid, particulate blowing agents are not easy (and sometimes difficult) to completely disperse in a liquid or paste to form plastisol with vinyl chloride homopolymers, copolymers, or their mixtures.[1] The powdered blowing agents contain particles in the range of 0.1 to 30 μm in diameter.[1] The particles tend to agglomerate due to moisture and static charges.[1] Different size of particles and formation of agglomerates are known to affect the rate of decomposition of the solid blowing agents.[1] The powdered blowing agent can be directly mixed with a liquid or paste, but such paste usually contains a substantial amount of agglomerated, undispersed particles, which produce voids and large gas pockets after the polymer is expanded.[1] This phenomenon is particularly pronounced with azodicarbonamide in vinyl plastisols, but it is also representative of other blowing agents.[1]

The common practice in the production of foamed vinyl products is to premix blowing agent(s) with a portion of plasticizer of plastisol formulation or to use commercially available premixes or masterbatches.[1] Such pastes usually contain 30 to 60 wt% of a blowing agent and are made by blending the blowing agent and a plasticizer in a mixer and milling the resultant paste using a paint mill, ball mill, or similar equipment.[1] The resultant premix should prevent the settling of a blowing agent.[1] The premix is then dispersed in plastisol.[1] An extra compounding step is a disadvantage of the method, which increases production cost.[1] The alternative, namely the use of commercial masterbatches, also increases production cost and restricts technology by the use of what is available rather than what is the best for a particular formulation.[1]

Addition of some inorganic substances (1-5 wt%), such as aluminum oxide, silica aluminate, hydrated magnesium silicate, calcium silicate, silica gel, fumed silica, magnesium oxide, titanium dioxide, dibasic calcium phosphate, tribasic calcium phosphate, sodium silica aluminate, potassium silica aluminate, calcium silica aluminate, diatomaceous silica, sodium silicate, potassium silicate and calcium silicate with dispersing aid helps to disperse blowing agent directly in the formulation.[1] This method is suitable for solid blowing agents, such as azodicarbonamide, p,p'-oxybis(benzene sulfonyl hydrazide), p-toluene sulfonyl hydrazide, and p-toluene sulfonyl semicarbazide.[1]

In the more contemporary literature, the discussion of dispersion of particulate blowing agents is not a part of open literature (patents or published papers), but infrequently, some information is given on particle size and distribution of azodicarbonamide. This suggests that the problem is mostly addressed on the level of manufacturing technology of a blowing agent.

It is known that when the particle size of a blowing agent decreases, its surface area increases, which improves its interaction with additives such as decomposition catalysts.[2]

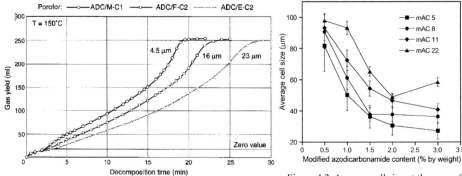

Figure 4.1. The influence of the particle size of different grades of azodicarbonamide on their decomposition rates. [Adapted, by permission, from Quinn, S, *Plast. Addit. Compound.*, **2001**, May, 16-21, 2001.]

Figure 4.2. Average cell size at the core of PVC/rice hull composites, foamed with various contents and particle sizes of modified azodicarbonamide. [Adapted, by permission, from Petchwattana, N; Covavisaruch, S, *Mater. Design*, **32**, 2844-50, 2011.]

Figure 4.1 shows the effect of decomposition time on gas yield for different grades of azodicarbonamide.[2] The finer the grade, the faster the decomposition rate.[2] The decomposition rate increases with temperature.[2] Once decomposition has been initiated, the reaction becomes exothermic and, as such, self-catalyzing.[2] Coarse grades do offer some advantages, such as powder flow, conveying performance, and reduction of dust.[2] Pre-dispersing is mostly relevant in the case of PVC plastisols, where the blowing agent is dispersed within a plasticizer.[2] The ratio of plasticizer to blowing agent is usually 1 to 1, but the correct viscosity is important.[2] If the viscosity is low, the premix is easier to disperse, but the blowing agent may settle to the base of the container during storage, meaning that agitation or stirring may be required before its use.[2]

Figure 4.2 shows that the cell size depends on the amount of azodicarbonamide and its particle size. The more azodicarbonamide, the smaller the cells; the smaller the size of the particles of azodicarbonamide, the smaller the cell sizes.[3]

It appears from the limited literature that studies on improvement of particle size are still continued. German company Tramaco GmbH launched some years ago Unicell D 200 LK to improve particle size of azodicarbonamide grade that it can provide a combination of high reactivity and optimum dispersion properties.[4]

There are still some cases in which the dispersion of solid additives impacts the results of foaming. Dispersion of exfoliated layered-silicate in the polystyrene block copolymer influenced cell size when supercritical carbon dioxide was used as a blowing agent.[5] Only exfoliated layered-silicate inhibited the cell expansion and had high nucleation efficiency during foaming.[5] The average cell diameter was reduced from 6 to 1.4 μm, and the cell density was increased from 7.6×10^9 to 5.0×10^{11} cells/cm^3.[5] The aggregated layered-silicate polystyrene did not show any effect on the cell morphology in PS foam.[5]

Foaming poly(3-hydroxybutyrate-co-3-hydroxyvalerate) was affected by organoclays in a continuous supercritical carbon dioxide-assisted extrusion process.[6] Preparation of premix by twin-screw extrusion and its dilution during the foaming process improved clay dispersion without extensive thermal degradation of PHBV.[6] A good clay dispersion

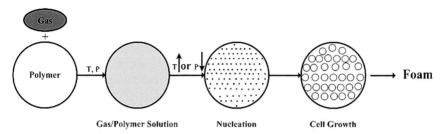

Figure 4.3. Foaming process by supercritical carbon dioxide. [Adapted, by permission, from Xu, Z-M; Jiang, X-L; Liu, T; Hu, G-H; Zhao, L; Zhu, Z-N; Yuan, W-K, *J. Supercritical Fluids*, **41**, 299-300, 2007.]

appears to favor homogeneous nucleation while limiting the coalescence of cells and hence renders foams having better homogeneity and higher porosity by up to 50%.[6] Similar observations were made for Cloisite 30B (nanoclay) in polylactide foamed by supercritical carbon dioxide.[9]

The solubility of blowing agents is in play in the case of gaseous and liquid blowing agents. Figure 4.3 outlines stages of the foaming process by the gaseous blowing agents.[7] There are four steps involved in a foaming process:[7]

1. gas dissolution – the dissolution of gas in a polymer to form a polymer/gas solution
2. cell nucleation – driven by thermodynamic instability as a result of an increase in temperature or a decrease in pressure; phase separation takes place between the polymer and gas to form cell nuclei
3. cell growth – a combination of mass and heat transfer
4. cell stabilization – the cell growth process is stopped by a natural or an imposed ending of the driving force for cell growth, and a cellular foam is obtained.

The saturation time, foaming temperature, saturation pressure, and depressurization rate have an effect on the foam structure and volume expansion ratio in polypropylene foamed by supercritical carbon dioxide.[7] The first three parameters control the dissolution of carbon dioxide in the polymer.[7] The upper limit of foaming was determined by the melt strength of polypropylene, while the lower limit was determined by the deformability of polypropylene (Figure 4.4).[7] A lower foaming temperature and a higher saturation pressure were more favorable for the production of a uniform foam than a higher foaming temperature and a lower saturation pressure.[7] The increased depressurization rate led to an increase in cell density.[7]

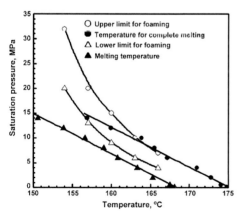

Figure 4.4. Upper and lower limits for the foaming temperature and corresponding saturation pressure. [Adapted, by permission, from Xu, Z-M; Jiang, X-L; Liu, T; Hu, G-H; Zhao, L; Zhu, Z-N; Yuan, W-K, *J. Supercritical Fluids*, **41**, 299-300, 2007.]

If the polymer properties, such as the glass transition temperature and the solubility of CO_2, are known, full control of the desired foam morphology can be obtained by a proper selection of temperature, pres-

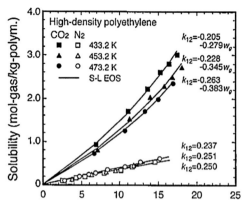

Figure 4.5. Solubility in HDPE of CO_2 and N_2. [Adapted, by permission, from Sauceau, M; Fages, J; Common, A; Nikitine, C; Rodier, E; *Prog. Polym. Sci.*, **36**, 749-66, 2011.]

sure, and depressurization rate.[8] Saturation at higher pressures followed by a fast depressurization leads to a less homogeneous foam morphology, whereas saturation at higher temperatures followed by a fast depressurization leads to larger average cell size and a lower density of the produced foam.[8]

The dissolution of gas leads to a decrease in viscosity of the liquid polymer, melting point, and glass transition temperature.[10] Figure 4.5 shows that carbon dioxide is more soluble in HDPE than nitrogen.[10] Solubility increases with pressure.[10] Carbon dioxide decreases the glass transition temperature and the viscosity of many polymers.[10] The foaming behavior of polystyrene blown with supercritical CO_2-N_2 blends indicates a high nucleating power of supercritical N_2 and a high foam expanding the ability of supercritical CO_2.[11] Combination of both produces a synergistic effect which can be used for regulation of the properties of the foam.[11] The synergy is based on the fact that the presence of supercritical CO_2 increases the solubility of supercritical N_2 in PS, so the concentration of dissolved supercritical N_2 becomes higher than the predicted from the simple mixing rule.[11] The additional supercritical N_2 further increases the cell nucleation performance.[11]

In a similar development, the addition of dimethyl ether to polystyrene not only decreased the pressure required to dissolve carbon dioxide in polystyrene, but plasticization was also observed. The plasticization effect, due to the presence of dimethyl ether, increased the carbon dioxide solubility *via* an increased free volume of polystyrene.[12]

The solubility of HFC-134a, butane, and isobutane in the molten low-density polyethylene has been measured with a volumetric method at pressures up to 3.4 MPa and the temperature in the range of 383.15-473.15K.[13] The solubilities for all liquids in LDPE increased with increasing pressures and decreased with increasing temperature.[13] At 413 K, the order of the solubilities in LDPE was butane > isobutane > HFC-134a, which was similar to the order of normal boiling temperatures of the solutes.[13]

The perturbed chain-statistical associating fluid theory and Sanchez–Lacombe equations of state are used to model the gas solubilities of binary systems composed of polystyrene and hydrofluorocarbons, and supercritical fluids.[14] Solubilities correlated using the perturbed chain-statistical associating fluid theory are in good agreement with the experimental data.[14]

The perturbed-chain version of the statistical fluid theory (PC-SAFT) equation of state was used to study the solubility of 5 blowing agents (CO_2, R134a, R142b, R152a, and R245fa) in 7 homopolymers (HDPE, PS, PMMA, PBMA, PLA, PEG, and PDMS) and 4 random copolymers (SMMA, PLGA, SAN, and poly(ethylene-co-1-octene)).[15] For the majority of polymer-blowing agent systems, the experimental solubility data were well represented by PC-SAFT predictions.[15] The results provided screening and optimizing industrial polymeric foaming processes.[15] Addition of fluorinated gas co-blowing agent enhanced the total solubility of blowing agent in the polymer matrix.[15]

The microcellular foaming of polyphenylsulfone and polysulfone was investigated in supercritical CO_2.[16] The expansion ratio, ranging from 1.10-2.45 for PPSU and 1.10-3.72 for PSU was not sufficient due to limited solubility of CO_2.[16] Porosity of foams was improved by co-blowing agents.[16] Ethanol had a more favorable interaction with CO_2, followed by water, acetone, and ethyl acetate.[16] Ethanol also significantly increased the interaction of CO_2 with polymer chains.[16] The addition of ethanol gave the highest expansion ratio; 5.02 for PPSU and 6.54 for PSU, and broadened the foaming temperature dramatically, decreasing it by 50°C for PPSU and 70°C for PSU.[16]

CO_2 solubility in TPU is one magnitude higher than that of N_2, but TPU foaming was enhanced with N_2 despite its lower solubility compared with CO_2 (cell density was higher with N_2 than it was with CO_2).[17] Diffusivity of both CO_2 and N_2 was promoted, within limits, by increased solubility.[17] The results differ from the results for polypropylene and polystyrene foaming because of the low diffusivity and the low plasticization effect of N_2 in TPU.[17]

Dissolution of both CO_2 and N_2 in PMMA obeyed Henry's law.[18] Solubility data could be fitted using the Sanchez-Lacombe equation of state.[18] For both gases, solubility has a determining effect on diffusivity (Figure 4.6).[18]

The interfacial tension of polylactide decreased with increased pressure.[19] At lower

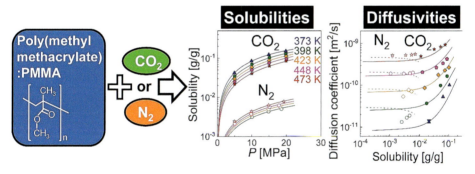

Figure 4.6. Solubility and diffusivity of CO_2 and N_2 in PMMA. [Adapted, by permission from Ushiki, I; Hayashi, S; Kihara, S-i; Takishima, S. *J. Supercritical Fluids*, **152**, 104565, 2019.]

pressure, the interfacial tension decreased with temperature, increasing at a higher pressure.[19] These opposing trends were attributed to two competing CO_2 mechanisms: hydraulic pressure and polymer swelling.[19]

PVC foam obtained by using supercritical CO_2 had a uniform closed-cell structure with thin cell walls (less than 5 μm).[20] At high solubility of supercritical CO_2, PVC foam samples had cell diameters of less than 20 μm and high density (10^8-10^{11} bubbles/cm^3).[20] PVC samples foamed at low solubility conditions displayed cells of large diameter (more than 140 μm) and lower cell density (10^6 bubbles/cm^3).[20]

REFERENCES

1. La, C R; **US3743605**, *Uniroyal Inc.*, Jul. 3, 1973.
2. Quinn, S, *Plast. Addit. Compound.*, **2001**, May, 16-21, 2001.
3. Petchwattana, N; Covavisaruch, S, *Mater. Design*, **32**, 2844-50, 2011.
4. *Addit. Polym.*, **2010**, July, 2, 2010.
5. Zhu, B; Zha, W; Yang, J; Zhang, C; Lee. L J, *Polymer*, **51**, 2177-84, 2010.
6. Le Moigne, N; Sauceau, M; Benyakhlef, M; Jemai, R; Benezet, J-C; Rodier, E; Lopez-Cuesta, J-M; Fages, J, J; Eur. Polym. J., **61**, 157-71, 2014.
7. Xu, Z-M; Jiang, X-L; Liu, T; Hu, G-H; Zhao, L; Zhu, Z-N; Yuan, W-K, *J. Supercritical Fluids*, **41**, 299-300, 2007.
8. Jacobs, L J M; Dannen, K C H; Kemmere, M F; Keurentjes, J T F, *Polymer*, **48**, 3771-80, 2007.
9. Keshtkar, M; Nofar, M; Park, C B; Carreau, P J, *Polymer*, **55**, 4077-90, 2014.
10. Sauceau, M; Fages, J; Common, A; Nikitine, C; Rodier, E; *Prog. Polym. Sci.*, **36**, 749-66, 2011.
11. Wong, A; Mark, L H; Hasan, M M; Park, C B, *J. Supercritical Fluids*, **90**, 35-43, 2014.
12. Mahmood, S H; Xin, C L; Gong, P; Lee, J H; Li, G; Park, C B, *Polymer*, **97**, 95-103, 2016.
13. Wang, M; Sato, Y; Iketani, T; Takishima, S; Masuoka, H; Watanabe, T; Fukasawa, Y, *Fluid Phase Equilibria*, **232**, 1-8, 2005.
14. Arce, P; Aznar, M, *J. Supercritical Fluids*, **45**, 134-45, 2008.
15. Feng, Z; Panuganti, S R; Chapman, W G, *Chem. Eng. Sci.*, **183**, 306-28, 2018.
16. Hu, D-D; Gu, Y; Zhao, L, *J. Supercritical Fluids*, **140**, 21-31, 2018.
17. Li, R; Lee, J H; Park, C B, *J. Supercritical Fluids*, **154**, 104623, 2019
18. Ushiki, I; Hayashi, S; Kihara, S-i; Takishima, S. *J. Supercritical Fluids*, **152**, 104565, 2019.
19. Nofar, M; Park, C B, **Polylactide Foams. Fundamentals, Manufacturing, and Applications**. *William Andrew*, 2018, pp. 35-56.
20. Chuaponpat, N; Areerat, S, *MaterialsToday, Proc.*, **5**, 11, 3, 23526-33, 2018.

5

PARAMETERS OF FOAM PRODUCTION

In this chapter, over 20 different influential factors are discussed which have an impact or characterize the outcome of foaming. There are various practical means used for regulation of the process conditions. Formulation components that can be used for adjustment of the foaming outcome are also discussed. For easier retrieval, all these elements of control are discussed in alphabetical order.

5.1 AMOUNT OF BLOWING AGENT

Figure 5.1 shows that foam density decreases when the amount of blowing agent is increased for all three blowing agents tested.[1] Blowing agent 1 is endothermic, whereas blowing agents 2 and 3 are exothermic.[1] The endothermic agent is composed of sodium bicarbonate, whereas two exothermic blowing agents are based on mixtures of components, including azodicarbonamide.[1] Figure 5.2 shows that the application of the same blowing agent gives a similar character of relationship, but the rates of density changes are affected by the polymer type.[1] The relationship is almost identical for polyolefins but differs in the case of PVC.[1] This is mostly due to the differences in densities of polymers because the average surface areas of pores are quite similar for all polymers.[1]

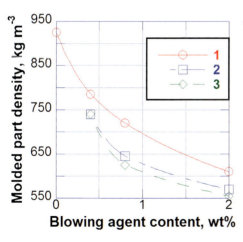

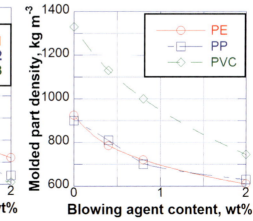

Figure 5.1. Molded part density vs. blowing agent contents for polyethylene containing variable amounts of the following blowing agents 1 – Hostatron P 1941, 2 – Hydrocerol PLC 751, 3 – Adcol blow x 1020. [Data from Garbacz, T; Palutkiewicz, P, *Cellular Polym.*, **34**, 4, 189-214, 2015.]

Figure 5.2. Molded part density vs. contents of blowing agent Hostatron P 1941 used with different polymers, as follows: 1 – polyethylene, 2 – polypropylene, 3 – poly(vinyl chloride). [Data from Garbacz, T; Palutkiewicz, P, *Cellular Polym.*, **34**, 4, 189-214, 2015.]

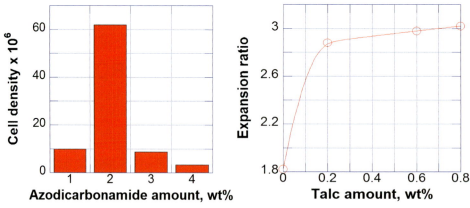

Figure 5.3. Cell density of HIPS foamed with variable amounts of azodicarbonamide. [Data from Zhang, Z X; Li, Y N; Xia, L; Ma, Z G; Xin, Z X; Lim, J K, *Appl. Phys. A*, **117**, 755-9, 2014.]

Figure 5.4. Expansion ratio of cellulose acetate vs. amount of nucleating agent (at constant concentration of blowing agent). [Data from Zepnik, S; Hendriks, S; Kabasci, S; Radusch, H-J, *J. Mater. Res.*, **28**, 17, 2394-2400, 2013.]

Figure 5.3 shows that cell density initially increases with an increase in blowing agent content but then decreases with a further increase of blowing agent concentration to 3 and 4 wt%.[2] Once the nucleated cells are fully grown, cell coalescence and/or collapse may take place with a further increase in the blowing agent content.[2] This causes a decrease in cell density.[2] The diffusion of small bubbles into the greater ones is promoted, and coarse cells are formed.[2] The cell sizes became bigger and distorted at 4 phr blowing agent.[2]

Figure 5.4 shows that the blowing agent is not the only component of formulation that can be useful in the regulation of expansion rate.[3] When the amount of blowing agent is kept constant (2 wt% of hydrofluorocarbon, HFO 1234ze) but the amount of talc is varied, the expansion ratio is affected.[3] Talc, which is a nucleating agent, on small addition, increases the expansion ratio very rapidly.[3] Further increase in concentration has a much smaller effect.[3]

The amounts of blowing agent and nucleating agents influence not only the number of cells, cell density, cell size, and expansion ratio, which are all related parameters, but also the thickness of the skin.[4] With an increase in blowing agent content, also gloss of the injection-molded part decreases, and color becomes lighter.[4]

During bubble growth, large bubbles are surrounded by some fresh small bubbles.[5] The interactions between adjacent large and small bubbles significantly affect polymer foamability.[5] There are two possible scenarios between adjacent large and small bubbles (Figure 5.5).[5] If a small-bubble pressure P_s is greater than that of large-bubble pressure P_b ($P_s > P_b$), the pressure gradient will drive the blowing gas from the small bubble to the large bubble, and the small bubble will shrink and eventually disappear.[5] This is an adverse phenomenon that is called bubble coalescence.[5] If $P_s < P_b$, the pressure gradient will drive the blowing agent from the large bubble to the small bubble so that the small bubble could further grow and eventually survive, which is favorable for polymer foaming.[5] Therefore, the fate of small bubbles is highly dependent on the bubble pressure.[5]

5.1 Amount of blowing agent

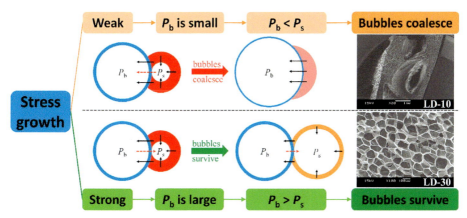

Figure 5.5. Schematic diagram of interactions between adjacent bubbles. [Adapted, by permission, from Zhang, H; Liu, T; Li, B; Xin, Z, *Polymer*, in press 123209, 2020.]

Bubble coalescence mechanism was evaluated for the melt foamability of polymer blends.[5] Linear low-density polyethylene has excellent mechanical properties, but poor foamability, and low-density polyethylene has good foamability.[5] The steep stress growth caused by LDPE addition dramatically increased the pressure in large bubbles and then effectively inhibited the coalescence of adjacent bubbles, thus improving the foamability of LLDPE/LDPE blends.[5] The addition of only 20 wt% of LDPE was sufficient to significantly improve the non-linear behavior under large deformations to the same level as for pure LDPE.[5]

In another study, the addition of carbon nanotubes to PS/LLDPE=50/50 blends led to the decrease of PS dispersed phase size and transformed morphology of blend from bicontinuous to sea-island structure, which was more conducive to preventing rapid desorp-

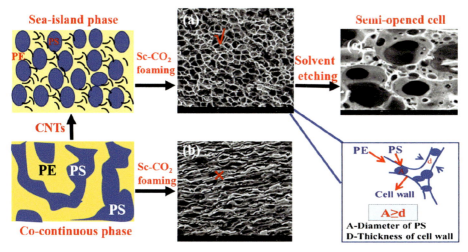

Figure 5.6. Effect of carbon nanotubes and morphology and blowing performance LLDPE/PS=50/50 blends. [Adapted, by permission, from Shi, Z; Zhang, S; Qiu, J, Tang, T, *Polymer*, **207**, 122896, 2020.]

tion and escape of CO_2 from the blend.[6] Figure 5.6 illustrates how it affected blowing outcome and why.[6]

In some processes, the use of high concentrations of foaming agent requires adaptation of equipment.[7] Production of lightweight, flexible foams is such an example.[7] It was not possible to use carbon dioxide for lightweight foams because high amounts of carbon dioxide lead to an uncontrolled foaming process.[7] The physical foaming technology for molded polyurethane foam has been developed at the Institute of Plastics Processing in Aachen, Germany in order to enable high amounts of carbon dioxide (up to 10 wt%) as a blowing agent.[7] A gas counter-pressure inside the mold cavity enables precise control of the physical foaming process.[7]

If large pores were required, a combination of supercritical carbon dioxide and ethyl lactate for foaming of lactide/caprolactone copolymer was used.[8] This combination had a plasticizing effect on copolymer, which was instrumental in the production of lightweight foam.[8] This is only one of the examples (others are scattered throughout this book), which show that synergistic mixtures of blowing agents or blowing agents and plasticizers (or fillers and other additives) permit conducting foaming processes that were not possible with a single component blowing agent.

The above examples show that there is much more to be expected from the concentration of blowing agent to be limited to the obvious effect of producing more gas when its amount is increased. Also, the composition and properties of the matrix have an important effect on the outcome of foam formation.

5.2 CLAMPING PRESSURE

The nanofoams produced from polyetherimide foamed using supercritical carbon dioxide had a void fraction of 25-64% and the average cell size of 40-100 nm.[9] Supercritical carbon dioxide at 20 MPa was used as a blowing agent.[9] The saturated sheets were foamed in a hot press. Sorption studies showed that at 20 MPa, PEI was able to absorb 10 wt% carbon dioxide.[9] An equilibrium concentration can be reached after 100 h at 45°C.[9] The foaming temperature was in the range of 165 to 210°C (glass transition temperature of PEI was 215°C) (Figure 5.7).[9] The clamping force controlled the nanostructure formation of the foam.[9] Figure 5.8 shows the morphology of cells obtained at different clamping pressures.[9]

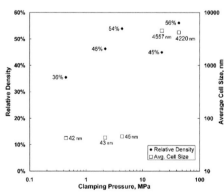

Figure 5.7. The effect of clamping pressure on the relative density of PEI nanofoams and their cell densities. [Adapted, by permission, from Aher, B; Olson, N M; Kumar, V, *J. Mater. Res.*, **28**, 17, 2366-73, 2013.]

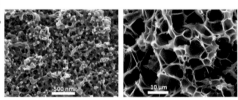

Figure 5.8. PEI foams after foaming for 3 min. at 195°C at clamping pressures of 0.44 MPa (left) and 22 MPa (right). [Adapted, by permission, from Aher, B; Olson, N M; Kumar, V, *J. Mater. Res.*, **28**, 17, 2366-73, 2013.]

5.3 DELAY TIME

Precision mold opening (also called the mold breathing or negative compression process) is gaining ground.[10,11] The mold is filled with melt containing a blowing agent.[10,11] A short delay time is allowed to help in the development of compact skin layers.[10,11] The mold is opened.[10,11] The pressure drop induced by the sudden increase in volume causes the foaming process.[10,11] Parts having compact skin layer and foamed core area with closed cells were produced.[10,11] A delay time of 1 s gave a higher amount of heat conducted from the mold wall and led to an increase in the skin layer thickness.[10,11] The delay time resulted in a thicker skin layer.[10,11]

5.4 DESORPTION TIME

The desorption time – the time the material saturated with a blowing agent (e.g., supercritical carbon dioxide) was left to desorb before it was subjected to expansion.[9] Typical desorption time is 10 to 60 min.[9] Higher rate of internal blistering can be expected at lower desorption times.[9] The internal blisters form around the midplane of the plastic sheet and expand perpendicularly to the clamp pressure.[9] The internal blisters join and separate foamed material into two halves attached only at the edges (at the most extreme cases).[9] All samples foamed after only less than 35 min. of desorption had internal blisters.[9] Samples foamed after more than 35 min. had no internal blisters, but they had thicker transition layers.[9] The skin was not affected by desorption time, remaining between 10 and 20 µm thick.[9] The transition layer grew steadily from 54 µm after 5 min of desorption to 326 µm after 75 min of desorption.[9] This effect was due to the diffusion of carbon dioxide out of the sample.[9] Figure 5.9 shows the effect of desorption time on relative density and thickness of the transition layer.[9]

PLA thin films were produced with a desorption time of 30 s.[12] Density of film was a function of desorption time.[12] The thickness of the solid skin at the surface can be increased by allowing for greater desorption times.[13]

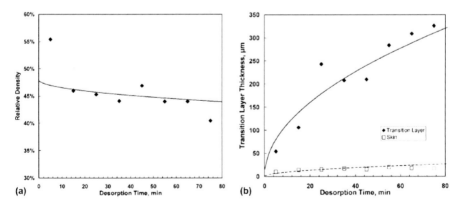

Figure 5.9. (a) Relative density of samples foamed for 3 minutes at 195°C, 2.2 MPa clamping pressure, at various desorption times, and (b) Thickness of microcellular transition layer and skin of samples foamed for 3 minutes at 195°C, 2.2 MPa clamping pressure, at various desorption times. The curves are based on a constant diffusion coefficient and 8.5% carbon dioxide concentration threshold for nanocellular foaming. [Adapted, by permission, from Aher, B; Olson, N M; Kumar, V, *J. Mater. Res.*, **28**, 17, 2366-73, 2013.]

It is difficult to obtain polypropylene foams with fine cellular structure and large volume expansion ratio using supercritical CO_2 because of low solubility and fast desorption rate of CO_2 in the polymer.[14] Consequently, PP foams usually exhibited a large cell size, small volume expansion ratio, and nonuniform cell size distribution.[14] The desorption coefficient of CO_2 in blends with PDMS was smaller (e.g., 2.06×10^{-10} for PP/PDMS blend as compared with 4.48×10^{-10} m^2/s for PP.[14] The desorption diffusion coefficient of CO_2 in PP/PDMS having M_w=4000 exhibited the lowest value due to the highest crystallinity of the polymer matrix (high crystallinity hindered desorption rate).[14]

5.5 DIE PRESSURE

The die pressure is kept higher than the solubility pressure of carbon dioxide that was injected into the polypropylene matrix in the extrusion process.[15] The die pressure depends on the die temperature.[15,16] It decreases with die temperature increasing.[15,16]

In the extrusion foaming of polystyrene, the pressure of carbon dioxide was kept at 6.7-7.6 MPa.[17] The die pressure was in the range of 8.3-10.3 MPa depending on the processing conditions.[17]

When producing a foam product using 5 wt% pentane, the die pressure of 140 bar was maintained.[18] Die pressures are maintained to keep the blowing agent in the solution.[18] A halo appearance at the die exit is desired, indicating that foaming occurs after the die exit.[18] This results in a uniform cell structure and uniform surface appearance.[18] If the die pressure is not sufficiently high to keep the blowing agent in solution, a condition known as pre-foaming occurs.[18] The extrudate begins foaming within the die body.[18] The resultant foam has a very poor surface appearance and poor cell size distribution.[18] Product made from such foam is inferior in quality, and it is generally rejected by a customer.[18]

In the polylactide-talc composites, the effective influence of talc on PLA crystallization kinetics dramatically increased the melt strength of PLA, reduced die pressure, and produced a less uniform foam morphology with decreased cell density.[19]

5.6 DIE TEMPERATURE

The void fraction in thermoplastic polyolefin foam blown with nitrogen was a function of cell density but not a function of the die temperature.[16] In polystyrene extrusion, higher die temperature increased the number of pores and changed their morphology to almost spherical.[20] Cell coalescence was also observed with increasing die temperature.[20] It appears from this that the die temperature increase causes an increase in cell growth and coalescence. The effect of die temperature on foam morphology is a result of the interplay of these two parameters acting in opposite directions.

Continuous foam extrusion of polyvinylidenefluoride was used in chemical microfoam formation with a mixture of azodicarbonamide, paraffin, and ZnO (used to decrease the decomposition temperature of azodicarbonamide).[21] Decreasing the die temperature from 135 to 130°C caused cell density to increase and cell size to decrease, while void fraction decreased from 58% to 39%.[21] This was due to the loss of melt strength upon increasing temperature of PVDF melt as it exited the die.[21] In the region of 125-130°C, the melt viscosity was low enough to allow cell formation and growth but high enough to prevent cell coalescence.[21]

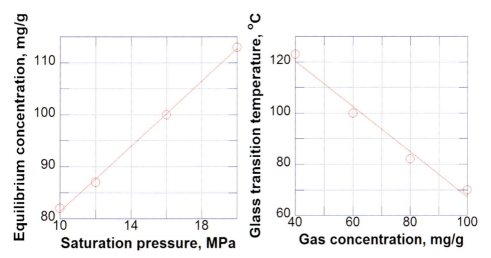

Figure 5.10. Equilibrium concentration of carbon dioxide in polycarbonate vs. saturation pressure. [Data from Ma, Z; Zhang, G; Yang, Q; Shi, X; Shi, A, *J. Cellular Plast.*, **50**, 1, 55-79, 2014.]

Figure 5.11. Glass transition temperature of polycarbonate vs. dissolved carbon dioxide concentration. [Data from Ma, Z; Zhang, G; Yang, Q; Shi, X; Shi, A, *J. Cellular Plast.*, **50**, 1, 55-79, 2014.]

In foam extrusion of low-density polyethylene with talc as nucleating agent and fluorocarbon as blowing agent, the bubble coalescence and the open cell fraction increased with increasing talc concentration and die temperature.[22] Especially, the die temperature has been considered as the main factor because viscosity and melt strength of the polymer rapidly dropped with temperature.[22]

Gas escape through thin walls decreases the amount of gas available for the growth of cells, and therefore, appropriate melt strength and freezing of a skin of foam by lowering the die temperature are necessary to prevent gas loss.[23]

5.7 GAS CONTENT

Figure 5.10 shows the effect of pressure on the equilibrium content of carbon dioxide in polycarbonate.[24] Figure 5.11 shows that the higher the concentration of carbon dioxide in polycarbonate, the lower its glass transition temperature.[24] This means that polymer under processing conditions containing supercritical carbon dioxide becomes softer and easier to be foamed.[24] Also, the viscosity of the polymer with dissolved carbon dioxide is decreased that results in its easier processing.[24]

In isotactic polypropylene, the nucleation energy barrier at high carbon dioxide concentration is lower, and the nucleation supersaturation degree can be achieved faster than at low carbon dioxide concentration.[25] This results in a higher nucleation rate and bubble density when carbon dioxide concentration is increased.[25] Because the bubble nucleation and growth are the two competitive stages during the foaming process, the high nucleation rate reduces the bubble growth rate, which leads to a smaller cell size under high carbon dioxide concentrations.[25]

The increase of PVDF-microcrystals and the decrease of CO_2 content in the polymer matrix significantly improved the strength of the polymer matrix, contributing to diminishing rupture and coalescence of cells.[26]

Considering the above information that nucleation and bubble growth are two competing processes, we ought to discuss findings for using mixtures of two gases: carbon dioxide and nitrogen. Out of the two, supercritical carbon dioxide has higher solubility and plasticization effect.[27] On the other hand, supercritical nitrogen exhibits better cell nucleating power than supercritical carbon dioxide.[27] The combination of 75% carbon dioxide and 25% nitrogen had the best foaming performance, as it yielded the highest cell density and cell growth rate over the widest processing window from 100 to 180°C.[27] This shows that using a combination of blowing gases, the effect of gas contents can be further enhanced. This is also true for additives, which increase the solubility of gases in polymers, such as additives having plasticizing action (e.g., dimethyl ether).[28]

When foaming LDPE with CO_2-N_2 mixture, the permeation of CO_2 and N_2 showed the same direction (both CO_2 and N_2 diffuse out), and the high permeability of CO_2 made CO_2 permeate out of the foam much faster than N_2, so that the gas composition (e.g., N_2 fraction) in the LDPE foam changed significantly with the decrease of the total gas content, thus resulting in an initial N_2 fraction to be much higher than the N_2 fraction used in the foaming process.[29]

5.8 GAS FLOW RATE

Figure 3.7 shows the experimental setup of extrusion equipment developed to test the effect of carbon dioxide flow rate on extrusion foaming of poly(ethylene terephthalate).[30] The gas input was varied from 0.1 to 10 ml/min at a gas pressure higher than 10 MPa.[30] Figures 5.12 and 5.13 show the effect of gas input on properties of the foam, such as expansion ratio (Figure 5.12) and average cell size and cell density (Figure 5.13).[30] The volume expansion ratio and cell density increase with an increase in gas input, whereas average cell size decreases when gas input is increased.[30]

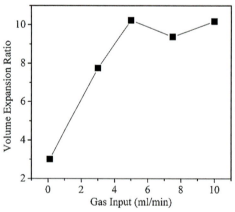

Figure 5.12. The influence of gas input on volume expansion ratio of PET foams. [Adapted, by permission, from Fan, C; Wan, C; Gao, F; Huang, C; Xi, Z; Xu, Z; Zhao, L; Liu, T, *J. Cellular Plast.*, **52**, 3, 277-98, 2016.]

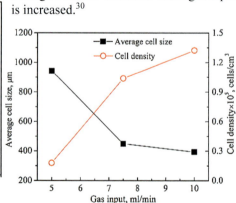

Figure 5.13. The influence of gas input on average cell size and cell density of PET foams. [Adapted, by permission, from Fan, C; Wan, C; Gao, F; Huang, C; Xi, Z; Xu, Z; Zhao, L; Liu, T, *J. Cellular Plast.*, **52**, 3, 277-98, 2016.]

5.9 GAS INJECTION LOCATION

Figure 3.8 shows the setup for nitrogen gas injection through any one of the four ports mounted on extruder setup for extrusion foaming of low-density polymer, and Figure 3.9 shows morphological outcomes of the selection of any of these ports.[31] The conclusion of studies is that longer residence time (use of the port furthest from extruder output side) increases interaction between gas and polymer matrix and produces foam with increased cell density.[31]

5.10 GAS SORPTION AND DESORPTION RATES

In the gas foaming process, sorption of the blowing agent – typically a gas, for example, CO_2 or N_2 – is the key step preceding foaming.[32] It is conducted at high, constant pressure for a sufficient period of time (sorption time is close to, but larger than the characteristic diffusion time, τ_d).[32] When pressure is then quickly released, the solubilized gas nucleates the bubbles, whose size and number depend on the gas concentration.[32]

Gas sorption of different polymers can be characterized using data obtained from sorption experiments in the form of sorption isotherms. The data quoted in Table 5.1 for selected polymers come from IUPAC-NIST data[33] and other sources, as indicated in the table. To make comparison between polymers easier, the data were extrapolated (if not directly available) to 100 kPa.

Table 5.1. Sorption data for carbon dioxide and nitrogen for various polymers at temperature=298K and pressure=100 kPa

Polymer	Concentration, mg/g CO_2	Concentration, mg/g N_2	Reference
Pressure=100 kPa; temperature=298K			
PMMA	8.5		33
Poly(styrene-co-acrylonitrile)	4.48		35
PVAc	2.5		34
PVC suspension)	3.38	0.08	33
Pressure=10.5 MPa; temperature=313K			
PCTFE	4.16		37
Crosslinked PDMS	10.19		37
PMMA	18.17		37
PLA	10		39
PP		11.87	38
HDPE		7.79	38
PS		6.31	38

There is a large number of data available in the open literature. Still, the data differ in units, the temperature of measurement, and the pressure. When one would like to compare the values for different polymers, only a few data points can be found in a similar range of measurement conditions, as presented in the table above. The sample size is not sufficient to draw any other conclusion than there are differences among polymers in the absorption of gases.

The desorption behavior can be described using the Fickian diffusion equation:[36]

$$M_d = M_\infty - \frac{4M_\infty}{l}\sqrt{\frac{D_d t_d}{\pi}} \qquad [5.1]$$

where:
- M_d — amount of gas dissolved in the sample at the desorption time t_d
- M_∞ — total sorbed amount of gas at saturation
- l — thickness of sample
- D_d — diffusivity of desorption
- t_d — desorption time

Figure 5.14 shows the effect of pressure on gas concentration in the polymer.[40] Figure 5.15 shows that a long time is required to attain the equilibrium concentration.[40] The higher the pressure, the shorter the time to equilibrium.[40] This means that gas diffusivity depends on pressure.[40] The early part of sorption (after normalization for sample thickness) shows a linear relationship between gas uptake and normalized time, indicating that diffusion follows Fick's law.[40] Long saturation time presents a challenge to the scale-up of the solid-state process in industrial applications.[40]

The gas saturation in amorphous gas-polymer systems is well described by the dual-mode gas sorption model in the steady-state, as follows:[40]

$$C(T_{sat}, P_{sat}) = k_D(T_{sat})P_{sat} + \frac{C'_H(T_{sat})b(T_{sat})P_{sat}}{[1 + b(T_{sat})P_{sat}]} \qquad [5.2]$$

where:
- $k_D(T_{sat})$ — Henry's gas law constant
- $C'_H(T_{sat})$ — capacity constant for Langmuir mode adsorption
- $b(T_{sat})$ — affinity constant of Langmuir mode sites
- P_{sat} — applied gas saturation pressure
- T_{sat} — saturation temperature.

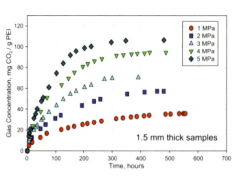

Figure 5.14. Plots of carbon dioxide uptake as a function of time at 21°C. The data is for 1.5 mm thick specimens. [Adapted, by permission, from Miller, D; Chatchaisucha, P; Kumar, V, *Polymer*, **50**, 23, 5576-84, 2009.]

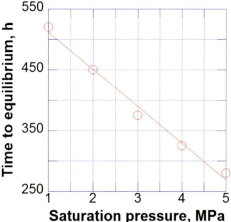

Figure 5.15. Time required for carbon dioxide to equilibrate in PEI vs. saturation pressure. [Data from Miller, D; Chatchaisucha, P; Kumar, V, *Polymer*, **50**, 23, 5576-84, 2009.]

5.11 Internal pressure after foaming

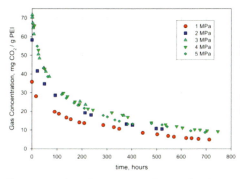
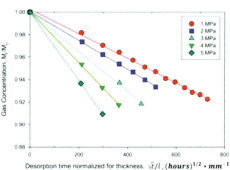

Figure 5.16. Plots of gas concentration as a function of desorption time at 21°C. [Adapted, by permission, from Miller, D; Chatchaisucha, P; Kumar, V, *Polymer*, **50**, 23, 5576-84, 2009.]

Figure 5.17. Plots of normalized gas concentration as a function of desorption time normalized for thickness. [Adapted, by permission, from Miller, D; Chatchaisucha, P; Kumar, V, *Polymer*, **50**, 23, 5576-84, 2009.]

This relationship shows that the relationship between saturation pressure and equilibrium concentration is linear for high saturation pressures and that the zero pressure intercept of this relationship gives the Langmuir capacity constant.[40] This relationship is confirmed by the data in Figure 5.15.[40]

Thymol acted as a molecular lubricant, which resulted in the increased free volume of the polymer matrix of polylactide and consequently higher gas sorption.[41] A prolonged exposure of PLA (24 h) to either scCO$_2$ only or scCO$_2$ with thymol has led to a 40-90% greater swelling extent.[41]

Figure 5.16 shows desorption time, which is also not short, but the majority of gas desorbs in the first minutes of the process.[29] Later, the kinetics of desorption slows down.[40] After normalization for sample thickness, the data in Figure 5.17 is obtained.[40] Figure 5.17 is useful for the calculation of diffusion coefficients.[40]

Similar observations and conclusions have been found from the studies of polycarbonate blown with supercritical carbon dioxide,[24] as well as from studies poly(ε-caprolactone).[42]

Gas saturation and desorption studies reveal that the crystal barriers hinder the diffusion of carbon dioxide in PPS/PES blends in semi-crystalline PPS component.[36] The equilibrium concentration and desorption diffusivity of carbon dioxide increase as the PES content increases.[36] Crystallites form obstacles in the pathways of gas diffusion.

The morphology of foams was controlled by the number of nucleation sites, the amount of CO$_2$ dissolved in polycarbonate, and the gas desorption rate.[43]

5.11 INTERNAL PRESSURE AFTER FOAMING

An increase in the chemical blowing agent content results in a greater amount of gas available for foaming.[44] As a consequence, the pressure after foaming (when the pressure is released and the material expands) increases.[44] The foam solidifies under significant pressure when greater amounts of azodicarbonamide are used (Figure 5.18).[44] The pressure was calculated using the ideal gas law equation, assuming that the volume occupied by the gas is that available for polymer expansion inside the mold and that the amount of gas is generated by all azodicarbonamide present in a sample.[44] It is estimated that the amount of

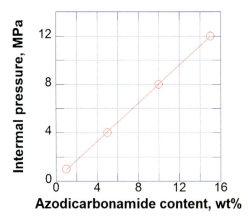

Figure 5.18. Internal pressure after foaming vs. concentration of azodicarbonamide. [Data from Saiz-Arroyo, C; Rodriguez-Perez, M A; Tirado, J; Lopez-Gil, A; de Saja, J A, *Polym. Int.*, **62**, 1324-33, 2013.]

gas generated by the blowing agent equals 210 g cm^{-3}.[44] The experimental results indicate that the external counter-pressure contributes to the stabilization of the cellular structure, reducing the cellular structure degeneration ratio.[44]

5.12 MOLD PRESSURE

The mold is filled under counter-pressure and then opened in one direction (Figure 5.19).[45] The sudden pressure drop initiates the foam nucleation process of the dissolved gas in the polymer melt.[45] The density reduction is achieved by the degree of the cavity opening.[45] In combination with a variotherm mold, the breathing mold technology permits control over the pressure and the melt temperature.[45] By this, the foaming conditions can be adjusted.[45] Figure 5.20 shows stages of the process and setup of the pressure-temperature regime.

In another technological development, the internal mold pressure is 70 to 90% of the saturation vapor pressure of dissolved carbon dioxide, so that about 10 to 30% of the dissolved carbon dioxide is released during the introduction into the mold.[46] Then, the internal mold pressure is reduced until the foamed reactive mixture fills the whole mold.[46]

In foam molding of reactive polyurethane systems, the low internal pressure maintained within the polymerization cavity helps polyisocyanurate reactive mixture fill the available space more and, therefore, reduce the required overpacking resulting in extra pressure on the press planes.[47] The internal pressure of the mold is controlled *via* a pipe connected to a 500-liter buffer tank that is connected to a medium-capacity vacuum pump (1500 1/min).[47]

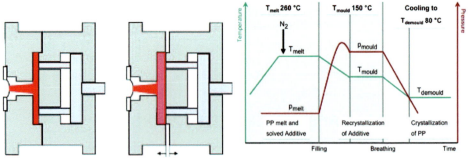

Figure 5.19. Schematic representation of the breathing mold. Left: after volumetric filling the complete cavity; right: after it is opened in one direction. The breathing is accompanied by a pressure drop and the melt is foamed. [Adapted, by permission, from Stumpf, M; Spoerrer, A; Schmidt, H-W; Altstaedt, V, *J. Cellular Plast.*, **47**, 6, 519-34, 2011.]

Figure 5.20. Pressure-temperature regime during foam injection molding. The temperature of the melt is decreasing after injection, while the pressure is increasing. The breathing of the mold initiates the iPP foaming. After cooling, the part is removed. [Adapted, by permission, from Stumpf, M; Spoerrer, A; Schmidt, H-W; Altstaedt, V, *J. Cellular Plast.*, **47**, 6, 519-34, 2011.]

The mold was completely filled with polymer/fiber/gas melt resulting in high in-mold pressure.[48] With high in-mold pressure, the nucleated gas was re-solved into the melt resulting in a single-phase polymer/fiber/gas melt-filled mold.[48] The cell nucleation started with increasing the mold volume, which was accompanied by a sudden pressure drop.[48] To initiate this foaming step after a certain delay time of the mold opening, the distance of the mold halves was increased to the final part dimensions *via* the precision opening.[48]

5.13 MOLD TEMPERATURE

Figure 5.20 discussed in the previous section shows the temperature graph in one of the processes. For different reasons, different temperature regimes are employed.

Mold temperature was regulated in the range of 30 and 80°C in the chemical blowing of polypropylene.[49] Foam obtained at higher temperature had thinner skin because of the slower heat transfer rate.[49] Cell size, cell density, and skin thickness were affected by the mold temperature in the injection molding of HDPE foamed with azodicarbonamide.[50,51] Foaming of polypropylene with supercritical carbon dioxide required a high concentration of carbon dioxide, high backpressure, low mold temperature, and low injection speed.[25] Comparison of the last three processes shows the difference between chemical and physical blowing. In chemical blowing, the higher mold temperature prolonged production of gas, whereas in physical molding lower mold temperature led to faster solidification of the matrix, which discontinued coalescence, causing enlargement of cells and formation of thicker skin.

Mold temperature was reported to critically influence crystallization behavior during the MIM process, and lower mold temperature could contribute to acceleration of crystallization rate and the decrease of crystal size.[52] High crystallization rate could enhance heterogeneous cell nucleation and simultaneously stable cell growth by improving melt strength, and tiny crystals could further promote cell nucleation by generating more local

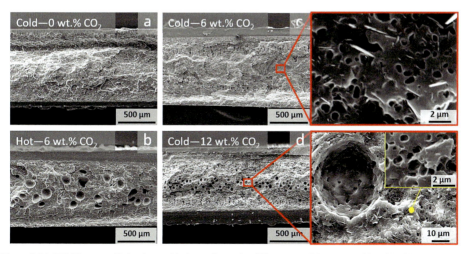

Figure 5.21. SEM images of injection molded samples under different conditions: (a) cold mold with 0 wt% CO_2, (b) hot mold with 6 wt% CO_2, (c) cold mold with 6 wt% CO_2, and (d) cold mold with 12 wt% CO_2. [Adapted, by permission, from Zhao, J; Qiao, Y; Park, C B, *Mater. Design*, **195**, 109051, 2020.]

stress or offering more nucleating sites, and refined crystals could also advance the mechanical properties of PP matrix.[52] Figure 5.21 displays SEM images of foamed PP/talc composites with different CO_2 contents and mold temperatures.[52] With hot mold, some microscale cells in the center layer of the foamed sample in Figure 5.21b were irregularly shaped and nonuniformly distributed across the foamed section.[52] Nanocellular structure, with a diameter of ~400 nm, was achieved throughout the entire foamed layer with 6 wt% CO_2 (Figure 5.21c) in samples foamed in a cold mold.[52] When gas content was further increased to 12 wt%, a bimodal cellular structure was fabricated, shown in Figure 5.21d.[52]

Comparisons of temperature profiles at three positions within polyisocyanurate rigid foams insulation panels (as a function of time) for mold temperatures of 25 and 55°C led to the conclusion that the temperature near the foam center was unaffected by the mold temperature for sufficiently thick samples because of the low thermal conductivity and low thermal diffusivity of the foam.[53]

It was demonstrated that increased mold temperature, use of gas counter-pressure, and reduced gas content benefited the surface appearance of the foam injection-molded parts.[54]

5.14 OPERATIONAL WINDOW

Successful production of foams requires proper vulcanization and appropriate control of blowing agent decomposition.[55] The result depends on the balance between both reactions. EPDM compound processed by extrusion without blowing agent has a wider processing window; that is, more time for homogenizing the compound in the extruder at higher temperatures, resulting in higher production rates.[55] But, more time is required to vulcanize the compound, thereby the need for a longer vulcanization tunnel (or lower tunnel speed).[55] In the case of the foamed profile, the processing window is narrower, the temperature in the extruder should be lower, and the production rate lower; however, the vulcanization tunnel may be shorter.[55]

Practical operational window lies between two extreme points in which the melt strength of the polymer is sufficiently low to allow bubble growth but high enough to prevent the total escape of carbon dioxide and cell collapse.[56]

5.15 PLASTISOL VISCOSITY

Rotational molding occurs at elevated temperature and involves the curing of plastisol and releasing of gases from the decomposition of chemical blowing agent.[57] Gelation and fusion processes and the development of the melt strength have to be synchronized with the evolution of the gas to enable the polymer matrix to withstand stresses involved and stabilize the foam structure.[57] This can be achieved by a proper selection of adequate type and concentration of plasticizer for plastisol so that the required balance of properties is timely obtained.[57] Better quality foams were obtained with plasticizers that have small linear molecules with good compatibility with PVC.[57]

5.16 SATURATION PRESSURE

Both saturation temperature and saturation pressure significantly influenced the cell morphology of poly(butylene succinate) foamed with supercritical carbon dioxide.[58] The

5.16 Saturation pressure

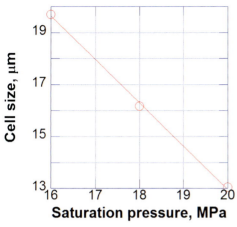

Figure 5.22. Cell size of poly(butylene succinate) vs. saturation pressure of supercritical carbon dioxide used for foaming. [Data from Wu, W; Cao, X; Lin, H; He, G; Wang, M, *J. Polym. Res.*, **22**, 177, 2015.]

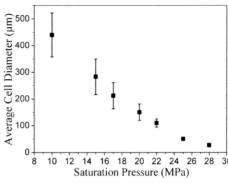

Figure 5.23. Cell diameter in polypropylene foam vs. saturation pressure of supercritical carbon dioxide. [Adapted, by permission, from Ding, J; Ma, W; Song, F; Zhong, Q, *J. Appl. Polym. Sci.*, **130**, 2877-85, 2013.]

higher the saturation pressure, the smaller the cell size (Figure 5.22), and the higher the volume expansion ratio.[58] Also, the lower depressurization rate contributed to smaller cell size.[58]

The batch foaming process is common because it can produce foams having small size cells and high cell density.[59] In pressure-induced phase separation process, the material is placed in a high-pressure vessel at polymer softening temperature, which has a narrow range for many polymers, especially polypropylene.[59] Variation in temperature may be sufficient to reverse the foaming process.[59] Also, the time to achieve saturation is usually very long in polypropylene (24 h).[59] The time to saturation can be reduced by decreasing the effect of crystalline structure.[59] This can be achieved by a two-stage process in which samples are saturated with supercritical carbon dioxide at temperatures sufficiently high to completely disrupt crystalline regions.[59] In the second step, the temperature is decreased to the foaming temperature, which is above the crystallization point.[59] The system is then depressurized to obtain PP foams.[59] Figure 5.23 shows the effect of saturation pressure on cell dimensions in polypropylene.[59] The time of production of polypropylene foams was reduced to 2.5 h with the above process.[59]

Polydimethylsiloxane was added to polypropylene to increase the solubility of carbon dioxide.[42] PDMS was playing the role of carbon dioxide reservoir, inducing increased nucleation and thus affecting the size of the bubble.[60] The saturation pressure was increased that caused the cell density to increase.[60]

When the saturation pressure of carbon dioxide increased from 18 MPa to 22 MPa in poly(ethylene terephthalate), the cell density increased more than 40 folds.[61] The amount of dissolved gas in the PET sheet increased with increasing saturation pressure, which led to greater supersaturation.[61] In homogeneous/heterogeneous nucleation, the greater the supersaturation, the greater the instability, which leads to more nucleation sites and smaller cell size.[61] The homogeneous nucleation theory predicts that when the magnitude of the pressure drop increases, the energy barrier to nucleation decreases, leading to more cells being nucleated within a given volume.[61]

The foaming behavior of pure ultra-high molecular weight polyethylene with supercritical carbon dioxide was evaluated for influence of foaming temperature and saturation pressure on the final foam structure.[62] Higher pressure led to lower cell density and larger average cell diameter in foaming during the heating stage due to the reduction of crystals and melt strength.[62] While during the cooling stage, higher saturation pressure led to higher cell density due to the increase in solubility of CO_2, and the cell density decreased as the pressure further increased due to cell coalescence.[62]

5.17 SATURATION TEMPERATURE

In the previous section, the effect of saturation temperature on shortening saturation time was discussed.[58]

The melting temperature of a semi-crystalline polymer can be depressed to a value below the operating saturation temperature.[63] The melting point depression permits a more uniform sorption of supercritical carbon dioxide by the polymeric material.[63] The increase of the saturation temperature in polylactide to 40°C induced the decrease of the amount of blowing agent dissolved in the polymeric material.[63]

If the polymer properties, such as the glass transition temperature and the solubility of carbon dioxide, are known, full control of the desired foam morphology can be obtained by choosing the correct combination of temperature, pressure, and depressurization rate.[64]

Too high a crystallinity will depress the foam's expansion ability, because the degree of crystallinity induced during blowing agent saturation needs to be controlled by selecting the proper saturation temperature and time.[65] When the saturation temperature was increased, the number of unmelted crystals was decreased.[65] The crystallization kinetics (i.e., nucleation and growth) of the PLA samples can be affected by saturation temperature, amount of dissolved blowing agent, and saturation time.[65]

In a solid-state foaming process of thermoplastic polymers, a combination of saturation pressure and saturation temperature determines the amount of physical blowing agent absorbed, and to a large extent, the subsequent foam structure.[66] The saturation temperature typically used is 20-30°C, and saturation pressure is 1-7 MPa.[66] At these conditions, carbon dioxide exists as a gas.[66] To achieve higher gas solubility and faster sorption, supercritical carbon dioxide is used (the saturation temperature above 31.1°C and the saturation pressure above 7.3 MPa).[66] The time needed to reach equilibrium concentration increases for carbon dioxide in polycarbonate as the saturation temperature decreases, from 12 h at 40°C to 72 h at -30°C.[66] Also, diffusivity decreases with the decreasing saturation temperature by nearly two orders of magnitude from 1.41×10^{-7} cm^2/s at 80°C down to 5.61×10^{-9} cm^2/s at -30 °C.[66] A lower saturation temperature results in a higher cell nucleation density (the cell nucleation densities have seven orders of magnitude increase from 10^8 cells/cm^3 at 40°C to 10^{15} cells/cm^3 at -30°C).[66]

The foam densities of random copolymers of polypropylene with respect to foaming temperature formed a Gaussian-like relationship.[67] Therefore, the same foam density can be reached using two different gas saturation temperatures.[67] At the same time, because of different processing temperatures, the polymer structure organization of the resulting foams, i.e., polycrystallinity, differed significantly.[67] Due to the Gaussian-like nature of these curves, a minimum foam density can be achieved for each resin at a singular processing temperature.[67]

The saturation temperature, pressure, and time influence the formation of hard segment crystals with different structures, which affect the TPU foaming behavior.[68] In the amorphous thermoplastics, saturation temperature affects CO_2 solubility and the melt strength of the polymer/gas mixture.[68] In semi-crystalline thermoplastics, crystallinity at various foaming temperatures may significantly influence the cell nucleation power and expandability of the foams.[68]

High saturation pressure (20 MPa) and low saturation temperature (-32°C) conditions were used to increase foaming gas dissolution.[69] After the PMMA sample was saturated, the pressure was released, and the sample was immersed in a thermal bath for foaming, resulting in cell sizes ten times smaller than the wavelength of visible light and very homogeneous cellular structures have been generated.[69]

5.18 SCREW REVOLUTION SPEED

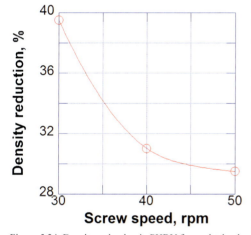

Figure 5.24. Density reduction in PHBV foam obtained with sodium bicarbonate vs. screw speed. [Data from Szegda, D; Duangphet, S; Song, J; Tarverdi, K, J. Cellular Plast., 50, 2, 145-62, 2014.]

Extrusion foaming of poly(3-hydroxybutyrate-co-3-hydroxyvalerate) with a chemical blowing agent based on sodium bicarbonate and citric acid and calcium carbonate nucleation agent was studied to evaluate the effect of screw revolution speed on the results of foaming.[70] The higher the screw speed, the lower the density reduction in the produced foam (Figure 5.24).[70] Higher screw speeds generate higher pressure drops at the die exit and this is expected to enhance cell nucleation and expansion.[70] But the higher screw revolutions also reduce the residence time of the material in the extruder barrel, especially when a low L/D ratio extruder is used.[70] Sodium bicarbonate is an endothermic blowing agent with slow decomposition rate over a wide range of temperatures.[70] The shorter residence time affected its decomposition.[70] Also, the shear-thinning at high screw speed reduced melt viscosity, increasing the probability of cell rupture during sheet foaming.[70]

The effect of the foaming temperature, screw revolution speed, gas input, and nucleating agent were investigated in extrusion foaming of poly(ethylene terephthalate) foamed with carbon dioxide.[30] In this foaming process, the polymer is first completely melted, then the carbon dioxide is injected into the mix.[30] The strong shear mixing effect of the well-designed screw causes big gas bubbles to be dispersed into a number of tiny bubbles.[30] The homogeneous melt containing gas is achieved through diffusion and dissolution.[30] At the die exit, the high-pressure drop induces thermodynamic instability, and the

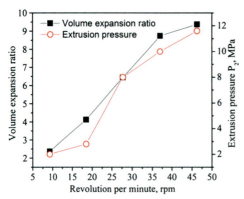

Figure 5.25. The influence of screw revolution speed on the extrusion pressure and the volume expansion ratio in processing of PET foams. [Adapted, by permission, from Fan, C; Wan, C; Gao, F; Huang, C; Xi, Z; Xu, Z; Zhao, L; Liu, T, *J. Cellular Plast.*, **52**, 3, 277-98, 2016.]

bubbles nucleate and grow.[30] On cooling, the bubbles gradually form a stable structure with pressure inside the bubble equal to the atmospheric pressure.[30] The extrusion pressure increases with increasing the screw revolution speed and then levels off at 11.6 MPa (Figure 5.25).[30] The thermodynamic instability is triggered by the rapid pressure drop in the die.[30] Increased extrusion pressure enhances the gas solubility in the polymer melt and the supersaturation of dissolved gas.[30] This increases the amount of the nucleation points in a homogeneous nucleation process.[30] The greater the number of nucleation points, the larger the cell density.[30] The larger foaming pressure drop results in the decline of the cell size.[30] The average cell size slightly decreases from 512 to 449 mm, and the cell density increases from 0.65×10^5 to 1.04×10^5 cells/cm^3 when the screw revolution speed increases from 37 to 46 r/min.[30]

These two practical examples show that the screw revolution rate is a useful parameter to control foam properties, but the actual effect also depends on other circumstances, in this particular case, on the type blowing agent.

5.19 SURFACE TENSION

Figure 5.26 shows that the surface tension decreases with an increase in both temperature and pressure.[59] Lowering surface tension helps to incorporate and disperse gas, whereas it reduces bubble lifetime and promotes coalescence. Further studies are necessary to assess the influence of this parameter on the foaming process.

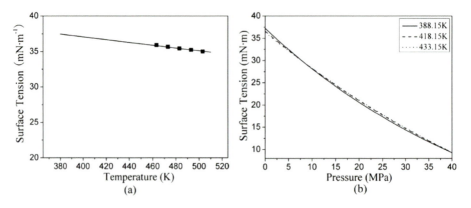

Figure 5.26. The surface tension of PP melt with temperature at (a) atmospheric pressure and (b) variable pressure. [Adapted, by permission, from Ding, J; Ma, W; Song, F; Zhong, Q, *J. Appl. Polym. Sci.*, **130**, 2877-85, 2013.]

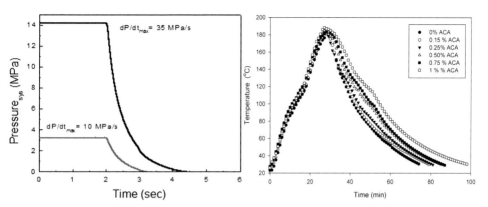

Figure 5.27. Pressure profiles in the batch foaming simulation chamber. [Adapted, by permission, from Kim, S G; Lee, J W S; Park, C B; Sain, M, *J. Appl. Polym. Sci.*, **118**, 1691-703, 2010.]

Figure 5.28. Air temperature profiles in mold containing polyethylene foamed with different concentrations of azodicarbonamide using biaxial rotational molding. [Adapted, by permission, from Moscoso-Sanchez, F J; Mendizabal, E; Jasso-Gastinel, C F; Ortega-Gudino, P; Robledo-Ortiz, J R; Gonzalez-Nunez, R; Rodrigue, D, *J. Cellular. Plast.*, **51**, 5-6, 489-503, 2015.]

A small bubble can be coalesced by a big bubble, but some critical size difference is required for coalescence to take place.[71] This size difference has been related to viscoelasticity and surface tension.[71]

5.20 TIME

In sections 5.4 and 5.10, time of sorption and desorption played an essential role. It is an important variable in equation 5.1. Many practical examples of studies are presented in these sections. Also, in section 5.9, the study shows that the residence time in the extruder has an essential influence on the dispersion and thus uniformity of produced foam. In many instances in this chapter, there was a discussion of the effect of pressure drop when material leaves extruder die. One should underline that it is not just pressure drop, which has an essential effect on thermodynamic instability and thus expansion of the gas-saturated melt, but it is the rate of drop of pressure that affects the process of cell formation and its size. Figure 5.27 shows pressure drop profiles as simulated in a batch chamber.[15] This pressure drop depends on the initial 15 in the extruder (or saturation pressure) but also depends on the construction of the die.[13] In this study, two dies were used with low (~30 MPa/s) and high (800 MPa/s) pressure drop rates.[15]

Cooling is also a time-related process – very essential for the development of foam properties.[72] From Figure 5.28, we can deconvolute another factor related to the exothermic chemical reaction of azodicarbonamide decomposition.[72] This process increases product temperature and requires more cooling time.[72]

5.21 TEMPERATURE

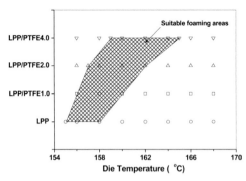

Figure 5.29. Foaming window for linear polypropylene and its blends with PTFE. [Adapted, by permission, from Wang, K; Wu, F; Zhai, W; Zheng, W, *J. Appl. Polym. Sci.*, **129**, 2253-60, 2013.]

Similar to time, the temperature is an essential parameter of the processes discussed in this book. Especially sections 5.6, 5.13, and 5.17 include extensive discussion of the effect of die, mold, and saturation temperatures on the effect of foaming. Figure 5.29 shows that foaming window (foams obtained in these ranges have a well-defined cell structure with very thin walls and uniform cell distribution) can be increased from 4 to 7°C when PTFE is added to the formulation.[73] PTFE particles were deformed into fine fibers under shear or extensional flows during the extrusion process, which significantly increased melt strength of PP, increased the range of temperatures in which foaming was successful (foaming window), and improved the quality of foam.[73]

Temperature restricts the selection of some blowing agents. The rate of gas evolution for a given chemical foaming or blowing agent is determined by the temperature-time relationship.[74] Applications of chemical foaming agents are generally divided into three areas: low, medium, and high-temperature processing polymers.[74] Organic foaming agents decompose at various temperatures.[74] As a practical limitation, blowing agents degrade at temperatures between about 150 and 200°C, and thus cannot be utilized in resins that melt above this temperature range.[74] Several solutions to this problem were found. The application of ammonium polyphosphate to melt-processable fluoropolymer is one of such solutions.[75] Polymer melting point is above 250°C, and extrusion temperature profile has temperatures in the range of 330-380°C.[75] The composition is useful for the production of electrical cables.[75]

Glass transition temperature and melting temperature of biodegradable poly(lactic acid) foams were improved by adding cellulose fiber, and decomposition temperature increased when blending it with poly(butylene succinate).[76] Also, temperatures higher than the glass transition of PLA allowed trapped gas from the decomposition of sodium bicarbonate to diffuse and to increase distances between molecules.[76] The gas expansion resulted in the formation of closed cells.[76]

5.22 VOID VOLUME

Void fraction is one of the parameters characterizing the result of foaming. Other similar parameters include cell size, cell density, the density of foamed materials, expansion ratio, etc. The void fraction is given by the following equation:[15]

$$Vf = \left(1 - \frac{1}{\Phi}\right) \times 100 \qquad [5.3]$$

5.22 Void volume

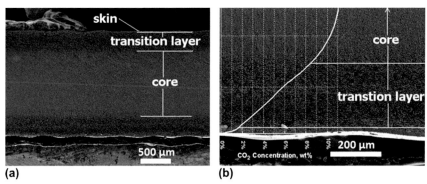

Figure 5.30. (a) SEM of nanocellular PEI sample cross-section showing internal blistering, and (b) SEM of the microcellular transition layer, with graph of carbon dioxide concentration at foaming. The sample was foamed for 3 minutes at 195°C, 2.2 MPa clamping pressure, after desorbing for 15 minutes. [Adapted, by permission, from Aher, B; Olson, N M; Kumar, V, *J. Mater. Res.*, **28**, 17, 2366-73, 2013.]

where:
    Vf   void fraction
    Φ   expansion ratio.

The cell density is given by the following equation:[15]

$$Cd = \left(\frac{nM^2}{A}\right)^{3/2} \times \Phi \qquad [5.4]$$

where:
    Cd   cell density
    n   number of bubbles in micrograph
    M   magnification factor in micrograph
    A   surface area of micrograph
    Φ   expansion ratio.

Both void fraction and cell density correlate very well, and they are a function of nitrogen content in thermoplastic polyolefin foam blown with nitrogen.[16] Void fraction, although easy to determine and calculate and handy in comparisons, is frequently misleading because foamed materials are not homogeneous throughout the thickness but have different structures in the skin, transitional layer, and core, as Figure 5.30 shows.[9]

A numerical model was developed to simulate the simultaneous bubble nucleation and growth during depressurization of thermoplastic polymers saturated with supercritical blowing agents.[77] The model has the ability to predict the formation of nanocellular foams, including cell size distribution within the foam, based on the specific process conditions and polymer properties.[77] Classical nucleation theory is used to predict the nucleation rate.[77] The mass, momentum, and species conservation equations are solved for each bubble, which makes the model capable of predicting bubble size distribution and bulk porosity.[77]

The objective of the study was to survey current foaming technologies to find out if voids can be incorporated into the matrix of a composite with a minimal negative impact on the mechanical properties.[78] Weight reduction of 20% is achievable for neat resins and of 10% for reinforced polymeric composites.[78] The introduction of voids reduces strength because the amount of load-carrying material is reduced.[78] But, in fiber composites, the main load carriers are the fibers, and matrix only transfers load and protects fibers from the environment.[78] The introduction of small, well-controlled bubbles into the matrix of a composite may actually improve the specific mechanical properties of a composite material.[78]

Unlike in typical foams having a void concentration in the range of 10-25% because of the constraints related to their mechanical performance, the ultrahigh void volume is used in polishing pads used in electronics.[79] In these materials, void volume fraction is as high as 70%.[79]

The void volume fraction of polyurethane foam (as any other foam) changes under uniaxial compressive load.[80] The higher the density of PUF, the higher the yield strength and the higher the rate of energy absorption due to the deformation and collapse of cells after the yield point.[80] Fracture occurs when energy absorption reaches a critical point.[80] When densification starts, deformation of the cell occurs perpendicular to the compression direction, and the deformed cells continuously change and form a pattern.[80]

REFERENCES

1 Garbacz, T; Palutkiewicz, P, *Cellular Polym.*, **34**, 4, 189-214, 2015.
2 Zhang, Z X; Li, Y N; Xia, L; Ma, Z G; Xin, Z X; Lim, J K, *Appl. Phys. A*, **117**, 755-9, 2014.
3 Zepnik, S; Hendriks, S; Kabasci, S; Radusch, H-J, *J. Mater. Res.*, **28**, 17, 2394-2400, 2013.
4 Bociaga, E; Palutkiewicz, P, *Cellular Polym.*, **32**, 5, 257-77, 2013.
5 Zhang, H; Liu, T; Li, B; Xin, Z, *Polymer*, in press 123209, 2020.
6 Shi, Z; Zhang, S; Qiu, J; Tang, T, *Polymer*, **207**, 122896, 2020.
7 Hopmann, C; Latz, S, *Polymer*, **56**, 29-36, 2015.
8 Salerno, A; Domingo, C, *J. Supercritical Fluids*, **84**, 195-204, 2013.
9 Aher, B; Olson, N M; Kumar, V, *J. Mater. Res.*, **28**, 17, 2366-73, 2013.
10 Heim, H-P; Tromm, M, *J. Cellular Plast.*, **52**, 3, 299-319, 2016.
11 Heim, H-P; Tromm, M, *Polymer*, **56**, 111-8, 2015.
12 Kumar, V; Lu, J C; Schirmer, H G; Miller, D, **WO2009036384**, *University of Washington*, Mar. 19, 2009.
13 Nadella, K; Miller, D; Kumar, V; Kuykendall, W; Probert, S, **US8357319**, *University of Washington*, Jan. 22, 2013.
14 Qiang, W; Zhao, L; Gao, X; Hu, D, *J. Supercritical Fluids*, **163**, 104888, 2020.
15 Kuboki, T, *J. Cellular Plast.*, **50**, 2, 113-28, 2014.
16 Kim, S G; Lee, J W S; Park, C B; Sain, M, *J. Appl. Polym. Sci.*, **118**, 1691-703, 2010.
17 Zhang, C; Zhu, B; Lee, L J, *Polymer*, **52**, 8, 1847-55, 2011.
18 Harfmann, W, **US20030138515**, *Genpak LLC*, Jul. 24, 2003.
19 Nofar, M; Park, C B, **Extrusion Foaming of PLA and Its Compounds. Polylactide Foams**, *William Andrew*, 2018, pp. 113-49.
20 Sauceau, M; Nikitine, C; Rodier, E; Fages, J, *J. Supercritical Fluids*, **43**, 2, 367-73, 2007.
21 Sameni, J; Jaffer, S A; Sain, M, *Adv. Ind. Eng. Polym. Res.*, **3**, 1, 35-45, 2020.
22 Tammaro, D; Di Maio, E, *Mater. Lett.*, **228**, 459-62, 2018.
23 Huang, P; Wu, F; Zheng, W, *Chem. Eng. J.*, **370**, 1322-30, 2019.
24 Ma, Z; Zhang, G; Yang, Q; Shi, X; Shi, A, *J. Cellular Plast.*, **50**, 1, 55-79, 2014.
25 Xi, Z; Chen, J; Liu, T; Zhao, L; Turng, L-S, *Chinese J. Chem. Eng.*, **24**, 180-9, 2016.
26 Shi, Z; Ma, X; Li, B, *Mater Design*, **195**, 109002, 2020.
27 Wong, A; Mark, L H; Hasan, M M; Park, C B, *J. Supercritical Fluids*, **90**, 35-43, 2014.
28 Mahmood, S H; Xin, C L; Gong, P; Lee, J H; Li, G; Park, C B, *Polymer*, **97**, 95-103, 2016.
29 Zhang, H; Liu, T; Xin, Z, *J. Supercritical Fluids*, **164**, 104930, 2020.
30 Fan, C; Wan, C; Gao, F; Huang, C; Xi, Z; Xu, Z; Zhao, L; Liu, T, *J. Cellular Plast.*, **52**, 3, 277-98, 2016.
31 Ghandi, A; Bhatnagar, N, *Polym. Plast. Technol. Eng.*, **54**, 1812-8, 2015.
32 Trofa, M; Di Maio, E; Maffettone, P L, *Chem. Eng. J.*, **362**, 812-7, 2019.
33 Paterson, R; Yampolskii, Y; Fogg, P G T; Bokarev, A, Bondar, V; Ilinich, O; Shishatskii, S; **IUPAC-NIST Solubility Data Series 70. Solubility of gases in Glassy Polymers**. *American Institute of Physics and American Chemical Society*, 1999.

References

34   Sato, Y; Takikawa, T; Takashima, S; Masuoka, H, *J. Supercritical Fluids*, **19**, 2, 187-98, 2001.
35   Balashova, I M; Danner, R P, *Fluid Phase Equilibria*, in press, 2016.
36   Ma, Z; Zhang, G; Shi, X; Yang, Q; Li, J; Liu, Y; Fan, X, *J. Appl. Polym. Sci.*, **132**, 42634, 2015.
37   Webb, K F; Teja, A S, *Fluid Phase Equilibria*, **158-160**, 1029-34, 1999.
38   Sato, Y; Fujiwara, K; Takikawa, T; Takishima, S; Masuoka, H, *Fluid Phase Equilibria*, **162**, 1-2, 261-76, 1999.
39   Li, B; Zhao, G; Gong, J, *Polym. Deg. Stab.*, **156**, 75-88, 2018.
40   Miller, D; Chatchaisucha, P; Kumar, V, *Polymer*, **50**, 23, 5576-84, 2009.
41   Kuska, R; Milovanovic, S; Ivanovic, J, *J. Supercritical Fluids*, **144**, 71-80, 2019.
42   Ogunsona, E; D'Souza, N A, *J. Cellular Plast.*, **51**, 3, 245-68, 2015.
43   Ho, Q B; Kontopoulou, M, *Polymer*, **198**, 122506, 2020.
44   Saiz-Arroyo, C; Rodriguez-Perez, M A; Tirado, J; Lopez-Gil, A; de Saja, J A, *Polym. Int.*, **62**, 1324-33, 2013.
45   Stumpf, M; Spoerrer, A; Schmidt, H-W; Altstaedt, V, *J. Cellular Plast.*, **47**, 6, 519-34, 2011.
46   Sulzbach, H-M, **US5834527**, *Maschinenfabrik Hennecke GmbH*, Nov. 10, 1998.
47   Bertucelli, L; Fantera, G; Golini, P, **WO2013174844**, *Dow Global Technologies LLC*, Nov. 28, 2013.
48   Lohr, C; Beck, B; Elsner, P, *Composite Structures*, **220**, 371-85, 2019.
49   Yousefian, H; Rodrigue, D, *J. Appl. Polym. Sci.*, **132**, 42845, 2015.
50   Tissandier, C; Gonzalez-Nunez, R; Rodrigue, D. *J. Cellular Plast.*, **50**, 5, 449-73, 2014.
51   Bociaga, E; Palutkiewicz, P, *Polym. Eng. Sci.*, **53**, 780-91, 2013.
52   Zhao, J; Qiao, Y; Park, C B, *Mater. Design*, **195**, 109051, 2020.
53   Reignier, J; Méchin, F; Sarbu, A, *Polym. Testing*, **93**, 106972, 2021.
54   Wang, G; Zhao, G; Park, CB, *Chem. Eng. J.*, **350**, 1-11, 2018.
55   Restrepo-Zapata, N C; Osswald, T A; Hernandez-Ortiz, J P, *Polym. Eng. Sci.*, **55**, 2073-88, 2014.
56   Sarver, J A; Hassler, J C; Kiran, E, *J. Supercritical Fluids*, **166**, 105015, 2020.
57   Zoller, A; Marcilla, A, *J. Appl. Polym. Sci.*, **121**, 1495-1505, 2011.
58   Wu, W; Cao, X; Lin, H; He, G; Wang, M, *J. Polym. Res.*, **22**, 177, 2015.
59   Ding, J; Ma, W; Song, F; Zhong, Q, *J. Appl. Polym. Sci.*, **130**, 2877-85, 2013.
60   Bing, L; Wu, Q; Zhou, N; Shi, B, *Int. J. Polym. Mater.*, **60**, 51-61, 2011.
61   Guang, R; Xiang, B; Xiao, Z; Li, Y; Lu, D; Song, G, *Eur. Polym. J.*, **42**, 5, 1022-32, 2006.
62   Liu, J; Qin, S; Gao, Y, *Polym. Testing*, **93**, 106974, 2021.
63   Salerno, A; Clerici, U; Domingo, C, *Eur. Polym. J.*, **51**, 1-11, 2014.
64   Jacobs, L J M; Danen, K C H; Kemmere, M F; Keurentjes, J T F, *Polymer*, **48**, 13, 3771-80, 2007.
65   Park, C B; Nofar, M, **WO2014158014**, *Synbra Technology BV*, Oct. 2, 2014.
66   Kumar, V; Guo, H, **WO2014210523**, *University of Washington*, Dec. 31, 2014.
67   Pin, J-M; Tuccitto, A V; Lee, P C, *Polymer*, in press, 123123, 2021.
68   Nofar, M; Batı, B; Jalali, A, *J. Supercritical Fluids*, **160**, 104816, 2020.
69   Ono, T; Wu, X; Furuya, T; Yoda, S, *J. Supercritical Fluids*, **149**, 26-33, 2019.
70   Szegda, D; Duangphet, S; Song, J; Tarverdi, K, *J. Cellular Plast.*, **50**, 2, 145-62, 2014.
71   Ge, Y; Liu, T, *Chem. Eng. Sci.*, **230**, 116213, 2021.
72   Moscoso-Sanchez, F J; Mendizabal, E; Jasso-Gastinel, C F; Ortega-Gudino, P; Robledo-Ortiz, J R; Gonzalez-Nunez, R; Rodrigue, D, *J. Cellular. Plast.*, **51**, 5-6, 489-503, 2015.
73   Wang, K; Wu, F; Zhai, W; Zheng, W, *J. Appl. Polym. Sci.*, **129**, 2253-60, 2013.
74   Kosin, J A; Mooney, G; Tarquini, M A; Garcia, R A, **US5009809**, *J. M. Huber Corporation*, Apr. 23, 1991.
75   Brix, S; Müller, F; Lankes, C; Gemmel, A; Pinto, O; Abeguile, M; Alric, J; Auvray, T; Kroushl, P, **EP2065155**, *Nexans*, Sep. 14, 2011.
76   Vorawongsagul, S; Pratumpong, P; Pechyen, C, *Food Packaging Shelf Life*, **27**, 100608, 2021.
77   Khan, I; Adrian, D; Costeux, S, *Chem. Eng. Sci.*, **138**, 634-45, 2015.
78   Rutz, B J; Berg, J C, *Adv. Colloid Interface Sci.*, **160**, 1-2, 56-75, 2010.
79   Fotou, G; Khanna, A; Vacassy, R, **US20150056895**, *Cabot Microelectronics Corporation*, Feb. 26, 2015.
80   Lee, E S; Goh, T S; Lee, C-S, Composites Part B: Eng., 163, 130-8, 2019.

6

FOAM STABILIZATION

Figure 6.1 shows the schematic interpretation of bubble dynamics during foam expansion. Initial growth is associated here with spherical bubbles.[1] When they expand (depending on the amount of gas either added or produced), they begin to fill the majority of space within the matrix, and their walls become thinner and more fragile.[1] At some point of an expansion, the walls of the bubbles may become so thin and stretched that they may rupture.[1] This depends on the extent of foaming (amount of the gas available in the mixture) but also on the melt strength of the polymer matrix.[1] When the wall between bubbles ruptures, the neighboring bubbles may merge and form one larger bubble, which will assume a non-spherical shape.[1] It is also possible that a large bubble may draw gas from smaller bubbles through tunnels formed from wall rupture caused by the lower internal pressure in the larger cell.[1] This may decrease the size or even annihilate the smaller bubbles.[1] If the blowing agent concentration is too large, the bubble wall ruptures can become so severe that a great amount of gas may escape, leading to the complete collapse of the foam structure.[1]

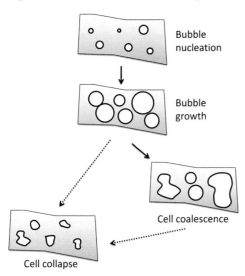

Figure 6.1. Bubble dynamics during the foam expansion. [Adapted, by permission, from Liao, Q; Billington, S; Frank, C W, *Polym. Eng. Sci.*, **52**, 1495-508, 2012.]

Figure 6.2 illustrates the above general model of the foam's morphological changes by the practical examples of SEM photographs. Pictures a, b, and c show mostly unrestricted growth of bubble; pictures d and e are characteristic of coalescence, and picture f shows the result of structure collapse when too much blowing agent was used.

When bubbles grow inside a polymer melt, the polymer matrix is forced to deform and flow.[2] The viscoelastic response of the polymer melt to the deformation opposes the bubble growth.[2] A momentum equation describes the dynamic process of bubble growth as follows:[2]

$$P_g - P_a - \rho\left(R\ddot{R} + \frac{3}{2}\dot{R}^2\right) + \frac{2\sigma}{R} + \frac{4\mu_s \dot{R}}{R} - 2\int_R^\infty \frac{\tau_{yy} - \tau_{\theta\theta}}{r} dr \qquad [6.1]$$

Foam Stabilization

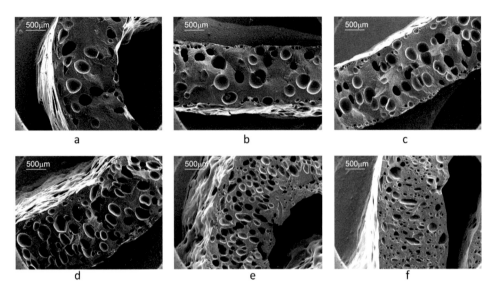

Figure 6.2. SEM images of the fractured cross-sectional areas of the extruded foams of blend of 80 wt% poly(3-hydroxybutyrate-co-3-hydroxyvalerate) and 20 wt% cellulose acetate butyrate with varying amounts of azodicarbonamide: (a) 0.5 phr; (b) 1.0 phr; (c) 1.5 phr; (d) 2.0 phr; (e) 2.5 phr; (f) 3.0 phr. [Adapted, by permission, from Liao, Q; Billington, S; Frank, C W, *Polym. Eng. Sci.*, **52**, 1495-508, 2012.]

where:
- P_g — pressure inside the bubble
- P_a — atmospheric pressure
- ρ — density of polymeric medium
- R — bubble radius
- $\dot{R}$ — first time derivative
- $\ddot{R}$ — second time derivative
- σ — surface tension
- μ_s — Newtonian viscosity of polymer matrix
- τ_{yy} — compressive normal stress term
- $\tau_{\theta\theta}$ — tensile normal stress term
- r — position of melt with respect to the center of bubble.

The left-hand side term is the differential gas pressure (the force for bubble nucleation and growth).[2] The right-hand side terms are effects that oppose bubble growth. These terms represent the effects of viscosity and viscoelastic normal stresses, respectively.[2]

In polymer foaming with a physical blowing agent, the three important mechanisms of bubble nucleation, bubble ripening, and bubble coalescence are recognized.[3] They determine the bubble density of the foamed samples.[3] Bubble nucleation is the formation of nuclei that result from gas supersaturation and depressurization or heating.[3] The bubble ripening is the growth of larger bubbles from smaller bubbles.[3] The bubble coalescence is caused by the rupture of the bubble wall when extensional force is applied.[3] The bubble collapse and bubble coalescence are important mechanisms of reduction of the final bubble density of foamed samples.[3] Figure 6.3 shows the data based on which the above statements have been made.[3] Also, here, the initial bubbles are spherical as in the chemically foamed material.[3] The bubbles grow increasingly closer to one another.[3] When a small bubble is close to a big one, the small one diminishes until it finally disappears.[3] The large

Foam Stabilization 77

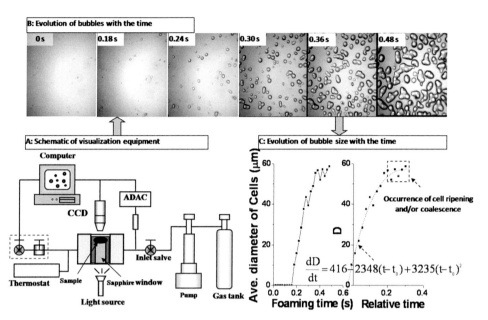

Figure 6.3. *In situ* visualization investigating the evolution of nucleated bubbles with time in microcellular foaming of poly(ethylene-co-octene) using supercritical carbon dioxide. [Adapted, by permission, from Zhai, W; Wang, J; Chen, N; Naguib, H E; Oark, C B, *Polym. Eng. Sci.*, **52**, 2078-89, 2012.]

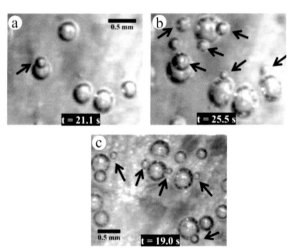

Figure 6.4. Formation of satellite bubbles around and on the surface of growing bubbles in high-pressure foam injection molding (a) and (b) PS+3 wt% CO_2 (injection speed = 80 cm³/s, packing pressure = 8 MPa, packing time = 1 s, T_{melt} = 230°C); (c) PS+0.5 wt% talc (injection speed = 50 cm³/s, packing pressure = 20 MPa, packing time = 1 s, 1 wt% CO_2). [Adapted, by permission, from Shaayegan, V; Wang, G; Park, C B, *Chem. Eng. Sci.*, **155**, 27-37, 2016.]

bubbles grow bigger, which is the stage of cell ripening.[3] If more time is allowed, the bubbles connect coalescing into larger bubbles.[3] There are many similarities between bubble formation in physical and chemical processes of foaming.[3] Figure 5.5 shows and discusses the mechanisms of coalescence in which internal pressures of the bubble are considered as important factors of bubble faith.[4]

Figure 6.4 shows still another mechanism of bubble formation.[5] The satellite bubbles form in the proximity of large bubbles.[5] Apparently, nucleation of the satellite bubbles occurs in regions that are gas-depleted due to the fast growth of former bubbles.[5] It is also possible that bub-

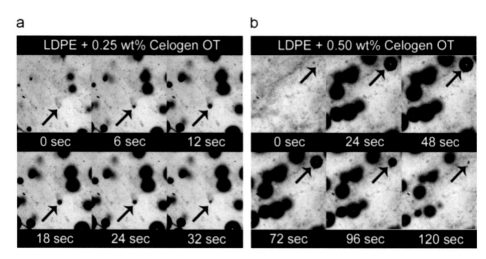

Figure 6.5. Bubble growth and collapse phenomena with different chemical blowing agent contents: (a) 0.25 wt% Celogen® OT and (b) 0.50 wt% Celogen® OT. [Adapted by permission, from Leung, S N; Wong, A; Guo, Q; Park, C B; Zong, J H, *Chem. Eng. Sci.*, **64**, 23, 4899-907, 2009.]

bles' expansion induces tensile stresses (or pressure fluctuations) in the melt around a growing bubble.[5] These may decrease pressure at regions close to the bubble surface.[5] The biaxial stretching of the melt around a growing bubble reduces the system's local pressure.[5] The growth of bubbles induces pressure fluctuations, such as extensional stresses, to their surroundings.[5] This reduces the local system pressure and, as a result, increases the supersaturation level of the system, facilitating the formation of new bubbles.[5]

The classical nucleation theory assumes that the critical radius of cell nucleation determines the fate of the bubble. The change in the critical radius during foaming has a strong impact on the stability of foamed cells.[6] The critical radius of cell nucleation is a function of the thermodynamic state.[6] It is determined by temperature, pressure, and the dissolved gas concentration in the polymer/gas solution.[6] These state variables change continuously during the foaming process, which causes a critical radius to vary.[6] The bubble stability depends on gas diffusivity (high diffusivity means more gas in the gas-rich region around the bubble and gas depletion in surroundings; this means fast growth and fast collapse), surface tension (low surface tension promotes bubble growth), and viscosity.[6] Figure 6.5 shows bubble faith, its growth, and collapse depending on the thermodynamical condition in the bubble closest environment.[6]

Enhanced cell nucleation and suppressed cell ripening and cell coalescence are needed to obtain a high-quality foam.[3] So far, no effective method exists to suppress cell ripening, but some suggestions are available for the prevention of cell coalescence.[3] The melt strength improvement is seen as a good method of slowing down cell coalescence.[3] The melt strength can be described as the resistance of the polymer melt to stretching. It is related to the molecular chain entanglements of the polymer and its resistance to untangling under strain. It is possible to improve the melt strength of polymer by long-chain branching and chemical or radiation crosslinking. Melt strength can also be improved by compounding methods, namely the addition of materials that can interact with a matrix to

Foam Stabilization

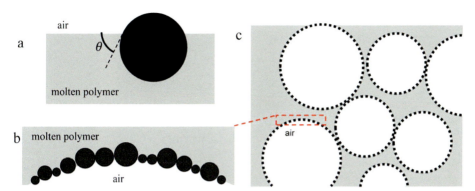

Figure 6.6. (a) A particle with surface energy lower than the polymer can be adsorbed at the interface between polymer and air making a contact angle of η. With sufficient adsorption, particles can endow cells with a shell (b), which protects against coalescence to create a stable foam (c). [Adapted, by permission, from Lobos, J; Lasella, S; Rodriguez-Perez, M A; Velankar, S S, *Polym. Eng. Sci.*, **56**, 9-17, 2015.]

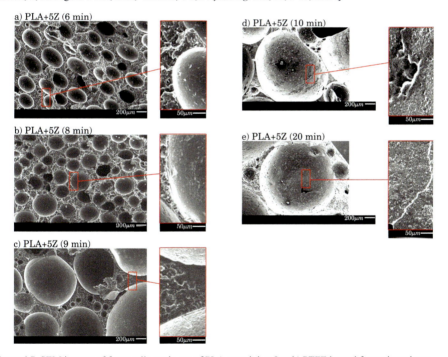

Figure 6.7. SEM images of foam cells made out of PLA containing 5 wt% PTFE heated for various times. [Adapted, by permission, from Lobos, J; Lasella, S; Rodriguez-Perez, M A; Velankar, S S, *Polym. Eng. Sci.*, **56**, 9-17, 2015.]

develop physical crosslinks by hydrogen bonding, physical or ionic attraction. These interactions may similarly act on the matrix as crosslinks or branches, but they also may influence the polymeric shell (namely, reinforce) of the bubble.

Figure 6.6 shows a potential mechanism by which the wall surface of the bubble is reinforced.[7] This method was adapted to poly(lactic acid), which has inferior melt

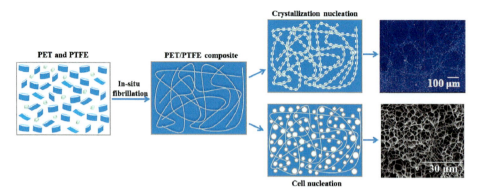

Figure 6.8. Effect of PTFE fibrils on morphology and stability of PET foams. [Adapted, by permission, from Jiang, C; Han, S; Wang, X, *Polymer*, **212**, 6, 123171, 2021.]

strength.[7] The addition of a few percent of a polytetrafluoroethylene improved the foaming process.[7] Because of their low surface energy, the PTFE particles adsorb on the foam bubbles' inner surface with a high surface coverage and endow the bubbles with an interfacial "shell" that prevents coalescence.[7] In successful applications, the particles must have low surface energy, and the polymer must have high surface tension.[7] Figure 6.7 shows that foams are stable (they can withstand a long heating time), and their stability is improved by PTFE particles that reside on the surface of bubbles.[7]

Microcellular poly(ethylene terephthalate) foams were fabricated with polytetrafluoroethylene fibrils.[8] PTFE network impeded mobility of PET chains and increased the crystal nucleation sites.[8] The crystallization temperature, crystallization rate, and viscoelastic behavior of PET were promoted effectively by PTFE fibrils.[8] The rheological percolation value was attained when PTFE content was 1 wt%.[8] During the foaming process, the presence of physical PTFE fibrils network structure promoted nucleation of crystallization and restricted the growth of cells (Figure 6.8).[8] PET foams with 1 wt% PTFE possessed the smallest cell diameter and the highest cell density.[8] The cell diameter with 1 wt% PTFE was reduced from 22 to 2 μm, and the cell density increased by three orders of magnitude from 10^8 to 10^{11} cells/cm^3.[8]

The molecular interaction, foaming ability, and foam stability of the mixtures of hydrophilic silica nanoparticles, a cationic hydrocarbon surfactant, and a nonionic short-chain fluorocarbon surfactant were studied.[9] The nanoparticle concentration had a considerable impact on surface activity, dynamic viscosity, conductivity, foaming ability, and foam stability.[9] The surface activity and dynamic viscosity of the foam dispersions increased rapidly with increasing nanoparticle concentration, but conductivity and foaming ability of the dispersions decreased with the addition of nanoparticles and further decreased with increasing nanoparticle concentration.[9] With nanoparticle concentration below 0.1%, foam drainage and coarsening were accelerated.[9] At a nanoparticle concentration above 0.1%, the addition of nanoparticles enhanced foam stability by decelerating foam drainage and foam coarsening.[9] The foam stability was further enhanced as the nanoparticle concentration increased.[9]

In another development, carbon nanotubes were used to improve the stability of poly(ethylene-co-octene) microfoam. Multiwall carbon nanotubes, MWNTs, bonded well

Foam Stabilization

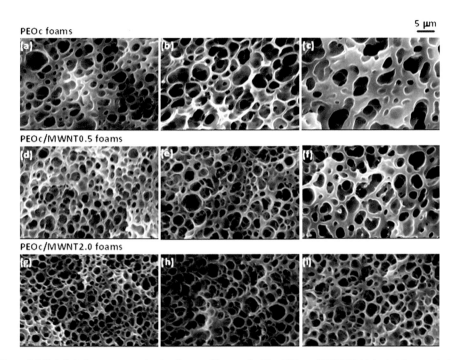

Figure 6.9. Poly(ethylene-co-octene) microfoams without and with addition of MWNT. [Adapted, by permission, from Zhai, W; Wang, J; Chen, N; Naguib, H E; Oark, C B, *Polym. Eng. Sci.*, **52**, 2078-89, 2012.]

with the PEO matrix.[3] This facilitated the orientation of MWNTs when shear and extensional forces were applied to the nanocomposite melts. MWNT is oriented in the cell wall.[3] Because of the strain hardening, MWNTs acted as self-reinforcing elements, protecting cells from destruction during their growth.[3] Figure 6.9 shows that MWNT improves the stability of PEO foam.[3] Figure 6.10 shows fiber-like materials having a size of 30-70 nm crossing the holes.[3] These holes are not the actual cells, but the broken parts of the cell wall that are the result of cell coalescence.[3] In other words, the fiber-like materials should be located in the cell wall.[3]

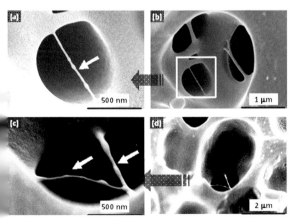

Figure 6.10. SEM micrographs of PEOc/MWNT0.5 foam (a, b) and PEOc/MWNT1.0 foam (c, d). These micrographs were used to show the distribution of MWNT in the cell structure. [Adapted, by permission, from Zhai, W; Wang, J; Chen, N; Naguib, H E; Oark, C B, *Polym. Eng. Sci.*, **52**, 2078-89, 2012.]

The foaming behavior of thermoplastic polyurethane is difficult to regulate due to its microphase-separation structure.[10] Organic montmorillonite was used in TPU foaming to regulate cell nucleation, growth, and stabilization.[10] Combination of hydrogen bonding and topological network formed by montmorillonite helped in optimizing TPU foam structure by simultaneous control of cell nucleation, growth, and stabilization.[10] The optimal montmorillonite loading was 1 wt%.[10]

Significantly higher melt viscosity and elasticity and reduced gas solubility of poly(3-hydroxybutyrate-co-3-hydroxyvalerate)/cellulose acetate butyrate blends contributed to retardation of bubble coalescence and collapse during foam expansion.[1] Blending with cellulose acetate butyrate increased melt viscosity and elasticity.[1] Blend having higher viscosity was able to withstand higher initial gas content.[1] The resulting blends had a more uniform cell size distribution.[1]

In PEEK/PEI blends, the crystallization rate was the main controlling parameter for the density reduction of foam.[7] PEEK/PEI=1:1 foams had the lowest crystallinity after foaming, and they had the best density reduction and improved morphological parameters.[11]

The siloxane-based amphiphiles are used as cell stabilizers at the polymer-gas interface during polyurethane foam blowing to prevent bubble coalescence.[12] They are used to generate a surface tension gradient at the interface.[12] The surfactants were used for tuning the pore size and morphology of the polyurethane foams.[12]

If polypropylene foaming composition exposure to ionizing radiation is too low, the cell stability is not maintained upon foaming.[13] If the exposure is too high, moldability of the foam compositions may be poor and/or the components may be degraded.[13] Stability of cells in polypropylene foam has been improved by the use of divinylbenzene as the crosslinking monomer.[14]

We were not able to report on the extensive developments, but some research findings already exist to pursue several avenues leading to the improvement of cell stability by modification of polymer structure, blending with other polymers, interfering with crystallinity, improving interaction by addition of special fillers (especially nanofillers), and reinforcing bubble shell with elongated fillers or powders of low surface energy that preferentially settle on the air surface of the bubble.

REFERENCES

1 Liao, Q; Billington, S; Frank, C W, *Polym. Eng. Sci.*, **52**, 1495-508, 2012.
2 Feng, J; Bertelo, C, *J. Rheol.*, **48**, 439-62, 2004.
3 Zhai, W; Wang, J; Chen, N; Naguib, H E; Oark, C B, *Polym. Eng. Sci.*, **52**, 2078-89, 2012.
4 Zhang, H; Liu, T; Li, B; Xin, Z, *Polymer*, in press 123209, 2020.
5 Shaayegan, V; Wang, G; Park, C B, *Chem. Eng. Sci.*, **155**, 27-37, 2016.
6 Leung, S N; Wong, A; Guo, Q; Park, C B; Zong, J H, *Chem. Eng. Sci.*, **64**, 23, 4899-907, 2009.
7 Lobos, J; Lasella, S; Rodriguez-Perez, M A; Velankar, S S, *Polym. Eng. Sci.*, **56**, 9-17, 2015.
8 Jiang, C; Han, S; Wang, X, *Polymer*, in press, 123171, 2021.
9 Sheng, Y; Xue, M; Liu, M, *Chem. Eng. Sci.*, **228**, 115977, 2020.
10 Lan, B; Li, P; Gong, P, *Polymer*, in press, 123159, 2021.
11 Cafiero, L; Iannace, S; Sorrentino, L, *Eur. Polym. J.*, **78**, 116-28, 2016.
12 Hasan, S M; Easley, A D; Browning Monroe, M B; Maitland, D J, *J. Colloid Interface Sci.*, **478**, 334-43, 2016.
13 Baldwin, J J; Sieradzki, P; Sharps, G V, **WO2012158203**, *Toray Plastics (America), Inc.*, Nov. 22, 2012.
14 Sieradzki, P; Baldwin, J J; Lippy, J S, **US9260577**, *Toray Plastics (America), Inc.*, Feb. 16, 2016.

7

FOAMING EFFICIENCY MEASURES

Several foaming efficiency measures are used in practice; some of them were already mentioned in the previous chapters. Here, we discuss them in alphabetical order to have their comparison in one place.

7.1 CELL SIZE

Specimens were immersed in liquid nitrogen for 10 min, fractured, and spurted with a palladium coating for 45 s. The average cell sizes and cell densities were obtained from the SEM micrographs by image analysis using the Image-Pro Plus 6.0 (Media Cybernetics, Silver Spring, MD).[3] The number average diameter of all cells on the micrograph was estimated using the following equation:[3]

$$d = \frac{\sum d_i n_i}{\sum n_i} \quad [7.1]$$

where:
- d average diameter
- d_i diameter of cell i-type
- n_i number of cells i-type

The average size of the cell was determined using a field emission scanning electron microscope.[1] The dimensions of fifty cells were measured and averaged.[1]

ASTM D3576 specifies the cell size measurement method for rigid cellular plastics, which discusses a sampling of thin and thick slices, the techniques of manual cell counting, and interpretation of results.[8]

The nanofoams have an average cell size in the range of 40-100 nm.[2] In microcellular foams, the bubble size is usually less than 10 μm.[3] Cells obtained by the extrusion foaming had sizes of 58-290 μm.[4] According to ASTM standard[5] microcellular urethane foams should have cell diameters of 1-10 μm.

7.2 CELL DENSITY

The cell density is determined by image analysis, as discussed in the previous section.[3] The following equation is used to calculate cell density:[3]

$$N_0 = \left(\frac{n}{A}\right)^{\frac{3}{2}} \quad [7.2]$$

where:
- N_0 cell density
- n   number of cells in area A
- A   area.

In microcellular foams, bubble densities are in the range of 10^8 to 10^9 cells cm^{-3},[3] as well as larger than 10^9 cells cm^{-3}.[9] Cells obtained by extrusion foaming had cell densities of 650-180,000 cells cm^{-3}.[4]

7.3 CELL WALL THICKNESS (AVERAGE)

The average cell wall thickness can be calculated from the following equation:[10]

$$\delta = d \times \left(\frac{1}{\sqrt{1 - \rho_f/\rho_p}} - 1 \right) \quad [7.3]$$

where:
- d   average cell size
- ρ_p density of unexpanded polymer composition
- ρ_f density of the foam

7.4 FOAM DENSITY

The foam density can be measured according to ASTM D792, which involves weighting polymer foam in water using a sinker. The result is calculated from the following equation:[7]

$$\rho_f = \frac{a}{a + w - b} \rho_{water} \quad [7.4]$$

where:
- a   apparent mass of specimen in the air without a sinker
- b   apparent mass of specimen and sinker completely immersed in water
- w   apparent mass of the totally immersed sinker.

According to ASTM D1622,[6] the term density should be understood as being the apparent overall density of the material to be used with skin; otherwise the density is known as the apparent core density if the skins are removed.[6] The density, according to D1622, is measured by the same method as described in the next sentence.[7] The foam density can be determined based on the precise weight (0.1 mg) and the dimensions (0.1 mm or +/−0.1% in ASTM D1622) of five samples.[7] Density of polyurethane foams can be determined according to ASTM standards D792 or D1622.[6]

The density of rigid and irregularly shaped molded cellular materials can be determined using ASTM standard D7710.[13] Determination can be done using one of the two methods: displacement method or bulk density method.

Polypropylene foam densities were produced in the range of 0.3-0.6 g cm^{-3}.[11] Microcellular urethane foam should have a minimum density of 160 kg m^{-3}.[5]

7.5 EXPANSION RATIO (BY VOLUME)

The volume expansion ratio of foam is calculated from the following equation:[12]

$$R_v = \frac{\rho_p}{\rho_f} \qquad [7.5]$$

where:
- R_v volume expansion ratio
- ρ_p density of unexpanded polymer composition
- ρ_f density of the foam.

The densities of foamed samples were determined from Archimedes' law by weighing the polymer foam in water with a sinker using an electronic analytical balance and Eq. 7.4 to calculate the density.[14]

7.6 OPEN CELL CONTENT

The percentage of open cells can be measured with an Eijkelkamp-Lange air pycnometer according to ASTM D6226. The following equation is used according to the ASTM standard:[15,16]

$$C = \frac{V_{sample} - V_{pycnometer}}{Vf \times V_{sample}} \qquad [7.6]$$

where:
- C open cell percentage
- V_{sample} the geometrical volume, (calculated from the specimen dimensions),
- $V_{pycnometer}$ volume measured with the pycnometer
- Vf see eq. 7.7

The test method is developed for determination of porosity in which the accessible cellular volume of a cellular plastic is determined by the application of the Boyle's Law, which states that the increase in the volume of a confined gas results in a proportionate decrease in pressure.[16] Standard contains a full description of equipment, its operation, and calculation of the results. Ultrapyc 1200e from Quantachrome Instruments is the most frequently used instrument for this determination.

ISO standard ISO 4590 contains two methods of determination of open-cell content: pressure variation and volume expansion.[17] The pressure variation method is similar to the method used in ASTM D6226.

7.7 VOID FRACTION

The void fraction is determined from the following equation:[15]

$$Vf = \left(1 - \frac{1}{R_v}\right) \times 100 \qquad [7.7]$$

The nanofoams have a void fraction in the range of 25-64%.[2]

REFERENCES

1. Lee, Y; Jang, M G; Choi, K H; Han, C; Kim, W N, *J. Appl. Polym. Sci.*, **133**, 43557, 2016.
2. Aher, B; Olson, N M; Kumar, V, *J. Mater. Res.*, **28**, 17, 2366-73, 2013.
3. Xi, Z; Chen, J; Liu, T; Zhao, L; Turng, L-S, *Chinese J. Chem. Eng.*, **24**, 180-9, 2016.
4. Liao, Q; Tsui, A; Billington, S; Frank, C W, *Polym. Eng. Sci.*, **52**, 1495-1508, 2012.
5. ASTM Standard D3489-17. Standard Test Methods for Microcellular Urethane Materials.
6. ASTM D1622-20. Standard Test Method for Apparent Density of Rigid Cellular Plastics.
7. Long, Y; Sun, F; Liu, C; Xie, X, *RSC Adv.*, **6**, 23726-36, 2016.
8. ASTM D3576-20. Standard Test Method for Cell Size of Rigid Cellular Plastics.
9. Ma, Z; Zhang, G; Yang, Q; Shi, X; Shi, A, *J. Cellular Plast.*, **50**, 1, 55-79, 2014.
10. Wang, X; Feng, N; Chang, S, *Polym. Compos.*, **34**, 849-59, 2013.
11. Saiz-Arroyo, C; Rodriguez-Perez, M A; Tirado, J; Lopez-Gil, A; de Saja, J A, *Polym. Int.*, **62**, 1324-33, 2013.
12. Fan, C; Wan, C; Gao, F; Huang, C; Xi, Z; Xu, Z; Zhao, L; Liu, T, *J. Cellular Plast.*, **52**, 3, 277-98, 2016.
13. ASTM D7710-14. Standard Test Method for Determination of Volume and Density of Rigid and Irregularly Shaped Molded Cellular Materials.
14. Yang, C; Zhang, Q; Wu, G, *Polym. Deg. Stab.*, in press, 109406, 2021.
15. Kuboki, T, *J. Cellular Plast.*, **50**, 2, 113-28, 2014.
16. ASTM D6226-15. Standard Test Method for Open Cell Content of Rigid Cellular Plastics.
17. ISO 4590-2016. Rigid cellular plastics — Determination of the volume percentage of open cells and of closed cells.

8

MORPHOLOGY OF FOAMS

Many aspects of foam morphology are discussed here, including bimodal morphology, cell density, cell size, cell wall thickness, closed cells, core, transfer and skin thicknesses, morphological features, and open cells. All these topics will be discussed in the above order.

8.1 BIMODAL MORPHOLOGY

Foams having bimodal structures have better thermal insulation properties and improved mechanical performance.[1] Similar to the application of fillers combination of large and small cells fills more space due to the geometrical packing advantage.

Many technological means have been exploited in attempts to develop bimodal morphology. The successful ones include:[2]
- blend phase morphology (the first bubble occurs in the less rigid domain, the second bubble occurs in the more rigid domain)[2,3]
- two-stage reduction of pressure (the second pressure drop induces secondary bubbles)[1]
- synergistic effect of temperature increase and depressurization[4]
- use of two blowing agents (e.g., water in the water-absorptive polymer as the second blowing agent in polystyrene foaming; a mixture of carbon dioxide and 2-ethyl hexanol (solubility difference in 2-ethyl hexanol and polystyrene was the reason for the formation of bimodal morphology); water and carbon dioxide[5])
- acute depression of carbon dioxide solubility in the polymer matrix during the foaming process[6]
- the time lag between heterogeneous (on crystal interface) and homogeneous nucleation (in the amorphous region)
- increasing the sorption pressure and depressurization rate and setting the foaming temperature close to the melting temperature of polypropylene (depressurization rate enabled to grow smaller bubbles in the crystalline region and larger bubbles in the amorphous region)
- blending PMMA having different molecular weights and using CO_2 as the blowing agent[6]
- use microcellular molding in combination with fast curing[7]
- use nanofillers in polymers foamed by supercritical carbon dioxide.[8] Nanofillers such as graphene, carbon nanotubes, and silica were used in thermoplastic polyurethane and polystyrene to apply heterogeneous nucleation effect that resulted in the formation of bimodal foam[8]
- the cellular structure of PP/NaCl foams depended on the mixing state of water and CO_2 in PP/NaCl composites[9]

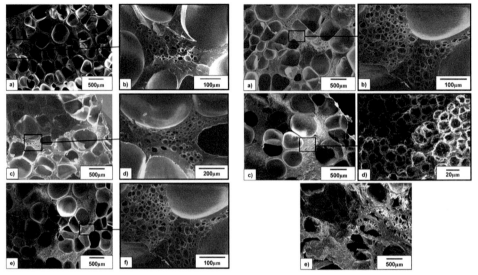

Figure 8.1. Different depressurization rates for the polymer–monomer system, polystyrene/methylmethacrylate 50:50 + 0.5 wt% diurethane dimethacrylate. (a and b) dp/dt = 1.5 bar s^{-1}, (c and d) dp/dt = 4.5 bar s^{-1}, and (e and f) dp / dt = 9 bar s^{-1}. [Adapted, by permission, from Kohlhoff, D; Nabil, A; Ohshima, M, *Polym. Adv. Technol.*, **23**, 1350-6, 2012.]

Figure 8.2. Different polymer–monomer systems foamed with the same depressurization rate dp/dt = 9 bar s^{-1}. (a and b) PS/MMA 50:50 + 0.5 wt% diurethane dimethacrylate, DUDMA, (c and d) PS/MMA 33:67 + 0.5 wt% DUDMA, and (e) PS/MMA 15:85 + 0.5 wt% DUDMA. [Adapted, by permission, from Kohlhoff, D; Nabil, A; Ohshima, M, *Polym. Adv. Technol.*, **23**, 1350-6, 2012.]

- bimodal open-porous scaffold architecture was obtained by synergistic control of temperature variation and a two-step depressurization in a supercritical carbon dioxide foaming process[10]

We show some examples of these technologies in the discussion below.

Exploiting polymer blend approach, the interpenetrating network of polystyrene and polymethylmethacrylate was used.[2] Saturation (60°C) with carbon dioxide and polymerization of methylmethacrylate dissolved in polystyrene (100°C) were performed in an autoclave followed by controlled depressurization to foam blend.[2] Figure 8.1 shows the effect of the depressurization rate on the formation of the bimodal structure.[2] The average diameter of the large bubbles was in the range of 200-400 μm, and that of the small bubbles was 10-30 μm.[2] The depressurization rates did not drastically affect the size of bubbles but affected the size of small bubble area.[2] The larger the depressurization rate, the larger the area of small bubbles.[2]

Figure 8.2 shows SEM micrographs of foams prepared from three PS/MMA solutions having different ratios, 50:50, 33:67, and 15:85.[2] The larger the concentration of methylmethacrylate, the larger the number of microbubbles, and the smaller the number of large bubbles.[2] The number density of large bubbles decreased, but their average diameter increased, and the foam density decreased from 0.26 to 0.23 g cm^{-3} because of the increase in the percentage of the microbubble area.[2]

In polypropylene/polystyrene blends, the large cells were mainly formed in the polystyrene phase, while the small ones were mostly formed in the polypropylene phase.[1] Fig-

8.1 Bimodal morphology

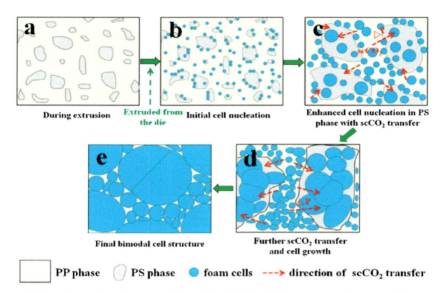

Figure 8.3. Formation of bimodal structure in PP/PS blend. [Adapted, by permission, from Wang, K; Pang, Y; Wu, F; Zhai, W; Zheng, W, *J. Supercritical Fluids*, **110**, 65-74, 2016.]

ure 8.3 shows the proposed mechanism of this formation process. Figure 8.3a shows the mixing and dispersion states of the polypropylene/polystyrene blend during the extrusion process with the dissolved supercritical carbon dioxide. Once the polymers/gas solution was extruded from the die, the cells were preferentially nucleated in the polypropylene-rich phase and at the polypropylene/polystyrene interface (Figure 8.3b). During the extrusion foaming process, polypropylene melt had a higher affinity for supercritical carbon

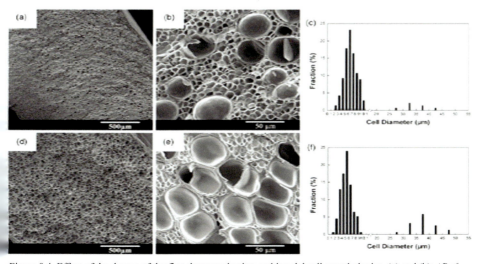

Figure 8.4. Effect of the degree of the first depressurization on bimodal cell morphologies. (a) and (b): ΔP=6 MPa, with magnifications of 100 and 1000, respectively. (d) and (e): ΔP=10 MPa, with magnifications of 100 and 1000, respectively. (c) and (f): statistical size distributions of cell diameters. [Adapted, by permission, from Ma, Z; Zhang, G; Yang, Q; Shi, X; Shi, A, *J. Cellular Plast.*, **50**, 1, 55-79, 2014.]

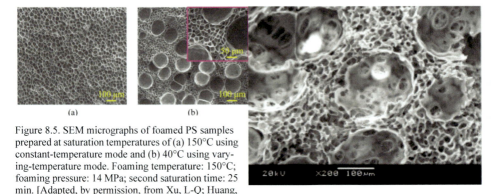

Figure 8.5. SEM micrographs of foamed PS samples prepared at saturation temperatures of (a) 150°C using constant-temperature mode and (b) 40°C using varying-temperature mode. Foaming temperature: 150°C; foaming pressure: 14 MPa; second saturation time: 25 min. [Adapted, by permission, from Xu, L-Q; Huang, H-X, J. Supercritical Fluids, 109, 177-85, 2016.]

Figure 8.6. Result of foaming at temperature of 140°C using foaming strategy explained in Figure 8.7. [Adapted, by permission, from Li, D-C; Liu, T; Zhao, L; Yuan, W-K, J. Supercritical Fluids, 60, 89-97, 2011.]

dioxide than polystyrene. This resulted in enhanced cell nucleation in the polypropylene phase. The cell nucleation was enhanced at the interphase regions because of the reduced energy barrier, as suggested by the classical heterogeneous nucleation theory. Because of the high concentration gradient of supercritical carbon dioxide across the polypropylene/polystyrene interface, the diffusion of supercritical carbon dioxide occurs from the polypropylene phase to the polystyrene phase (Figure 8.3c). The cell nucleation in the interfacial regions and in the polystyrene phase is enhanced. The excess supercritical carbon dioxide may also promote the growth of polystyrene cells. Because of the persistent pressure difference between the polypropylene phase and the polystyrene phase, the large cells (in the polystyrene phase) become even larger, whereas the growth of small cells (in the polypropylene phase) is hindered (Figure 8.3d). After the temperature falls below the crystallization temperature of polypropylene, the bimodal cell structure is stabilized (Figure 8.3e).

A two-step batch depressurization process is applied to produce bimodal polycarbonate foams.[3] These foams exhibit significantly improved tensile properties compared to the unimodal foams.[3] The extensional stress-induced nucleation promotes the generation of nanocellular structures around the expanding larger cells if the foaming temperature is increased to 160°C.[3] Figure 8.4 shows the effect of depressurization in the first step on the formation of the bimodal morphology.[3] A higher degree of depressurization results in a greater number of large cells and fewer small cells.[3] No bimodal cell structure is formed when the degree of the first depressurization is higher than 10 MPa or lower than 6 MPa.[3] When the foaming temperature is raised to 160°C, nanocellular structures are generated by a stress-induced nucleation mechanism.[3]

By manipulating the foaming parameters (foaming temperature, saturation temperature, second saturation time, and foaming pressure), the bimodal cell structure in the polystyrene foams is obtained with a wide range of cell diameters (50.0-713.7 μm) and cell densities (2.1×10^3-3.6×10^6 cells/cm^3), and a wide range of foam densities (0.09–0.73 g/cm^3).[4] Figure 8.5 shows the effect of constant and variable temperature on the development of bimodal morphology.[4]

8.1 Bimodal morphology

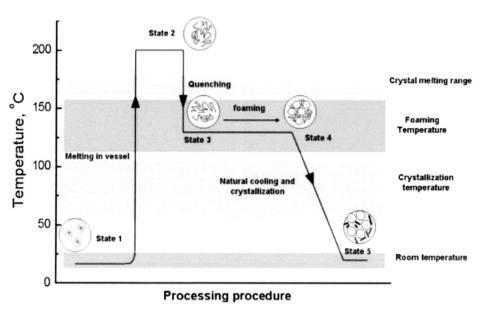

Figure 8.7. Foaming strategy. [Adapted, by permission, from Li, D-C; Liu, T; Zhao, L; Yuan, W-K, *J. Supercritical Fluids*, **60**, 89-97, 2011.]

The bimodal cell morphology was obtained by the extrusion foaming process using carbon dioxide and water as co-blowing agents.[5] Also, two particulate additives were used as nucleation agents.[5] Nanoclay decreased the water foaming time; therefore, both carbon dioxide and water could cause simultaneous blowing.[5] Activated carbon increased the carbon dioxide nucleation rate with little effect on the carbon dioxide foaming time.[5] The bimodal foams had much better compressive properties and slightly better thermal insulation.[5]

The compressed carbon dioxide has a strong plasticization effect on the isotactic polypropylene matrix.[11] It retards the formation of critical size nuclei, which effectively postpones the crystallization peak to a lower temperature region.[11] Because of the acute depression of carbon dioxide solubility in the isotactic polypropylene matrix during the foaming process, the isotactic polypropylene foams with the bimodal cell structure were fabricated (Figure 8.6).[11] A new foaming method to fabricate isotactic polypropylene foams using the unmodified linear isotactic polypropylene and supercritical carbon dioxide as the foaming agent was used.[11] The upper and lower temperature limits of foaming were 155 and 105°C.[11] They were determined by the melt strength and crystallization temperature of the isotactic polypropylene in the presence of supercritical carbon dioxide.[11] Figure 8.7 shows the schematic diagram of the process involved.[11] As shown in Figure 8.7, the material undergoes several states, as follows:[11]

- the iPP specimen is placed in the high-pressure vessel (state 1), and a selected amount of carbon dioxide is charged
- the vessel is heated to a temperature of 200°C (the saturation temperature) and kept at this temperature for 20 min to completely melt iPP crystals and dissolve

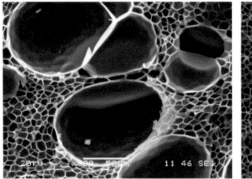 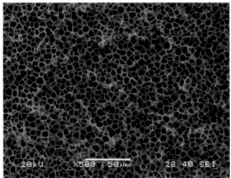

Figure 8.8. SEM images of the cell morphologies with bimodal cell structure (left) and uniform cell structure. (right) A two-step depressurization process: saturation at 20 MPa and 100°C. The pressure was released from 20 to 15 MPa, held at the latter pressure for 60 min, and followed by further depressurization to the ambient pressure. The scale bar is 50 μm. (right) A one-step depressurization process: saturation at 15 MPa and 100°C. The scale bar is 50 μm. [Adapted, by permission, from Bao, J-B; Liu, T; Zhao, L; Hu, G-H, *J. Supercritical Fluids*, **55**, 3, 1104-14, 2011.]

carbon dioxide in the sample (state 2). The 200°C permits complete melting of the iPP matrix, and saturation pressure of 30 MPa allows to achieve high solubility of carbon dioxide and high depressurization rate.

- the high-pressure vessel is quenched at a lower temperature (the foaming temperature), and retained for another 25 min to make sure the whole system will reach a stable state (state 3). During the cooling process, the carbon dioxide pressure decreases to reach stable foaming pressure. At state 3, compressed carbon dioxide postpones the crystallization of iPP and keeps the iPP melt from crystallizing at the lower foaming temperature. The lower temperature also increases the melt strength and enhances the foamability of polymer melts.
- the valve is rapidly opened to release the carbon dioxide to induce cell nucleation and bubble growth (state 4). The maximum depressurization rate was between 300 and 400 MPa/s.
- After a natural cooling process (state 5), the vessel is opened, and the foamed samples are taken out for analysis.

PMMA grades having different molecular weights are compatible.[6] They differ in glass transition temperature and melt flow index, in addition to the differences in molecular weight.[6] Bimodal nanocellular foam with a foam density as low as 0.249 g/cm^3 was successfully generated by blending 60 wt% PMMA-L (M_w=58,000) with 40 wt% PMMA-H (M_w=81,000).[6] The foam structure was controlled by the composition of the blend, without changing processing conditions.[6]

The microcellular injection molding method assisted by a fast cooling produced lightweight, bimodal nanoporous polypropylene foams with high toughness.[7] Mold with a thin cavity was designed to achieve rapid cooling during microcellular injection molding process, and to further accelerate crystallization and refine crystals.[7] With a carbon dioxide content of 12 wt%, a bimodal cellular structure was fabricated (see Figure 5.21).[7] In the case of bimodal foam, its microcellular structure was better than that of microfoam with a cell size of 51 μm and cell density of 3.1×10^5 cells/cm^3, and its nanocellular struc-

8.1 Bimodal morphology

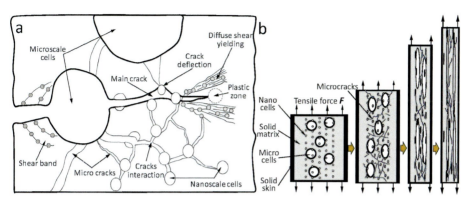

Figure 8.9. Schematic illustration of (a) the deformation process of bimodal foams during tensile testing and (b) the crack toughening mechanism by micro-nano bimodal cellular structure. [Adapted, by permission, from Zhao, J; Qiao, Y; Wang, G; Park, C B, *Mater. Design*, **195**, 109051, 2020.]

ture with a cell size of 490 nm and cell density of 5.4×10^{10} cells/cm^3.[7] Very high gas saturation (12 wt%) not only promoted cell nucleation (formation of nanoscale cells), but also supplied driving force for cell growth, leading to the partial cell coalescence and formation of microscale cells.[7] The bimodal cellular foams had increased toughness that was 327% higher than that of microcellular foams and up to 53% higher than that of solid.[7] The effect of nano and/or bimodal cellular structures on the tensile performance is illustrated in Figure 8.9 based on the toughening mechanism theories.[7] Nano cells enhance the mechanical performance of PP foams *via* matrix crazing and initiating and controlling craze growth.[7] Cells, similar to rubber particles, act as stress concentrators.[7] Nano cells also control craze growth or craze path.[7] For example, a craze could be terminated by encountering a further cell.[7] A large number of small crazes is formed in the nanocellular area, and a small number of large crazes are formed in the solid samples.[7] The occurrence of dense crazing throughout a comparatively large area of nanocellular PP foams absorbs a considerable amount of energy under applied tensile stress, which aids in toughening.[7]

The presence of NaCl increased the adsorption of water by PP/NaCl composites.d Both CO_2 and adsorbed water acted as blowing agents during CO_2 batch foaming.[9] Figure 8.10 illustrates the effect of particle size of salt on the morphology of foam.[9] When the size of NaCl particles is large (Figure 8.10a), only a few particles are present, and cells formed by NaCl as nucleation site and absorbed water rapidly grow to form large cells with a diameter of 895.9±81.2 μm.[9] Small cells of the diameter of 18.9±3.5 μm are formed by carbon dioxide.[9] When the size of NaCl particles is very small (Figure 8.10c), the cellular structure originating from heterogeneous nucleation of NaCl particles has the cell diameter of 48.1±5.1 μm.[9]

Based on the above data, it seems that the two-step depressurization is the most effective (and simplest) method of the creation of the bimodal structure. The mechanism of this process is still to be studied in detail, but the holding stage between the two depressurization steps seems to be the key controlling factor of the bimodal cell structure.[12] During the holding stage, larger cells further grow in size, and smaller ones further decrease in size until complete collapse, by gas diffusion from small cells to large ones.[12] The volume ratio of the large cells to the small ones can be tuned both by the holding time

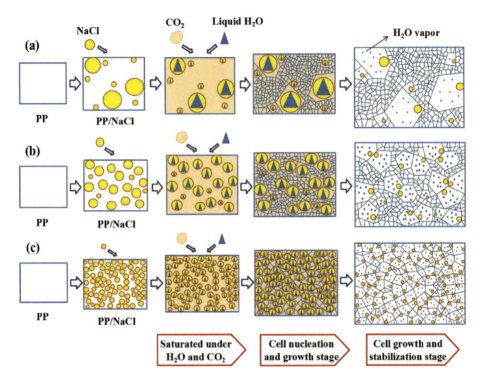

Figure 8.10. Schematic diagram for foaming process of PP/NaCl composites with different size of NaCl particles: (a) NaCl particles with large size; (b) NaCl particles with middle size; (c) NaCl particles with small size. [Adapted, by permission, from Liu, Z; Qiu, J; Tang, T, *Polymer*, **194**, 122406, 2020.]

and the degree of depressurization in the first step.[12] The depressurization rate and the temperatures of the first and the second steps also influence the cell structure.[12] Figure 8.8 shows an example of such a timed process leading to the development of bimodal structure in a polystyrene matrix.[12] Two contrasting pictures of two-step (left) and one step (right) show obvious benefits of the two-step process in the creation of the bimodal structure.[12]

8.2 CELL DENSITY

The method of determination of cell density has been outlined in chapter 7. The previous section shows that the numerical result can sometimes be misleading, especially when the foam has bimodal morphology. Table 8.1 shows the ranges of cell densities obtained in the experimental works.

Table 8.1. Cell density

Polymer	Blowing agent	Cell density, cells per cm^3	Comment	Ref.
PPSU	CO_2	1.2E12-3E15	saturation pressure	24
Non-isocyanate polyurethane	CO_2	1.5E10-1.72E11	increased cell density decreased thermal conductivity	13
PP/EPDM	N_2	4.5E9-1.4E10	organoclay nucleation affected CD	22
PP	ADM	9.5E7-1.5E9	effect of ADM density & content	21
PC	CO_2	5E7-6.1E9	graphene content increase	28
PC	N_2	1.5-8.2E6	cell coalescence reduced CD	20
POM	ADM	2-9E6	CD increases impact strength	26
PMMA/MWCN	CO_2	1E4-1E10	CD improves electromagnetic absorption	25
PCL	CO_2	7.6E4-2.3E6	addition of starch increased CD	18
PP	ADM	6E4-1E7	NCC increased CD 5-37x	15
PS	CO_2	2.7-25.2E4	screw rotation speed increase decreases CD	27
PP	water	3.7E3-1.8E5	talc nucleation	23
LMDPE	ADM	3.6E3-1.8E4	increased with ADM decreased	17
PVAl	water	1E3		19
LDPE	N_2	7E1-3E4	temperature, screw rotation	16
PLA	ADM	2E1-1E3	addition of PTFE increased CD	14
UHMWPE	CO_2	0.3-6.55E6	effect of temperature	29
Phenolic resin	n-pentane	2.9-13.3E6	modification by lignin	30
PBS	CO_2	8.9E6-4E7	modification with CTN	31
TPU	CO_2	1.2E12	addition of nanographite flakes	32

ADM – azodicarbonamide; CD – cell density, NCC – nanocrystalline cellulose

8.3 CELL MORPHOLOGY

Microcellular PS/PE foam obtained by foaming with supercritical carbon dioxide with and without the addition of SEBS copolymer was studied by x-ray microtomography.[33] The reason for the selection of this technique was that the application of SEM to microtomed slices, which were fractured in liquid nitrogen, has given a distorted picture of morphology.[33] The foam without SEBS (Figure 8.11A) has some large irregular bubbles, and the homogeneity of the cells is inferior.[33] The cell morphology of the PS/PE foams was improved with the increased addition of SEBS.[33] The cells of the PS/PE foams with SEBS are nearly spherical and surrounded by thick walls.[33] A smaller mean cell diameter and higher cell density result from the addition of SEBS.[33] The triblock copolymer SEBS acts as a compatibilizer of PS/PE blends, which improves cell morphology.[33]

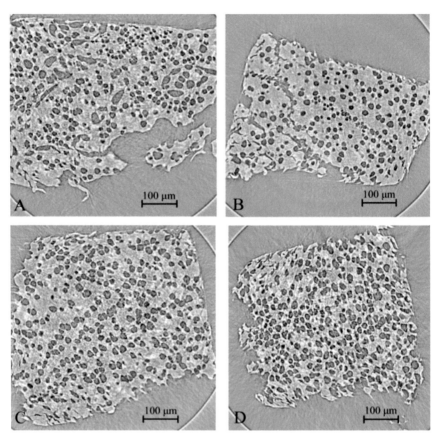

Figure 8.11. Reconstructed slices of PS/PE (70:30) foams with different SEBS content: 0 wt% (A), 2 wt% (B), 5 wt% (C) and 10 wt% (D). [Adapted, by permission, from Xing, Z; Wang, M; Du, G; Xiao, T; Liu, W; Qiang, D; Wu, G, *J. Supercritical Fluids*, 82, 50-5, 2013.]

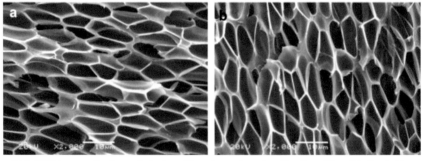

Figure 8.12. SEM images of cell morphologies with various orientation directions. The white bar is 10 μm. [Adapted, by permission, from Bao, J-B; Liu, T; Zhao, L; Hu, G-H; Miao, X; Li, X, *Polymer*, **53**, 25, 5982-93, 2012.]

Compatible LDPE/HDPE blend generated uniform cell morphology.[35] LDPE/PP blend with sea-island micro-structure also generated uniform cell morphology.[35] But, poor

8.3 Cell morphology

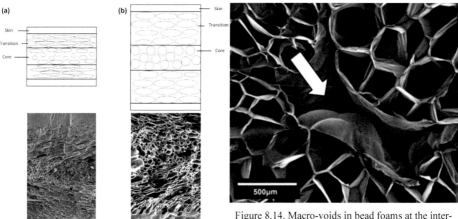

Figure 8.13. SEM micrographs and schematic structures of acoustic foams made (a) without mold opening and (b) with 1.5 mm length of mold opening. [Adapted, by permission, from Jahani, D; Ameli, A; Jung, P U; Barzegari, M R; Park, C B; Naguib, H, *Mater. Design*, **53**, 20-8, 2014.]

Figure 8.14. Macro-voids in bead foams at the intersection of several beads. [Adapted, by permission, from Raps, D; Hossieny, N; Park, C B; Altstädt, V, *Polymer*, **56**, 5-19, 2015.]

foaming of LDPE/PP with co-continuous structure was attributed to a higher carbon dioxide diffusion.[35]

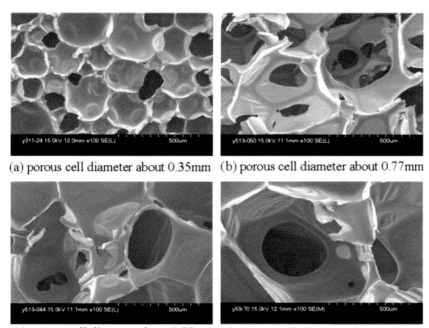

(a) porous cell diameter about 0.35mm (b) porous cell diameter about 0.77mm

(c) porous cell diameter about 0.77mm (d) porous cell diameter about 1.05mm

Figure 8.15. Open, porous cell morphologies of different pore cell diameters. [Adapted, by permission, from Zhang, C; Li, J; Hu, Z; Zhu, F; Huang, Y, *Mater. Design*, **41**, 319-25, 2012.]

Oriented foaming of polystyrene with supercritical carbon dioxide was performed in special molds.[34] With the restriction of the walls of the mold, the expanded sample was controlled to grow in only one direction (a-axis, b-axis, or c-axis) with the other two directions confined.[34] This produced foams with cells oriented in one direction, as can be seen in Figure 8.12.[34] If bubbles are oriented perpendicular to the impact direction, materials has higher impact strength.[34]

A commercially available foam injection-molding machine was enhanced with a mold opening technique to produce polypropylene open-cell acoustic foams.[36] Figure 8.13 shows the difference in the morphology of polypropylene foams obtained without and with mold opening.[36] A foamed structure with an open-cell content of 67% and an expansion ratio of 4.6 was obtained when the mold was opened by 4.5 mm.[36] The injection-molded foams with a cavity and a high open-cell content presented remarkable acoustic properties: a peak absorption coefficient of 0.95 was observed for a foam with a 73% open-cell content and a 9 mm cavity.[36]

Good cohesion between the beads and low content of macro-voids (marked with an arrow in Figure 8.14) is necessary to ensure favorable mechanical properties.[37]

Figure 8.15 shows the open, porous cell morphologies with different cell sizes obtained from polyurethane with water used as a blowing agent.[38] The white part corresponds to the pore shape, whereas the dark part was related to the open holes.[38] Interconnected pore cells in the appropriately chosen porous media can help to enhance their sound absorption performance.[38] Although there is a decrease in the strength of interconnected cells (92.6%) compared to the closed-cell with an interconnected cell ratio of 88.6%, the absorption coefficient of the interconnected cell foam was 0.66 at a low frequency of 250-600 Hz with an increase of 100% compared to the closed-cell ones.[38]

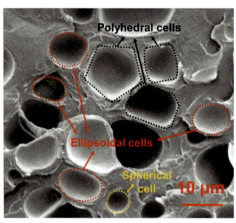

Figure 8.16. Microscopic topography of microcellular foam with void porosity of 37.8% and average cell size of 10.5 μm. [Adapted, by permission, from Zhu, Y; Luo, G; Zhang, L, *Compos. Sci. Technol.*, **192**, 108110, 2020.]

The ellipsoidal cells of PMMA foams had a superior compression performance compared to the spherical and the polyhedral cells (Figure 8.16); the compressive strength increased by 8.2%.[39]

8.4 CELL SIZE

The method of determination of cell size has been outlined in Chapter 7. Section 8.1 shows that the numerical results can be sometimes misleading when the foam has bimodal morphology. In many instances, it is more appropriate to determine cell size distribution, even though the results are more difficult to compare. Table 8.2 shows the ranges of cell density obtained in numerous experimental works.

8.4 Cell size

Table 8.2. Cell size.

Polymer	Blowing agent	Cell size, nm	Comments	Ref.
PPSU	CO_2	21-1220	saturation pressure 6-30 MPa	24
PP/TBS	water	0.2-1.8E3	thermoplastic starch supplies water	23
PS	cyclohexane	0.2-3.2E3	freeze-drying of solvent	47
PP/EPDM	N_2	1.4-2.8E3	nanoclay nucleated	22
PC	CO_2	5.5-19E3	bimodal and unimodal cell structures	3
PS	CO_2	7-20E3	foaming temperature	48
PC/graphene	CO_2	11-60E3 30-210E3	2nd step in the bimodal technology 1st step	49
iPP	CO_2	30-53E3 35-60E3	surface layer core	44
PLA	N_2 and CO_2	30-110E3		43
PC	N_2	42-163E3	mold opening; open cell foams	20
PS/SBS	CO_2	60-120E3	SBS and $CaCO_3$ influenced cell size	41
LPP	CO_2	100-300E3	PTFE particles	46
PEI	CO_2	100-800E3	die pressure affect cell size	45
PUR	HFC-365mfc	155-228E3	perfluoroalkane decreased thermal conductivity of foam	40
PE	ADM	231-623E3	closed/open cells; azodicarbonamide	17
PET	CO_2	265E3	narrow cell size distribution in presence of 0.1 wt% nanosilica	42
PVAl	water	450E3	melamine phosphate	19
PLA	ADM	750-2250E3	PTFE powder stabilization	14
PMMA	CO_2	2-4E3	CNT reinforced	50
PMI		300E3	creep behavior evaluation	51
PMMA	CO_2	42-58	high pressure/rapid depressurization	52
PP/carbon black	CO_2	120-320E3	steam chest molding	53
rigid PVC	ADM	600-800E3	thermal insulation foam	54

Figure 8.17 shows the effect of nanosilica nucleation on the cell size distribution of PET foams.[42] The cell size distribution of the PET foam without nano-SiO_2 has a wide range while the PET foam with 0.1 wt% nano-SiO_2 has narrow cell size distribution.[42] This indicates that the cell structure of the PET foam was improved with the addition of 0.1 wt% nano-SiO_2.[432]

Figure 8.18 shows an example of bimodal cell size distribution obtained by the cooperative action of two blowing agents: azodicarbonamide and water.[55]

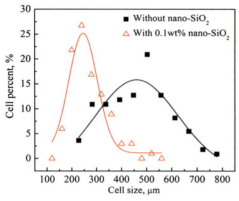

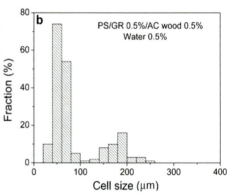

Figure 8.17. The influence of nano-SiO$_2$ on cell size distribution of PET foams. [Adapted, by permission, from Fan, C; Wan, C; Gao, F; Huang, C; Xi, Z; Xu, Z; Zhao, L; Liu, T, *J. Cellular Plast.*, **52**, 3, 277-98, 2016.]

Figure 8.18. Cell size distribution in bimodal polystyrene foam blown with azodicarbonamide and water as the co-blowing agent. [Adapted, by permission, from Zhang, C; Zhu, B; Lee, L J, *Polymer*, **52**, 8, 1847-55, 2011.]

8.5 CELL WALL THICKNESS

Figures 8.19 and 8.20 help us in stressing the importance of knowledge of the wall thickness.[56] Figure 8.19a explicitly indicates that cracks grow by scaling cell walls. Figure 8.19b shows that if there is a stronger element on the crack's way (e.g., the intersection of 3 cell walls that is substantially thicker than cell wall), the crack changes direction to avoid a less convenient (stronger) pathway. Figure 8.20 shows that foam failure is a composite

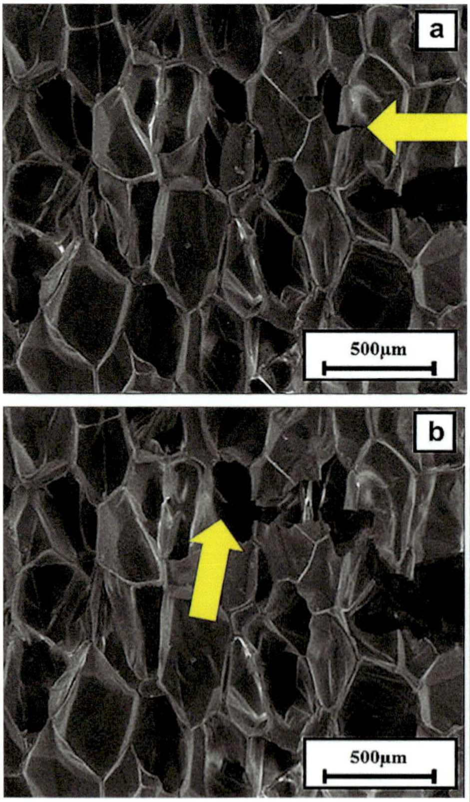

Figure 8.19. The growth of cracks in PES foam. [Adapted, by permission, from Saenz, E E; Carlsson, L A; Karlsson, A M, *Eng. Fract. Mech.*, **101**, 23-32, 2013.]

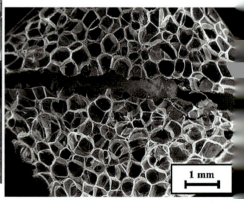

Figure 8.20. Final fracture of PVC foam. [Adapted, by permission, from Saenz, E E; Carlsson, L A; Karlsson, A M, *Eng. Fract. Mech.*, **101**, 23-32, 2013.]

8.5 Cell wall thickness

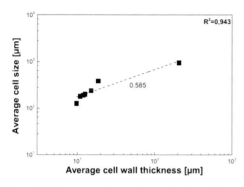

Figure 8.21. Average cell size of extrusion-foamed thermoplastic cellulose ester foamed using HFO 1234ze as a function of the average cell wall thickness. [Adapted, by permission, from Zepnik, S; Hendriks, S; Kabasci, S; Radusch, H-J, *J. Mater. Res.*, **28**, 17, 2394-400, 2013.]

result of individual wall failures. The crack growth in the PES foams occurred through the center of the cells.[56]

Figure 8.21 shows that the larger the average cell, the thicker the average wall.[57] The cell walls in this relationship are calculated from porosity, and they are not actual cell sizes but averages obtained from calculations similar to results which can be obtained using equation 7.3.

Figure 8.22 shows that walls, even in the same cell and also along their length, do not have the same thickness.[58] In addition, cells have different sizes, and walls surrounding them have a thickness not always

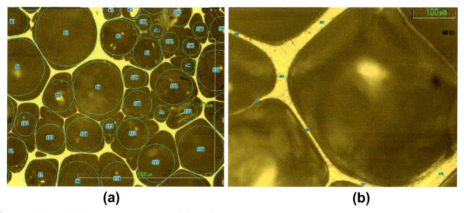

Figure 8.22. (a) Cell size measurement, and (b) cell wall thickness measurement. [Adapted, by permission, from Chen, Y; Das, R; Battley, M, *Int. J. Solids Structures*, **52**, 150-64, 2015.]

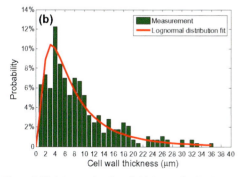

Figure 8.23. Measured cell wall thickness distribution and its probability distribution fit. [Adapted, by permission, from Chen, Y; Das, R; Battley, M, *Int. J. Solids Structures*, **52**, 150-64, 2015.]

relative to the dimensions of a cell. It can be therefore predicted that variation in cell thickness will have an influence on correlations with other properties of foams. And indeed, it is found that when the cell size and cell wall thickness are assumed to be uniform (taken as averages) in the models, the Kelvin, Weaire-Phelan and Laguerre models overpredict the stiffness of the foam.[58] But, Young's modulus and shear modulus predicted by the Laguerre models incorporating measured foam cell size and cell wall thickness distributions (Figure 8.23) agree well with the experimental

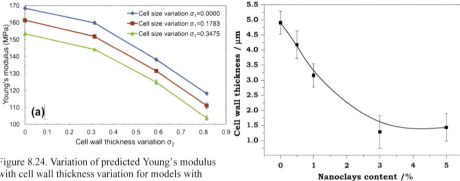

Figure 8.24. Variation of predicted Young's modulus with cell wall thickness variation for models with different cell size variations. [Adapted, by permission, from Chen, Y; Das, R; Battley, M, *Int. J. Solids Structures*, **52**, 150-64, 2015.]

Figure 8.25. Cell wall thickness variation as a function of the nanoclay content. [Adapted, by permission, from Pardo-Alonso, S; Solorzano, E; Brabant, L; Vanderniepen, P; Dierick, M; Van Hoorebeke, L; Rodriguez-Perez, M A, *Eur. Polym. J.*, **49**, 999-1006, 2013.]

data.[58] This emphasizes the fact that the integration of realistic cell wall and cell size variations is vital for foam modeling.[58] Figure 8.24 shows the cell wall thickness distribution. Property prediction depends on cell wall thickness variation.[58]

Figure 8.25 shows a significant reduction in the cell wall thickness when nanoclays are added.[59] Foams with smaller cells show a redistribution of polymer mass from cell walls to struts.[59] This is, on the one hand, caused by the cell size reduction and, on the other hand, due to the strut volume fraction increase.[59] The longer time between blowing and gelling gives more time to capillary forces to enhance drainage in the cell walls and then increase the strut volumetric fraction.[59] The cell wall thickness influences mechanical properties through a higher strut volume fraction.[59] The dimensions of cell walls have a clear influence on the radiative and conductive properties of foams.[59]

LDPE has substantially higher melt strength than LLDPE.[60] With increasing saturation pressure of blowing agent (CO_2), the cell density increased and cell size decreased resulting in a decrease in cell wall thickness, which may affect the formation of open cells.[60] With decreasing cell wall thickness, the open-cell content of LDPE foam was little changed, while the open-cell content of LDPE/LLDPE 90/10 foam increased significantly.[60]

8.6 CLOSED CELL

Foams can be classified as open-cell or closed-cell.[61] The open-cell foams have a network of cell struts but no cell walls.[61] The cells of closed-cell foams are surrounded by thin cell walls and are sealed off from neighboring cells.[61] In modeling of closed-cell foams tetrakaidecahedron is used (it is also used in open-cell models).[61] A tetrakaidecahedron (polyhedron) was introduced by Kelvin (for this reason, models based on it are called Kelvin's models).[61] It consists of 14 faces, of which six are flat squares and eight are slightly curved hexagons.[61] The tetrakaidecahedrons can be packed in a body-centered cubic lattice such as in the Voronoi cell (from another popular model).[61]

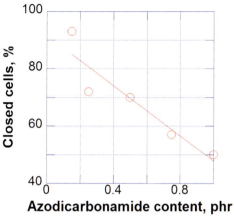

Figure 8.26. Closed-cell content in polyethylene foam vs. azodicarbonamide content. [Data from Moscoso-Sanchez, F J; Mendizabal, E; Jasso-Gastinel, C F; Ortega-Gudino, P; Robledo-Ortiz, J R; Gonzalez-Nunez, R; Rodrigue, D, *J. Cellular Plast.*, **51**, 5-6, 489-503, 2015.]

In polyurethanes, the open-cell foams are most conveniently formed using a one-step reaction involving the production of carbon dioxide when isocyanate reacts with water used in the formulation.[62] In contrast, closed-cell foams are usually formed using physical blowing agents.[62] Use of chain extender (e.g., 1,4-butanediol or malonic acid) helps to increase close-cell fraction.[62] The reduction of the hydrogen-bonded urea aggregation can suppress the cell-opening process, and the addition of the chain extender accelerates the molecular weight build-up in the reacting system and thus increases viscosity, decreasing the tendency of cells to rupture.[62]

Closed-cell foams can be obtained using different blowing agents such as gas obtained from dispersion and dissolution of supercritical gases, a gas formed during foaming of polyurethanes by reaction with water, decomposition of solid blowing agents (e.g., azodicarbonamide), etc. Figure 8.26 shows the effect of the amount of azodicarbonamide on the percentage of closed-cells formed in polyethylene during rotational molding.[17] The general observation from this figure suggests that the smaller the amount of gas produced, the higher the probability of forming closed cells.[17] With *in situ* formations of the gaseous product, the rate of its formation determined the probability of closed-cell formation.[17] This means that prime influence can be expected from the amount of blowing agent, the temperature of foaming, and external pressure.[17] Also, rheological properties of the matrix have influence since they are responsible for preventing coalescence and collapse of cells formed during the decomposition of blowing agent.[17]

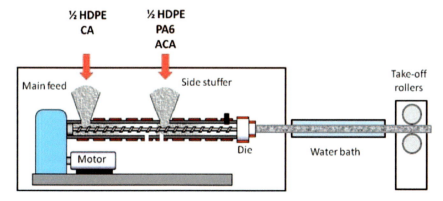

Figure 8.27. Schematic diagram of extrusion set-up used for the production of closed-cell HDPE/PA6 foams. [Adapted, by permission, from Reyes-Lozano, C A; Ortega Gudino, P; Gonzalez-Nunez, R; Rodrigue, D, *J. Cellular Plast.*, **47**, 2, 153-72, 2010.]

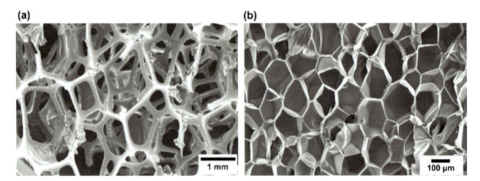

Figure 8.28. (a) Open-cell, (b) closed-cell cellular structure of polymeric foams. Note: Closed-cell foams can have between 1% and 10% open-cell content. [Adapted, by permission, from Okolieocha, C; Raps, D; Subramaniam, K; Altstädt, V, *Eur. Polym. J.*, **73**, 500-19, 2015.]

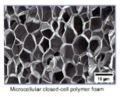 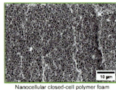

Figure 8.29. Macro- (left) and nanofoams (right). [Adapted, by permission, from Okolieocha, C; Raps, D; Subramaniam, K; Altstädt, V, *Eur. Polym. J.*, **73**, 500-19, 2015.]

Figure 8.27 shows technological means required to solve the problem of the rheological performance of different grades of polyethylene and their blends with polyamide that closed-cell morphology could be produced.[63] The solution was obtained from proper compatibilization of the blend, which had sufficient melt strength to prevent coalesce of gas bubbles.[63]

Figure 8.28 shows that it is very easy to distinguish between closed and open-cell foams based on their morphological features.[64] Also, it is very easy to differentiate between macro- and nanofoams (Figure 8.29).

8.7 CORE & SKIN THICKNESS

Heat dissipates readily from the surface layer, causing a major temperature gradient, which delays the reaction rate in the top layer.[65] This lower temperature on the surface reduces liberation of CO_2 at the top layer and in the proximity to the surface with air increases diffusion of any gas formed, which contribute to a compact layer with increased skin thickness.[65]

Figure 8.30 shows the structure of the injection-molded, foamed part, which includes 3 layers. In the skin, practically no foam exist, some bubbles are present in the transition layer, and, fully consistent with processing technology, bubbles are formed in the core, which determines most properties of the part, especially because it consists most of the part's thickness.[20] Figure 8.31 shows the extruded polypropylene foam.[15] It is pertinent that the skin in the extruded foam has a quite uniform thickness, but transitional and core layers do not.[15] Also, the thickness of the core section is much smaller than in the case of the injection molded part.[15]

8.7 Core & skin thickness 105

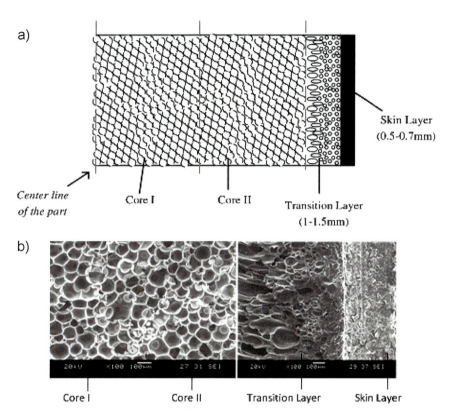

Figure 8.30. a) A schematic diagram showing the overall structure of the foam of injection-molded sample using mold opening, and b) representative SEM micrographs, obtained at temperature of 300°C, showing the skin layer and the transition and core regions of the injection-molded samples using mold opening. [Adapted, by permission, from Jahani, D; Ameli, A; Saniei, M; Ding, W; Park, C B; Naguib, H E, *Macromol. Mater. Eng.*, **300**, 48-56, 2015.]

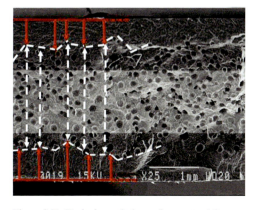

Figure 8.31. Typical morphology of a structural foam: two unfoamed skins (solid lines) enclosing a foamed core (dashed lines). [Adapted, by permission, from Yousefian, H; Rodrigue, D, *J. Appl. Polym. Sci.*, **132**, 42845, 2015.]

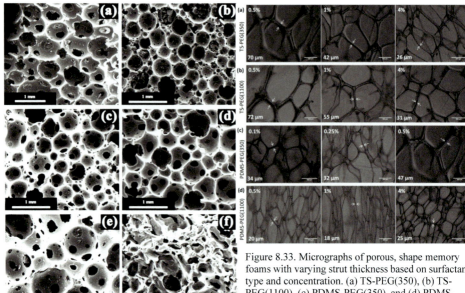

Figure 8.32. SEM micrographs of the graphene-reinforced carbon composite foams prepared at various GO concentrations a 0.0 wt%, b 0.25 wt%, c 0.5 wt%, d 0.75 wt%, e 1.00 wt% and f 1.25 wt%. [Adapted, by permission, from Narasimman, R; Vijayan, S; Prabhakaran, K, *J. Mater. Sci.*, **50**, 8018-28, 2015.]

Figure 8.33. Micrographs of porous, shape memory foams with varying strut thickness based on surfactant type and concentration. (a) TS-PEG(350), (b) TS-PEG(1100), (c) PDMS-PEG(350), and (d) PDMS-PEG(1100). TS – 1,1,1,3,5,5,5-heptamethyltrisiloxane, PEG – polyethylene glycol, PDMS -– polydimethylsiloxane. Adapted, by permission, from Hasan, S M; Easley, A D; Browning Monroe, M B; Maitland, D J, *J. Colloid Interface Sci.*, **478**, 334-43, 2016.]

8.8 MORPHOLOGICAL FEATURES OF FOAMS

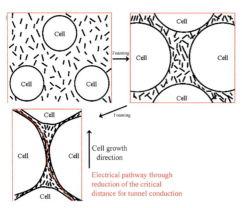

Figure 8.34. Scheme of improved electrical conduction by tunnel mechanism due to reduction of the critical distance between graphene nanoplatelets with foaming. [Adapted, by permission, from Antunes, M; Gedler, G; Abbasi, H; Velasco, J I, *Mater. Today: Proc.*, **3**, 2, S233-39, 2016.]

Figure 8.32 shows carbon foam having almost spherical cells, which are interconnected by circular or oval-shaped pores.[48] The size and number of interconnecting pores increase with an increase in graphene oxide concentration, but when graphene concentration is 1.25 wt%, bubbles collapse, which results in the formation of irregular cells.[66]

Figure 8.33 shows strut morphologies and their thicknesses, which depend on the type and concentration of surfactant used in PU porous, shape-memory foams.[67]

Figure 8.34 shows how foam morphology helps in the improvement of electric conductivity.[68] The electric conductivity of a polymer containing con-

8.8 Morphological features of foams

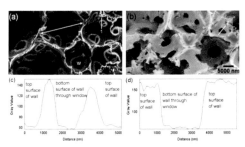

Figure 8.35. (a) Helium ion microscopy image taken with an acceleration voltage of 27.8 kV, (b) SEM image at 1 kV acceleration voltage, (c) intensity profile along line 1 in (a), and (d) intensity profile taken along line 2 in (b). [Adapted, by permission, from Rodenburg, C; Viswanathan, P; Jepson, M A E; Liu, X; Battaglia, G, *Ultramicroscopy*, **139**, 13-19, 2014.]

ductive particles (in this case, graphene nanoplatelets) depends on the distances between neighboring particles, which affect tunnel conductivity.[68] The foam bubbles take space in the polymer matrix by which form pathways with the reduced critical distance between particles.[68]

Helium ion microscopy was used to examine uncoated polymer foams (Figure 8.35).[69] A methods for the measurement of wall thickness variations and wall roughness measurements have been developed, based on the modeling of helium ion transmission.[69] The results indicate that within the walls of the void structure, there are small features with height variations of ~30 nm and wall thickness variations from ~100 nm to 340 nm in the regions surrounding interconnecting cells.[69]

On addition of malonic acid salt (chain extender), the viscosity build-up was accelerated during the reactive processing of polyurethane foam.[62] The resulting changes in the

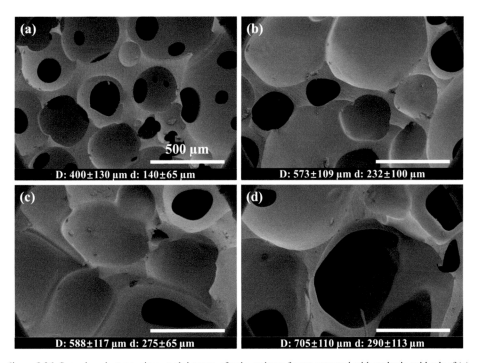

Figure 8.36. Scanning electron micrograph images of polyurethane foams prepared with malonic acid salt of (a) 0%, density 0.193±0.007 (b) 1.5%, density 0.108±0.008 (c) 3%, density 0.128±0.005 and (d) 6%, density 0.182±0.009. D: average cell diameter (μm), d: average pore diameter (μm). [Adapted, by permission, from Zhao, W; Nolan, B; Bermudez, H; van Walsem, J, *Polymer*, **193**, 122344, 2020.]

reaction kinetics lead to foams with a higher degree of closed-cell structure (Figure 8.36).[62]

Highly regular, honeycomb structure PP foam materials were prepared by supercritical CO_2 foaming in the presence of β-nucleating agent (N,N'-dicyclohexyl-2, 6-naphthalenedicarboxamide).[70] Figure 8.37a shows the preparation diagram.[70] The obtained honeycomb closed-cell structure not only has excellent mechanical properties, but also good thermal insulation (26.4 mW/mK) and reusable performance (compressed foam had a thermal conductivity of 37.1 mW/mK).[70] The polygonal honeycomb cells greatly increased the curved path of heat propagation and reduced heat transfer efficiency and enhanced mechanical properties.[70] The hexagon closed cells (Figure 8.37c) were full of air without circulation, which greatly decreased thermal conduction.[70] This polygonal structure simultaneously endowed foam with outstanding compression resistance, which indicated good reusability (contradicts some earlier discussed observations).[70]

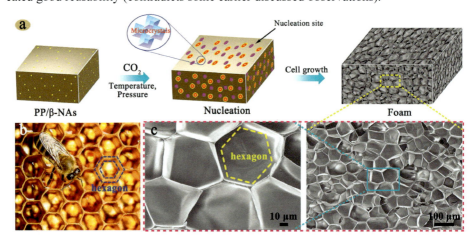

Figure 8.37. Schematic diagram of the preparation of honeycomb structure PP foams. (b) Photograph of honeycomb. (c) SEM images showing the microstructure of the PP foam. [Adapted, by permission, from Yang, C; Zhang, Q; Wu, G, *Polym. Deg. Stab.*, in press, 109406, 2021.]

8.9 OPEN CELL

Contrasting examples of open-cell foams were given in Section 8.6, and many morphological features were already discussed throughout this chapter. Below are a few examples of how these foams can be made.

Open-cell foams made out of natural rubber are used for sound absorption.[71] Open-cell foams have higher efficiency than closed-cell foams in absorbing sound energy.[71] Potassium oleate can be used as a blowing agent to create open-cell foam from natural rubber.[71] The addition of natural fibers results in a decrease in the average cell size and an increase of sound absorption coefficient.[71]

Expansion ratios of 1 to 8 and open-cell content up to 85% were obtained using foam injection molding of polycarbonate foamed with nitrogen.[20] Their sound wave transmission loss was increased by 250% compared with the unfoamed formulation.[20]

The molecular structure of isocyanate has a strong effect on the formation of interconnecting pores in polyurethane foam.[72] The addition of modified isocyanate containing uretonimine linkages improves sound absorption.[72] The optimal amount of uretonimine linkages for good sound absorption is 4.3x10^5 mol/g, giving the highest number of partially open pores.[72]

The bimodal structures are also suitable for acoustic absorption applications providing they have an open structure.[2] The open structure in the bimodal foams can be obtained by several methods, including[2]
- high temperature differences between the surface and core of an extrudate
- use of mixed blowing agents
- polymer blends with different crystallization temperatures
- non-homogeneity of the polymer blend phase morphology.

The second bubble nucleation forms pores that interconnect cells formed during the first stage.[2]

Water can be used as a blowing agent in the production of open-cell morphology in polypropylene foams.[23] Considering that the polypropylene is hydrophobic, the addition of thermoplastic starch as a water carrier permits effective operation of this process.[23] Up to 95.6% open cells can be produced in this process.[23] The resultant material is suitable for oil cleanup applications.[23]

REFERENCES

1   Wang, K; Pang, Y; Wu, F; Zhai, W; Zheng, W, *J. Supercritical Fluids*, **110**, 65-74, 2016.
2   Kohlhoff, D; Nabil, A; Ohshima, M, *Polym. Adv. Technol.*, **23**, 1350-6, 2012.
3   Ma, Z; Zhang, G; Yang, Q; Shi, X; Shi, A, *J. Cellular Plast.*, **50**, 1, 55-79, 2014.
4   Xu, L-Q; Huang, H-X, *J. Supercritical Fluids*, **109**, 177-85, 2016.
5   Zhang, C; Zhu, B; Li, D; Lee, J, *Polymer*, **53**, 12, 2453-42, 2012.
6   Yeh, S-K; Demewoz, N M; Kurniawan, V, *Polym. Testing*, **93**, 107004, 2021.
7   Zhao, J; Qiao, Y; Wang, G; Park, C B, *Mater. Design*, **195**, 109051, 2020.
8   Chen, Y; Weng, C; Yang, J, *J. Supercritical Fluids*, **147**, 107-15, 2019.
9   Liu, Z; Qiu, J; Tang, T, *Polymer*, **194**, 122406, 2020.
10 Ju, J; Gu, Z; Kuang, T, *Int. J. Biol. Macromol.*, **147**, 1164-73, 2020.
11 Li, D-C; Liu, T; Zhao, L; Yuan, W-K, *J. Supercritical Fluids*, **60**, 89-97, 2011.
12 Bao, J-B; Liu, T; Zhao, L; Hu, G-H, *J. Supercritical Fluids*, **55**, 3, 1104-14, 2011.
13 Grignard, B; Thomassin, J-M; Gennen, S; Poussard, L; Bonnaud, L; Raquez, J-M; Dubois, P; Tran, M-P; Park, C B; Jerome, C; Detrembleur, C, *Green Chem.*, **18**, 2206-15, 2016.
14 Lobos, J; Iasella, S; Rodriguez-Perez, M A; Velankar, S S, *Polym. Eng. Sci.*, **56**, 9-17, 2016.
15 Yousefian, H; Rodrigue, D, *J. Appl. Polym. Sci.*, **132**, 42845, 2015.
16 Gandhi, A; Bhatnagar, N, *Polym. Plast. Technol. Eng.*, **54**, 1812-8, 2015.
17 Moscoso-Sanchez, F J; Mendizabal, E; Jasso-Gastinel, C F; Ortega-Gudino, P; Robledo-Ortiz, J R; Gonzalez-Nunez, R; Rodrigue, D, *J. Cellular Plast.*, **51**, 5-6, 489-503, 2015.
18 Ogunsona, E; D'Souza, N A, *J. Cellular Plast.*, **51**, 3, 245-68, 2015.
19 Guo, D; Bai, S; Wang, Q, *J. Cellular Plast.*, **51**, 2, 145-63, 2015.
20 Jahani, D; Ameli, A; Saniei, M; Ding, W; Park, C B; Naguib, H E, *Macromol. Mater. Eng.*, **300**, 48-56, 2015.
21 Saiz-Arroyo, C; Rodriguez-Perez, M A; Tirado, J; Lopez-Gil, A; de Saja, J A, *Polym. Int.*, **62**, 1324-33, 2013.
22 Keramati, M; Ghasemi, I; Karrabi, M; Azizi, H, *Polym. J.*, **44**, 433-8, 2012.
23 Xu, M Z; Bian, J J; Han, C Y; Dong, L S, *Macromol. Mater. Eng.*, **302**, 149-59, 2016.
24 Bernardo, V; Martín-de León, J; Rodríguez-Pérez, M A, *Mater. Lett.*, **178**, 155-158. 2016.
25 Soltani Alkuh, M; Famili, M H N; Shirvan, M M M; Moeini, M H, *Mater. Design*, **100**, 73-83, 2016.
26 Mantaranon, N; Chirachanchai, S, *Polymer*, **96**, 54-62, 2016.
27 Peng, X-F; Liu, L-Y; Chen, B-Y; Mi, H-Y; Jing, X, *Polym. Testing*, **52**, 225-33, 2016.
28 Gedler, G; Antunes, M; Velasco, J I, *Compos. Part B: Eng.*, **93**, 143-52, 2016.
29 Liu, J; Qin, S; Gao, Y, *Polym. Testing*, **93**, 106974, 2021.
30 Gao, C; Li, M; Zhuang, W, *Composites Part B: Eng.*, **205**, 108530, 2021.

31 Huang, A; Lin, J; Peng, X, *Compos. Sci. Technol.*, **201**, 108519, 2021.
32 Li, X; Wang, G; Zhang, A, *Polym. Testing*, **93**, 106891, 2021.
33 Xing, Z; Wang, M; Du, G; Xiao, T; Liu, W; Qiang, D; Wu, G, *J. Supercritical Fluids*, **82**, 50-5, 2013.
34 Bao, J-B; Liu, T; Zhao, L; Hu, G-H; Miao, X; Li, X, *Polymer*, **53**, 25, 5982-93, 2012.
35 Wan, C; Sun, G; Gao, F; Liu, T; Esseghir, M; Zhao, L; Yuan, W, *J. Supercritical Fluids*, **120**, 421-31, 2017.
36 Jahani, D; Ameli, A; Jung, P U; Barzegari, M R; Park, C B; Naguib, H, *Mater. Design*, **53**, 20-8, 2014.
37 Raps, D; Hossieny, N; Park, C B; Altstädt, V, *Polymer*, **56**, 5-19, 2015.
38 Zhang, C; Li, J; Hu, Z; Zhu, F; Huang, Y, *Mater. Design*, **41**, 319-25, 2012.
39 Zhu, Y; Luo, G; Zhang, L, *Compos. Sci. Technol.*, **192**, 108110, 2020.
40 Lee, Y; Jang, M G; Choi, K H; Han, C; Kim, W N, *J. Appl. Polym. Sci.*, **133**, 43557, 2016.
41 Jing, X; Peng, X-F; Mi, H-Y; Wang, Y-S; Zhang, S; Chen, B-Y; Zhou, H-M; Mou, W-J, *J. Appl. Polym. Sci.*, **133**, 43508, 2016.
42 Fan, C; Wan, C; Gao, F; Huang, C; Xi, Z; Xu, Z; Zhao, L; Liu, T, *J. Cellular Plast.*, **52**, 3, 277-98, 2016.
43 Geissler, B; Feuchter, M; Laske, S; Fasching, M; Holzer, C; Langecker, G R, *J. Cellular Plast.*, **52**, 1, 15-35, 2016.
44 Xi, Z; Chen, J; Liu, T; Zhao, L; Turng, L-S, *Chinese J. Chem. Eng.*, **24**, 180-9, 2016.
45 Aktas, S; Gevgilili, H; Kucuk, I; Sunol, A; Kalyon, D M, *Polym. Eng. Sci.*, **54**, 2064-74, 2014.
46 Wang, K; Wu, F; Zhai, W; Zheng, W, *J. Appl. Polym. Sci.*, **129**, 2253-60, 2013.
47 Vonka, M; Nistor, A; Rygl, A; Toulec, M; Kosek, J, *Chem. Eng. J.*, **284**, 357-71, 2016.
48 Shirvan, M M M; Famili, M H N; Alkuh, M S; Golbang, A, *J. Supercritical Fluids*, **112**, 143-152, 2016.
49 Gedler, G; Antunes, M; Borca-Tasciuc, T; Velasco, J I; Ozisik, R, *Eur. Polym. J.*, **75**, 190-9, 2016.
50 Zhou, D; Yuan, H; Xiong, Y; Luo, G; Shen, Q, *Compos. Sci. Technol.*, **203**, 108614, 2021.
51 Zhong, J; Yang, C; Ma, W; Zhang, Z, *Polym. Testing*, **93**, 106893, 2021.
52 Ono, T; Wu, X; Horiuchi, S; Furuya, T; Yoda, S, *J. Supercritical Fluids*, **165**, 104963, 2020.
53 Li, Y; Lan, X; Wu, F; Zheng, W, *Compos. Commun.*, **22**, 100508, 2020.
54 You, J; Xing, H; Xue, J; Jiang, Z; Tang, T, *Compos. Sci. Technol.*, in press, 108566, 2021.
55 Zhang, C; Zhu, B; Lee, L J, *Polymer*, **52**, 8, 1847-55, 2011.
56 Saenz, E E; Carlsson, L A; Karlsson, A M, *Eng. Fract. Mech.*, **101**, 23-32, 2013.
57 Zepnik, S; Hendriks, S; Kabasci, S; Radusch, H-J, *J. Mater. Res.*, **28**, 17, 2394-400, 2013.
58 Chen, Y; Das, R; Battley, M, *Int. J. Solids Structures*, **52**, 150-64, 2015.
59 Pardo-Alonso, S; Solorzano, E; Brabant, L; Vanderniepen, P; Dierick, M; Van Hoorebeke, L; Rodriguez-Perez, M A, *Eur. Polym. J.*, **49**, 999-1006, 2013.
60 Zhang, H; Liu, T; Li, B; Xin, Z, *J. Supercritical Fluids*, **163**, 104883, 2020.
61 Fahlbush, N-C; Grenestedt, J L; Becker, W, *Int. J. Solids Stuctures*, **97-98**, 417-430, 2016.
62 Zhao, W; Nolan, B; Bermudez, H; van Walsem, J, *Polymer*, **193**, 122344, 2020.
63 Reyes-Lozano, C A; Ortega Gudino, P; Gonzalez-Nunez, R; Rodrigue, D, *J. Cellular Plast.*, **47**, 2, 153-72, 2010.
64 Okolieocha, C; Raps, D; Subramaniam, K; Altstädt, V, *Eur. Polym. J.*, **73**, 500-19, 2015.
65 Mukherjee, M; Gurusamy-Thangavelu, S A; Mandal, A B, *Appl. Surf. Sci.*, **499**, 143966, 2020.
66 Narasimman, R; Vijayan, S; Prabhakaran, K, *J. Mater. Sci.*, **50**, 8018-28, 2015.
67 Hasan, S M; Easley, A D; Browning Monroe, M B; Maitland, D J, *J. Colloid Interface Sci.*, **478**, 334-43, 2016.
68 Antunes, M; Gedler, G; Abbasi, H; Velasco, J I, *Mater. Today: Proc.*, **3**, 2, S233-39, 2016.
69 Rodenburg, C; Viswanathan, P; Jepson, M A E; Liu, X; Battaglia, G, *Ultramicroscopy*, **139**, 13-19, 2014.
70 Yang, C; Zhang, Q; Wu, G, *Polym. Deg. Stab.*, in press, 109406, 2021.
71 Tomyangkul, S; Pongmuksuwan, P; Harnnarongchai, W; Chaochanchaikul, K, *J. Reinf. Plast. Compos.*, **35**, 8, 672-81, 2016.
72 Sung, G; Kim, S K; Kim, J W; Kim, J H, *Polym. Testing*, **53**, 156-64, 2016.

9

FOAMING IN DIFFERENT PROCESSING METHODS

9.1 BLOWN FILM EXTRUSION

A coextrusion method is used for making a blown-film liner having a textured surface (provides improved soil gripping properties) and smooth, untextured edges (improve the integrity of the joint between adjacent liner sheets).[1] The production is carried out using a blown-film extrusion technique for producing a three-layer product.[1] The outer textured layers contain a blowing agent that is activated during extrusion to create the texture.[1] The smooth surface areas comprise only the center layer with no textured layer.[1]

It can be challenging to maintain a good cell structure in blown film extrusion because there is a tendency to distort the foam cells during the blowing process.[2] The developed method has an average aspect ratio of the cells of less than 5:1 and an average maximum dimension of the cells of less than 150 μm.[2] The 0.5 wt% nitrogen was used for the production of the film from MDPE.[2]

Blown film HDPE-MWNT nanocomposites had better mechanical performance due to improved orientation and disentanglement of MWNTs during processing.[3] At the same time, higher blow-up ratios led to the destruction of conductive pathways formed by nanotubes, resulting in nanocomposite with lower electrical conductivities because of increased distance between nanotubes that exceeded the maximum critical distance for electron hopping (~1.8 nm).[3]

9.2 CALENDERING

Films can be produced by calendering PVC mix containing 0.1-3 wt% blowing agent.[4] The first stage of the process is conducted at a temperature below the activation temperature of the blowing agent (150-180°C).[4] The second stage of the process is conducted at a temperature above the activation temperature of the blowing agent (210-250°C).[4]

A cushioned floor covering has a backing material formed from a polymeric waste material and a blowing agent.[5] The blowing agent may be activated either before or after the backing material is adhered to the floor covering.[5] The polymeric waste material contains up to 40 wt% aliphatic polyamide.[5]

9.3 CLAY EXFOLIATION

A blowing agent was used to enhance clay exfoliation in polymer nanocomposites.[6] A dimethyl dehydrogenated tallow ammonium chloride modified montmorillonite (Nanomer 144P) was treated with azodicarbonamide:ZnO=1:1 or sodium bicarbonate before being used in polypropylene or polystyrene formulations.[6] The enhancement of thermal stability,

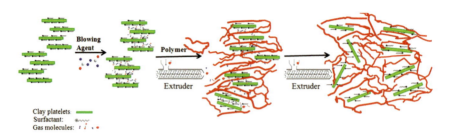

Figure 9.1. Intercalation/exfoliation process in polystyrene/blowing agent-organoclay or polypropylene/polypropylene maleic anhydride (compatibilizer)/blowing agent-organoclay nanocomposites. [Adapted, by permission, from Istrate, O M; Chen, B, *Polym. Int.*, **63**, 2008-16, 2014.]

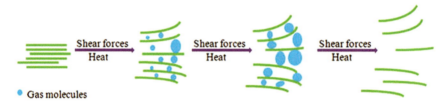

Figure 9.2. Mechanism of exfoliation of blowing agent-treated organoclay during an extrusion process. [Adapted, by permission, from Istrate, O M; Chen, B, *Polym. Int.*, **63**, 2008-16, 2014.]

stiffness strength, ductility, and toughness were attributed to the action of blowing agent in clay interlayers.[6] Blowing agent degraded during melt compounding, generated gases, and pushed clay layers apart, facilitating clay exfoliation under the shear forces.[6] Figures 9.1 and 9.2 show the mechanisms of exfoliation and reinforcement, leading to the improvement of properties of composites.[6] Sodium bicarbonate, having a lower degradation temperature, was preferred due to the lower processing temperature that was beneficial for preserving the structure and properties of the clay surfactant and to the ease of its decomposition into gases during processing.[6]

Tensile mechanical properties, such as tensile modulus and tensile strength, of both solid and foamed formulations of PE/clay improve with the addition of clays.[7] However, despite having a higher clay exfoliation degree in the foamed materials, the improvements associated with incorporation of the layered particles are more significant in the solid samples.[7]

9.4 COMPRESSION MOLDING

The biodegradable poly(butylene succinate) foam was produced by compression molding using dicumyl peroxide as a crosslinking agent and trimethylolpropane trimethacrylate as a curing coagent.[8] Zinc oxide/zinc stearate was used to reduce the thermal decomposition temperature of azodicarbonamide to balance it with the vulcanization temperature of the polymer.[8] The formulation produced closed-cell foams.[8] Figure 9.3 shows a schematic diagram of the process, including some of the process parameters.[8]

Linear medium-density polyethylene was used to produce foam using a chemical blowing agent (Celogen 754A; azodicarbonamide-based) by compression molding with the application of temperature gradient to the molding sheet (different temperatures of the

9.4 Compression molding

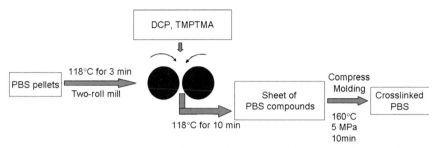

Figure 9.3. Process of production of crosslinked poly(butylene succinate) foam. [Adapted, by permission, from Li, G; Qi, R; Lu, J; Hu, X; Luo, Y; Jiang, P, *J. Appl. Polym. Sci.*, **127**, 3586-94, 2013.]

top and bottom plates).[9] Temperature gradient produced a density gradient in the foam.[9] The asymmetric density profile of the material caused that flexural strength and apparent modules depended on the side to which the load was applied.[9] Azodicarbonamide was also used for blowing chlorinated polyethylene rubber/ethylene vinyl acetate blends in the compression molding process.[10] The expansion ratio and void fraction increased with an increase in the EVA component.[10] Azodicarbonamide-based Porofor ADC/M-C1 was used in the production of medium-density polypropylene foams.[11] Improved compression molding process permits independent control of density and cellular structure.[11] The process comprises three stages, as follows:[11]
1. all materials are compounded in a co-rotating twin-screw extruder with the production of pellets
2. the pellets are compression molded in a two-hot plate press
3. the foaming process is performed in self-expandable molds, which can control the expansion of the material.

ZnO was used to lower the decomposition temperature of azodicarbonamide in the production of polylactide foams.[12] Casico-based foams are produced from the base developed by Borealis containing 30 wt% calcium carbonate, 5 wt% polydimethylsiloxane, and acrylate containing copolymer (EBA).[13] The Casico-based foams are produced using an azodicarbonamide-based blowing agent.[13]

The foam blocks, made out of polyolefins having a closed-cell structure, produced by compression molding, have a sub-atmospheric, non-constant gas pressure after production.[14] The increase in pressure with time is a very slow process.[14] The process was not completed in 200 days after production.[14] The cooling of the foam during production affects the pressure distribution within the foam.[14]

By controlling the foaming parameters, such as blowing agent concentration, foaming temperature, pressure drop, and pressure drop rate, it was possible to regulate the cellular structure of foams.[15] The foams were from markedly isotropic-like cellular structures to foams containing highly-elongated cells in the vertical foam growth direction (honeycomb-like cell orientation).[15] These structural differences helped to produce foams having thermal conductivity anisotropy.[15]

An injection compression molding system includes a mold having a fixed first half and a displaceable second half.[16] The foaming of the melt once it is injected into the tool acts to pressurize the melt against the cooled walls of the mold, therefore enhancing the heat transfer between the tool wall and the molten plastic.[16] The material expands to fill all the areas of the tool.[16]

UHMWPE can be processed by compression molding into a sheet, block, and precision parts, and it is one of the most suitable polymers for compression molding with high filler content.[17] The degree of crystallinity of the UHMWPE compression-molded sample was as high as 70%.[17] Crystals act as heterogeneous nucleation points for bubbles maximizing cell density.[17] The foam structure was affected by the combined action of crystals and pressure.[17]

Vapor-grown carbon fiber-based polyurethane foam was produced by compression molding.[18] With increasing compression, the electrical conductivity of the foam was improved by orders of magnitude, and the microwave shielding performance was also improved.[18] The hot compression improved the electrical interconnections of the nanoscale carbon nanofibers in the foam.[18]

9.5 DEPRESSURIZATION

A two-step batch depressurization process was applied to produce bimodal polycarbonate foams having both small and large cells. The resultant foams exhibit significantly improved tensile properties compared to the unimodal foams. A high-pressure vessel capable of operation at high temperature (200°C) and high pressure (30 MPa) was automatically controlled.[19] The polycarbonate was placed in the high-pressure vessel and exposed to supercritical carbon dioxide at 20 MPa and 60°C for 6 h until complete saturation.[19] Then, the vessel was depressurized to an intermediate pressure (14 or 10 MPa) within 10 s and then maintained at the intermediate pressure for 1 h.[19] The saturation temperature was kept constant at 60°C during the holding time.[19] The second depressurization to atmospheric pressure was conducted (also in 10 s).[19] The material was immediately immersed in the glycerol bath to get foamed. Bimodal foams were obtained in this way.[19]

PMMA foams with a cell size of 42-58 nm and a porosity of 27-58% were obtained by two-step foaming with ultra-high saturation pressure (> 50 MPa) and rapid depressurization (> 1.25 GP/s).[20] The cell growth was inhibited because the temperature during depressurization was less than glass transition temperature, and, as a result, a high cell nucleation density occurred at high saturation pressure, namely, by taking advantage of a higher solubility (higher saturation pressure).[20]

A new foaming apparatus is able to depressurize vessel from 100 to 0.1 MPa in 50 ms (Figure 9.4).[21] Sub-micron cellular

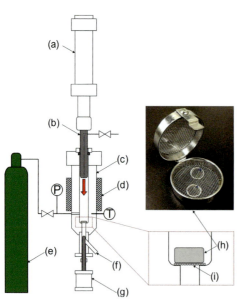

Figure 9.4. Schematic drawing of foaming process apparatus: (a) oil hydraulic cylinder; (b) piston; (c) high pressure vessel; (d) heater; (e) CO2; (f) Y-type piston valve; (g) pneumatic cylinder; (h) metal basket containing polymer samples: (i) metal mesh. The inserted picture shows the metal basket containing two polymer samples with different molecular weights. [Adapted, by permission, from Ono, T; Wu, X; Furuya, T; Yoda, S, *J. Supercritical Fluids*, **149**, 26-33, 2019.

structures of PMMA with 10^{12} to 10^{14} cells/cm^3 were obtained.[21] Faster decompression was needed to take advantage of a high degree of supersaturation.[21]

At a low depressurization rate, the CO_2 absorbed was slowly removed, decreasing the size of the cells and increasing their density in microcellular scaffolds obtained from poly(lactic-co-glycolic) acid.[22] This trend is contrary to what is found in literature, where a shorter depressurization time causes the formation of smaller cells.[22]

In order to avoid a bimodal cell distribution in rapid depressurization foaming of rigid polyurethane using supercritical CO_2 as a blowing agent, a new strategy called the two-step pressurization process was developed with lower CO_2 saturation pressure and higher curing pressure.[23] According to the classical nucleation theory, the higher the depressurization rate, the lower the free energy barrier of bubble nucleation.[23] When the depressurization was very low (24 or 54 MPa/s), the PU foams did not meet the definition of microcellular foams of cell size less than 10 µm and cell density larger than 10^9/cm^3.[23] When depressurization was increased to 389 MPa/s, the average cell size, and cell density were 7.1 µm and 10^{11}/cm^3, respectively, which indicated that microcellular rigid PU foams were only produced at very high depressurization rate.[23]

The bimodal foam was also obtained from polymethylmethacrylate and polystyrene blend.[24] Methyl methacrylate monomer was dissolved in polystyrene under supercritical carbon dioxide at a temperature of 60°C and a pressure of 8 MPa, and the polymerization of methyl methacrylate was conducted at 100°C and 8 MPa carbon dioxide, with a cross-linking agent present in polystyrene.[24] The blend was then foamed by depressurizing the carbon dioxide.[24] The smaller cells ranging from 10-30 µm in diameter were located in the wall of the large cells having 200-400 µm in diameter.[24]

9.6 EXTRUSION

Figure 9.5 shows a schematic diagram of the extrusion foaming line.[25] Extruder is equipped with a gas injection system and a static mixer with a Sulzer mixture unit, aimed at enhancing gas dispersion and dissolution in the polymer melt, is installed at the end of the extruder screw.[25] The line was used for foam extrusion of polycarbonate using carbon dioxide at the gas pressure higher than 10 MPa.[25] A low die temperature, the increased screw revolution speed, and gas input improve the volume expansion ratio and cell structure of the PET foams.[25]

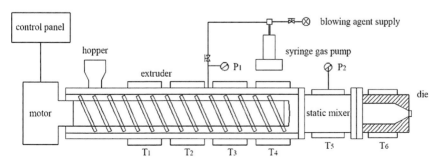

Figure 9.5. Extrusion foaming equipment. [Adapted, by permission, from Fan, C; Wan, C; Gao, F; Huang, C; Xi, ; Xu, Z; Zhao, L; Liu, T, *J. Cellular Plast.*, **52**, 3, 277-98, 2016.]

Prior to processing polylactate, all components were dried for at least 4 h at 80°C to avoid any degradation caused by hydrolysis.[26] A static mixer was used to increase the residence time for the formation of a single-phase and improvement of melt homogeneity.[26] At lower nucleating agent contents, nitrogen leads to smaller cell sizes compared to supercritical carbon dioxide.[26] At the higher nucleating agent content, the quality of the foam sheets was worse with nitrogen.[26] This is explained by the higher degassing pressure of nitrogen compared to carbon dioxide.[26] The higher degassing pressure causes an earlier phase separation inside the die.[26]

The addition of the second polymer in foam extrusion enhances cell nucleation during foaming because of the heterogeneous nucleation boundaries created by the interphase between the two components.[27]

The blowing agent injection location in the extrusion barrel affects the residence time inside the barrel, which influences the foam microstructure.[28] The higher gas residence time gives foams with smaller cells, higher expansion ratio, and higher cell density.[28]

A process using dry ice as a physical blowing agent has been developed at the Institute of Plastics Processing, Aachen University, Germany.[29] The process has been implemented for foam injection molding and the foam extrusion processes.[29] In foam extrusion, dry ice in the form of cylindrical pellets is dosed *via* the hopper using a modified screw feeder that is commonly applied for additives and masterbatches.[29] A downpipe is used to pass the pellets into the feeding zone to reduce heat conduction between the dry ice pellets and the plastic granules (Figure 9.6).[29]

Foaming of plasticized starch was performed by adding water as a natural blowing agent and by increasing the die temperature.[30] Both the expansion ratio and the porosity increase with increasing die temperature.[30] Addition of more water permits reduction of the foaming temperature.[30]

Extrusion foaming of poly(3-hydroxybutyrate-co-3-hydroxyvalerate) using sodium bicarbonate and citric acid, and calcium carbonate as nucleation agents.[31] Use of negative gradient temperature profile (100/180/170/160/150/150°C) helped to minimize the thermal degradation and maintain melt strength required to stabilize the cell structure.[31] High

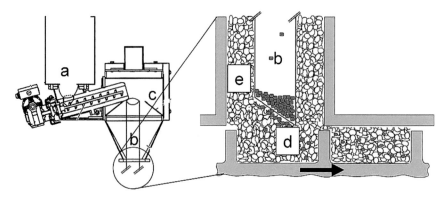

Figure 9.6. The dry ice pellets are conveyed by the screw feeder (a) and dropped in a downpipe (b) inside the tumbler mixer (c). In the feeding zone (d), the dry ice pellets are carried into the barrel with the flow of plastic granules (e). [Adapted, by permission, from Hopmann, C; Hendriks, S; Feldges, R, *Polym. Eng. Sci.*, **55**, 2851-8 2015.]

9.6 Extrusion

Figure 9.7. Die buildup associated with extrusion of the PHBV under super-cooled conditions. The outer layer is the crystallized PHBV and the core is a purge polymer. [Adapted, by permission, from Szegda, D; Duangphet, S; Song, J; Tarverdi, K, *J. Cellular Plast.*, **50**, 2, 145-62, 2014.]

level of supercooling, however, can result in a buildup of the material in the die (Figure 9.7).[31]

A single screw extruder with a length to diameter ratio (L/D) of 40:1 was used for foaming of cellulose acetate using HFO 1234ze as a low global warming blowing agent.[32] The extruder was equipped with a mixing screw optimized for the foam extrusion process.[32] The last 11 D of the extruder length was temperature-controlled with oil to cool the polymer melt.[32] The blowing agent was compressed and injected into the extruder barrel through a pressure hole at 16 D using a metering system equipped with a diaphragm pump.[32]

A tandem extrusion system (Figure 9.8) was used for foaming of linear polypropylene with the addition of polytetrafluoroethylene.[33] The first extruder is used for plasticizing the polymer resin, and the second extruder provides mixing and cooling to completely dissolve the carbon dioxide in the polymer melt.[33]

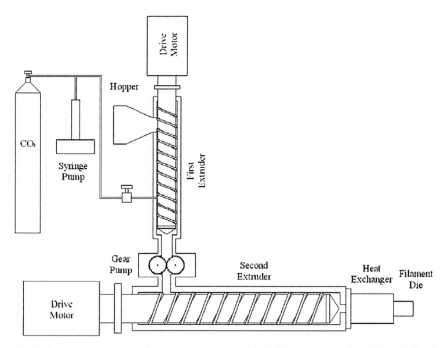

Figure 9.8. Schematic diagram of tandem extrusion system. [Adapted, by permission, from Wang, K; Wu, F; Zhai, W; Zheng, W, *J. Appl. Polym. Sci.*, **129**, 2253-60, 2013.]

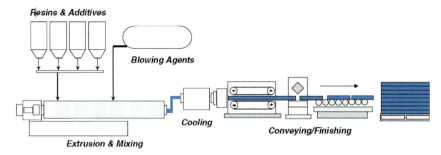

Figure 9.9.Stryrofoam production line. [Adapted, by permission, from Fox, R; David Frankowski, D; Jeff Alcott, J; Dan Beaudoin, D; Lawrence Hood, L, *J. Cellular Plast.*, **49**, 4, 335-49, 2013.]

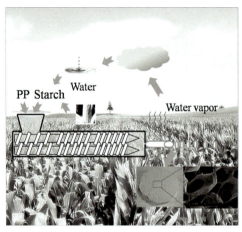

Figure 9.10. Production of closed-cell foam. [Adapted, by permission, from Xu, M Z; Bian, J J; Han, C Y; Dong, L S, *Macromol. Mater. Eng.*, **301**, 149-59, 2016.]

Figure 9.9 shows the line for the continuous production of styrofoam.[34] It includes melting of the polymer resin, introduction of physical additives, mixing with blowing agent under pressure followed by controlled cooling and expansion to generate the resultant foam structure.[34]

The extruder barrel's temperature profile for the production of foam from a blend of poly(3-hydroxybutyrate-co-3-hydroxyvalerate) with cellulose acetate butyrate using azodicarbonamide was set as follows: zones 1 to 3, 140°C to 170°C to 160°C, and zone 4, which is at the die, 140°C.[35] The die temperature was chosen so that no significant crystallization of the polymer melt would occur before exposure to the air.[35]

Double-layer extrusion coating was used to manufacture cables and conduits using PVC and azodicarbonamide-based blowing agent.[36] The coextrusion line contained an angular extrusion head.[36]

Figure 9.10 symbolizes the direction in which foam production is heading – namely towards incorporation, if not all, at least the largest possible fraction of raw materials from renewable resources.[37] One option of reducing the impact on nature is to produce foams from natural polymers such as starch or cellulose.[38] To pursue this direction, foams from hydroxypropyl methylcellulose are manufactured with water as a blowing agent by continuous extrusion.[38]

Figure 9.11 shows an illustration of an assembly for visualization of the developments in the material in the extruder, including bubble formation.[39] The visualization system combines data on extrusion foaming behavior with simultaneous measurement of temperature and pressure profiles.[39]

A polymeric nanofoam production method has been developed using a continuous extrusion process by introducing carbon dioxide at a concentration above the solubility in

9.7 Free foaming

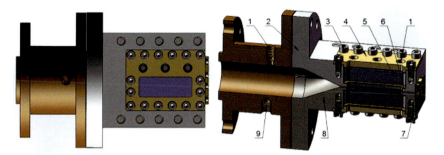

Figure 9.11. Assembly illustration of the visualization die: 1. temperature & pressure sensors 2. upper die 3. fasten screws 4. quartz window clamp 5. quartz window 6. flexible graphite gaskets 7. replaceable die lip 8. lower die 9. thermocouple. [Adapted, by permission, from Peng, X-F; Liu, L-Y; Chen, B-Y; Mi, H-Y, Jing, X, *Polym. Testing*, **52**, 225-33, 2016.]

the polymer melt, cooling the polymer melt without increasing the pressure to achieve conditions where all carbon dioxide is soluble in the polymer composition and then extruding the polymer composition and carbon dioxide mixture through an extrusion die so as to experience a pressure drop of at least 5 MPa at a rate of at least 10 MPa/s and allowing the polymer composition to expand into a polymeric nanofoam.[40] The thermal conductivity of foam is reduced by almost half upon reducing a foam cell size from one micrometer to 300 nm, and it is further reduced by almost 2/3 upon reducing the cell size from one micrometer to below 100 nm.[40]

Properties of polyvinylidenefluoride extruded foam were strongly dependent on melt viscosity controlled by die temperature.[41] In the region of 125-130°C, the melt viscosity was low enough to allow cell formation and growth but high enough to prevent cell coalescence.[41] Above 130°C, the cell size and void fraction drastically increased, and cell density was decreased due to low melt viscosity for cell growth and coalescence.[41]

9.7 FREE FOAMING

The classical case of free foaming is that of polyurethane foams, including the commonly used spray foams. The method of manufacture of polyurethane foam is based on mixing components of the foam and allowing the foam to grow freely. Depending on the method of blowing (reaction with water and/or use of liquid blowing agents) and the rate of gas release from the blowing agent (in addition to the rheological properties of polymer mix), either open-cell or closed-cell is formed. The free foaming technology requires that proportions of components are mixed in a proper ratio, which can be accomplished either by simple weighing or sophisticated dosing equipment. The other requirement is fast and good mixing of components, which assures uniformity of foam. The foaming mass can then be spread on a surface (e.g., spraying), injected into a mold, or can simply fill any space in a container in which it was placed. The only remaining task is to make sure that the static electricity is removed, which is formed during foaming because of the movement of material, which has insulating properties. The formulation of foam requires the use of materials that do not cause global warming, which became the major issue several decades ago.[42]

Free foaming of other polymers is a more complicated process. It can be well illustrated by the example of free foaming of a blend of natural rubber and ethylene vinyl acetate.[43] Ethylene-vinyl acetate pellets and natural rubber are first melt-mixed in an internal mixer at a temperature of 70°C and rotor speed of 60 rpm, followed by the addition of the fillers and other additives according to ingredients and mixing procedure required by the technological process.[43] The obtained ethylene vinyl acetate/natural rubber blend was compounded in this example with azodicarbonamide, dicumyl peroxide, and trimethylolpropane trimethylacrylate using a two-roll mill at 70°C for 12 min.[43] The compounds are then hot-pressed to a 90% cure in a mold with a hydraulic pressure at 160°C cure temperature.[43] The cure time of the rubber blend is determined by an oscillating disk rheometer at a test temperature of 160°C.[43] Finally, the pressure is removed, and the precured material is allowed to expand.[43] From this description, it is evident that there are 3 basic steps in this process, including mixing (uniform mixing produces a uniform composition of the material, which is a prerequisite for a good uniform foaming; especially important is a good dispersion of blowing agent), vulcanization/curing of polymeric material (this type of process is usually conducted at the temperature lower than the degradation temperature of blowing agent or, similar to this technology, under increased pressure that does not permit formed gas to expand), and final stage is a free expansion. The alternative to this process will be to thoroughly mix all the ingredients, cure polymer at a temperature lower than the temperature of decomposition of blowing agent under atmospheric pressure, increasing temperature to allow the blowing agent to expand, followed by cooling. In many processes, cure and expansion can be synchronized that both processes occur simultaneously.

In addition to these two cases, we have a case of products that foam when released from a pressurized container. These are mainly froth spray insulation foam packaged in high-pressure gas cylinders similar to the propane tanks, polyurethane foam packaged in small metal, pressurized containers, numerous cosmetic products (the most popular shaving gel), and fire-fighting foaming compositions. In the self-foaming shaving gel, in addition to typically 15-20 different cosmetic ingredients, the self-foaming agent is usually selected from volatile hydrocarbons and halogenated volatile hydrocarbons, having a boiling point ranging from -20 to 40°C (usually isopentane and isobutane).[44]

Finally, it should be mentioned that all methods used in the bakery are designed for free foaming, and these are the most ancient technological process used today with a very similar composition to the ones used thousands years ago.

9.8 INJECTION MOLDING

In applications using solid blowing agents, all components of the formulation are blended, for example in a twin-screw extruder, with blend either pelletized for use in the solid form or directly injected into the cavity of mold in which the composition is exposed to the temperature higher than the decomposition temperature of blowing agent, held at this temperature until expansion continues, and cooled down to the temperature at which part can be safely removed from the mold. In the case of polylactide using azodicarbonamide blending was done at 150°C and foaming at 200°C.[45] The foaming temperature can be lowered by the addition of catalysts of decomposition. Figure 9.12 shows that holding time is an essential parameter.[45] The cellular structure at the shorter times is very uniform

9.8 Injection molding

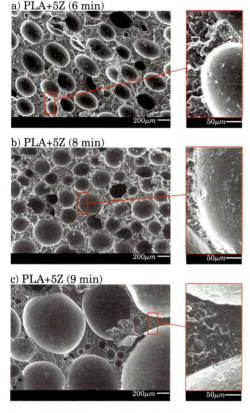

Figure 9.12. SEM images of the cells of the poly(lactic acid) composition foamed for various times. [Adapted, by permission, from Lobos, J; Iasella, S; Rodriguez-Perez, M A; Velankar, S S, *Polym. Eng. Sci.*, **56**, 9-17, 2015.]

when bubble nucleation has just started, and the cells are 200 µm in size (Figures 9.12a and 9.12b).[45] Figure 9.12c shows a bi-disperse population of bubbles with some of the bubbles still having a size of 200 µm (same as after 8 minutes), and some being over 600 µm.[45] Small bubbles vanish, and large bubbles grow at their expense.[45]

The injection molding of compounds foamed by physical blowing agents has one essential difference as compared with chemical blowing agents: mixing equipment and injection molding equipment are integral parts of the line. For example, foaming of polypropylene with supercritical carbon dioxide is realized in a twin-screw extruder in tandem with an injection molding machine.[46] In the extruder, the polymer is premixed with additives, and supercritical carbon dioxide, and the homogeneous mix is injected through a dosing system (MuCell SCF) into the cavity of the injection molding machine (Arburg Allrounder 320).[46]

Trexel supplies MuCell injection molding technology to the industry.[47] Supercritical fluid (N_2 or CO_2) is introduced into the polymer through injectors mounted on the plasticizing barrel, mixing section homogenizes fluid with polymeric compound creating a single-phase liquid containing polymer melt and supercritical fluid.[47] After an injection to a mold, cells begin to nucleate because they are exposed to lower pressure in the mold cavity.[47] Closed-cell structure with a skin develops uniformly.[47] The controlled cell growth ceases once the mold cavity is filled.[47] The technology was successfully tried with 27 polymers and their blends for the production of a large number of articles.[47] Figure 9.13 shows an example of the MuCell line.[47] Many variations are possible, based on injection molding machines from leading manufacturers (e.g., Milacron, Engel, Krauss Maffei, Arburg, etc.) retrofitted with Trexel's equipment and technology.[47] It is also possible to retrofit existing injection molding equipment with modular MuCell upgrade with installation by qualified Trexel technicians.[47]

Glass fiber reinforced-polypropylene foams obtained by IQ Foam® technology exhibited thicker solid surface layers and lower cell density than that of MuCell® ones, but the consequently higher resistant area, and thus, slightly higher mechanical properties.[48] The new IQ Foam® technology produced foamed parts with properties comparable to that

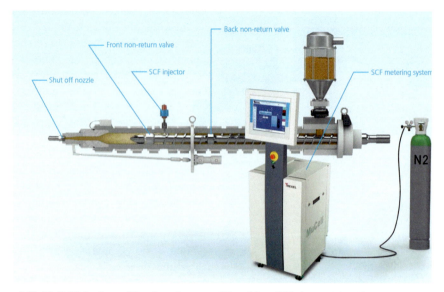

Figure 9.13. MuCell injection molding line. Courtesy of Trexel, Inc., Wilmington, MA, USA.

of MuCell® process, offering additional benefits such as cost-effectiveness, easy to use, and machine-independence.[48]

Essential in physical foaming in the injection molding machine is the proper application of nucleating agents, considering that cycle time is usually very short (15-20 sec.).

A mold opening technique and gas counter-pressure were used for the production of highly expanded (8 times expansion; 85% open-cell content), open-cell foams from rigid polycarbonate foamed with nitrogen.[49] The degree of mold opening, the melt temperature, and the injection flow rate were the most influential parameters for a high open-cell content.[49]

Breathing mold technology has to be employed to achieve a density reduction of 50% during foam injection molding.[50] The mold is completely filled under counter-pressure and then opened in one direction.[50] By opening the cavity (Figure 5.19), the sudden pressure drop initiates the foam nucleation process of the dissolved gas in the polymer melt.[50] The exact density reduction is achieved by the degree of the cavity opening.

Polylactide foam having a high expansion and fine cells was manufactured by injection foam molding technique in spite of intrinsically low melt strength of polylactide.[51] To overcome the inferior foaming characteristics of PLA, nano-fibrils of polytetrafluoroethylene were added to increase the crystallization rate.[51] High-pressure injection foam molding process combined with mold-opening technique permitted the production of high-quality foam.[51]

Figure 9.14 shows the sequence of the pull and foam method, which might be considered as the further development of breathing mold technology.[52] The technology enables the production of foam injection molded components with different foaming ratios.[52] The pull and foam method is based on the idea of equipping thin-walled components with locally foamed thick-volume elements – i.e., increasing stiffness, creating spac-

9.8 Injection molding

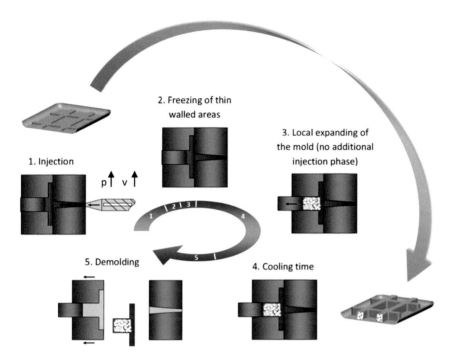

Figure 9.14. Process sequence of the pull and foam method. [Adapted, by permission, from Heim, H-P; Tromm, M, *J. Cellular Plast.*, **52**, 3, 299-319, 2016.]

ers, or joining surfaces.[52] The first processing step is a conventional injection molding – the cavity is volumetrically filled with melt containing blowing agent (1).[52] A high injection pressure applied or a high injection speed helps to counteract instant foaming.[52] An optional short delay time enables the development of the skin layers and the thin-walled areas; more specifically, areas that should have minimal foaming (2).[52] Pulling cores in the closed mold leads to a local enlargement of the cavity volume.[52] The correlating pressure drop initiates the actual foaming process in the enlarged cavity volume (3).[52] The remaining cooling time and demolding are again the same as in the standard injection molding process (4+5).[52]

A method of improving the appearance of foaming injection molding product comprises the following steps:[53]
1. closing the moving mold and the fixed mold and setting a clamping force on the closed mold, wherein a mold cavity is formed between the moving mold and the fixed mold
2. inflating the mold cavity with high pressure and high-temperature gas until the pressure in the mold cavity reaches 2-25 MPa and the temperature of the high pressure and high-temperature gas is between 60-200°C
3. injecting molten resin that contains foaming agent into the mold cavity while continuously inflating high pressure and high-temperature gas
4. after injection is complete, stopping inflating high pressure and high-temperature gas and simultaneously releasing pressure (the step of releasing pressure includes the step of opening the mold).

The method is suitable for chemical foaming injection molding of the thin-walled product, wherein the initial wall thickness before foaming is between 1-2 mm, and the foaming ratio of the product is 0.5-1.5.[53] The resin is a polyolefin, and the foaming agent is either sodium bicarbonate or an azo compound.[53] The foaming agent has a concentration of 0.5-4 wt%.[53]

Parts produced by foam injection molding sometimes display poor surface quality, and the desired quality is only achieved after additional post-processing.[54] The formation of streaks on the surface, as well as a relatively high roughness depth of 20-100 μm, occurs.[54] These effects can be attributed to the tearing and collapse of individual foam pores on the tool surface during the injection.[54] They freeze on the tool wall, creating a commonly known swirled pattern.[54] When the polymer melt cools down quickly on the tool wall, an unfoamed layer forms on the component surface because the low viscosity keeps the blowing agent from expanding.[54]

Controlling the cell nucleation rate of the polymer/gas solution through material formulation and gas concentration, microcellular injection molded parts free of surface defects (swirl-free foamed plastic parts) were achieved.[55] Surface imperfections such as lack of brightness, gas flow marks, and rough surfaces were eliminated without any additional equipment or equipment modification.[55] The reduction of the degree of supersaturation causes an increase in the activation energy for cell nucleation.[55] This reduces the cell nucleation rate and retards cell nucleation during the mold-filling stage, preventing bubble formation on the melt front of the polymer/gas solution and permitting to produce of swirl-free microcellular injection molded parts.[55]

Microcellular injection molding is widely applied in many industries due to its excellent flexibility and capabilities, good scalability and environmental benefits, low cost, and high efficiency to directly mold components with complex structures.[56] In the case of polypropylene foams, cell nucleation occurs at high temperatures without crystal forming, and thus homogeneous cell nucleation takes place, which dramatically decreases the number of cell nuclei, leading to inferior cellular structure, having large cell size and low cell density.[56] This reduces the mechanical performance of the molded PP foam.[56] Also, PP has a linear molecular chain structure, which leads to its low viscoelasticity and poor melt strength.[56] Thus, severe cell coalescence and collapse are usually caused at the relatively high foaming temperature in the MIM procedure of PP.[56] Mold with a thin cavity was designed to obtain a rapid cooling rate in the MIM process, which could considerably accelerate the crystallization and produce refined crystals.[56] The crystal was reduced by increased cooling rate, and nano cell nucleation was happening.[56]

9.9 MICROWAVE HEATING

Microwave heating is a highly efficient volumetric heating system.[57] Microwaves are electromagnetic waves having a frequency from 0.3 to 300 GHz and wavelengths of 1 m to 1 mm.[57] They travel at the velocity of light and are comprised of oscillating electric and magnetic fields.[57] The heat is generated within the material as opposed to originating from external heating sources.[57] Heating time can be reduced to less than 1% compared to conventional techniques. Expansion of PS beads (2.5 times the initial volume) was achieved.[57]

Many thermoplastic polymers are semi-transparent to microwave radiation, and they require additives such as filler to increase absorption of microwaves.[58] Composites of thermoplastic polyurethane and carbon black were foamed using azodicarbonamide and microwave radiation.[58] The addition of carbon black improves the uniformity of cell structure with anisotropy index close to unity.[58]

Natural rubber foam was produced using microwave-assisted processing and azodicarbonamide as the blowing agent.[59] Good foam structure was produced by sequential heating in the microwave and convection heating.[59] Consecutive use of microwave and conventional heating, having different heating mechanisms, gave out uniform heat distribution and allowed curing and foaming processes to occur at a suitable period, thus eliminating blister formation on the foam skin and larger cell size in the middle section of the foam.[59]

Starch blocks were foamed by microwave heating of thermoformed sheets using water as a blowing agent.[60] The foam stiffness and strength were higher than if produced from pellets.[60] The reinforcement with natural fiber-containing fillers reduced the cell size and increased the strength.[60]

A foamed article was made by infusing an article of thermoplastic polyurethane with a supercritical fluid (N_2 or CO_2), removing the article from the supercritical fluid and foaming.[61] The foaming methods include immersion in a heated fluid, infrared, or microwave radiation heating.[61] The method is used to expand midsoles.[61] In microwave heating, the material is exposed to an electromagnetic wave that causes the molecules in the material to oscillate, thereby generating heat.[61]

Nano-hydroxyapatite-reinforced polycaprolactone composite foam was made using microwave energy for rapid fabrication.[62] NaCl acted as microwave absorber and the porogen agent.[62] The interconnected porosity was induced by leaching NaCl particles from composite in an ultrasonication bath.[62]

9.10 ROTATIONAL MOLDING

LLDPE reinforced with wood flour was foamed by azodicarbonamide type blowing agent during rotational molding.[63] Premixes can be prepared by dryblending from raw materials with good control of particle size (possible segregation in rotomolding equipment).[63] Wood particles do not require drying.[63] Composite foam containing 20 wt% wood flour, foamed with 0.4 wt% azodicarbonamide, had the best performance.[63]

Foamed linear medium density polyethylene parts were prepared by rotational molding in biaxial mode, using azodicarbonamide.[64] Zinc oxide was used as a catalyst of decomposition for previously dried components.[64] The mold was preheated to 280°C, and the molding cycle was 20 min.[64] The total cycle time was increased with the blowing agent content as a result of the insulating effect of gas bubbles generated, reducing thermal conductivity across the part thickness.[64] The 6% foam contents were found to be the optimal level to obtain sufficient melt flow index and good impact strength.[65]

Rheological properties of the polypropylene resin, processing conditions, and choice of composition of the chemical blowing agent were found to play important roles in obtaining high-quality polypropylene foam structures by rotational molding.[66] A control agent needs to be used to lower the effective decomposition temperature of the blowing agent closer to the sintering temperature.[66] Such agent is zinc oxide when using azodicarbonamide or inorganic bicarbonate.[66]

9.11 SOLID-STATE FOAMING

Supercritical carbon dioxide at 20 MPa was used as a blowing agent for polyetherimide.[67] Foaming was done in a hot press to ensure flatness (Figure 9.15).[67] The samples were first desorbed at room temperature and atmospheric pressure for a set amount of time prior to foaming (5-75 min).[67] Samples were placed one at a time in the middle of the hot press platens, which were preheated to the desired foaming temperature (165-210°C).[67] PEI can absorb about 10 wt% of carbon dioxide at 20 MPa.[67] A 1-mm thick specimen can reach an equilibrium concentration in 100 h at 45°C.[67] The nanofoams had a bulk porosity in the range of 25-64% and the average cell size in the range of 40-100 nm.[67] The critical process parameters are desorption time (minimum 35 min to avoid blisters), gas concentration, foaming time (the longer the time, the larger the particles), foaming pressure, and foaming temperature.[67]

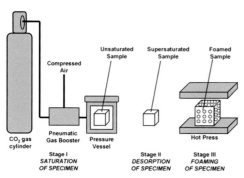

Figure 9.15. Diagram of solid-state PEI foaming process. [Adapted, by permission, from Aher, B; Olson, N M; Kumar, V, *J. Mater. Res.*, **28**, 17, 2366-73, 2013.]

The presence of a small amount of polyethersulfone as a dispersed phase in poly(ethylene 2,6-naphthalate)-based blends acted as blowing agent reservoir and allowed to extend the processing temperature range for obtaining low-density foams produced with supercritical carbon dioxide as a blowing agent.[68] The carbon dioxide sorption occurred in a high-pressure vessel operated at 8.0 MPa, and 50°C for 72 h.[68] After gas sorption, samples were foamed by immersion in silicon oil at the desired foaming temperature (160-220°C) for 10 s and quenching in cold water.[68] The minimum density of foams made out of PEN rich blends, prepared at 180°C, was 0.2 g/cm³.[68]

A solid-state nanofoaming method (close to or less than 100 nm) has been developed by saturating thermoplastic polymers with liquid carbon dioxide at low saturation temperatures below room temperature.[69] After saturation, the material was heated to above the glass transition temperature of the polymer.[69] The method has been applied to numerous polymers.[69] A carbon dioxide concentration for producing an average cell size of 100 nm or less may be 27.5 wt% or 31 wt% or greater if the polymer is polymethylmethacrylate.[69] Different concentrations are appropriate for different polymers.[69] The saturation is performed at a temperature of 0°C or less and a pressure of 5 MPa or less.[69] The low-temperature saturation step is advantageous for various reasons.[69] It permits to reach the very high concentrations needed for the creation of nanofoams.[69] Also, lower temperatures permit saturation to take place at lower pressures, such as 5 MPa or less.[69]

Bimodal PMMA foam was fabricated by solid-state foaming using CO_2 as a blowing agent.[70] The foam structure was controlled by changing the composition of the blend with

9.11 Solid-state foaming

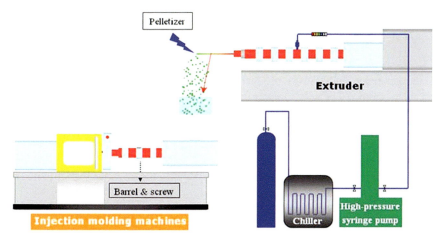

Figure 9.16. Supercritical injection molding foaming technology, SIFT, extrusion/injection molding process. [Adapted, by permission, from Sun, X; Turng, L-S, *Polym. Eng. Sci.*, **54**, 899-913, 2014.]

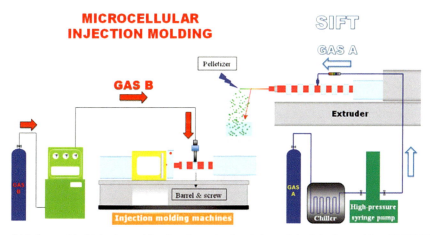

Figure 9.17. Supercritical injection molding foaming technology/microcellular injection molding, SIFT/MIM, combined process using co-blowing agents [Adapted, by permission, from Sun, X; Turng, L-S, *Polym. Eng. Sci.*, **54**, 899-913, 2014.]

out changing the processing conditions.[70] Bimodal foam was formed by two mechanisms:[70]
- micron-sized cells were generated during pressure release, with ultramicro cells and nanocells obtained during post-foaming
- both micron-sized cells and nanocells were formed during post-foaming.

Generation of micron-sized cells in a nanocellular structure resulted in a 29% reduction in relative density.[70]

9.12 SUPERCRITICAL FLUID-LADEN PELLET INJECTION MOLDING FOAMING TECHNOLOGY

Two processes discussed here are SIFT (Figure 9.16) and SIFT/MIM (Figure 9.17).[71] In SIFT, the gas-laden pellets were produced in a single-screw extruder using a high precision syringe pump.[71] The solution was extruded through a 12-orifice filament die.[71] Once the strands came out of the die; they were immersed in a cold water bath to prevent foaming and a loss of gas from the strands.[71] Strands were blow-dried and pelletized, and pellets were oven-dried for 1 h at 30°C to remove moisture.[71] The gas-laden pellets can be taken to injection molding machines to produce foamed parts.[71] In the MIM process another gas is injected into the mixture and blended into the polymer-gas solution, which is then extruded to produce foam.[71] Carbon dioxide is more suitable for SIFT because of easier metering and dosing and a longer gas-laden pellet shelf life.[71] Nitrogen was used as a co-blowing gas in SIFT/MIM system.[71] MIM improves foam morphology compared with using either blowing agent alone.[71] SIFT/MIM was successfully applied to LDPE, PP, and HIPS.[71]

9.13 THERMOFORMING

A solid-state process utilizes gas impregnation to enhance the thermoforming of thermoplastic material.[72] The gas may have plasticizing properties which contribute to thermoforming.[57] Foaming the polymer prior to or during thermoforming can also be performed.[72] Foaming may proceed spontaneously upon decompression from gas

(a)

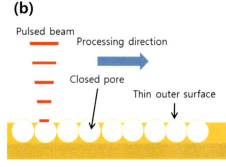
(b)

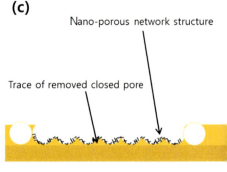
(c)

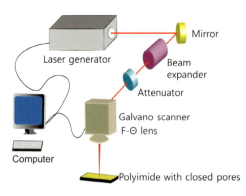

Figure 9.18. Schematic diagram of the laser system. [Adapted, by permission, from Ma, Y-W; Jeong, M Y; Lee, S-M; Shin, B S, *J. Korean Phys. Soc.*, **68**, 5, 668-73, 2016.]

Figure 9.19. Fabrication process for the high-density nano-porous network structure: (a) pre-curing at 130°C, (b) curing at 205°C before laser processing, and (c) elimination of closed pores by using an UV laser. [Adapted, by permission, from Ma, Y-W; Jeong, M Y; Lee, S-M; Shin, B S, *J. Korean Phys. Soc.*, **68**, 5, 668-73, 2016.]

pressure, or foaming may be enhanced by heating the polymer sheet near to or above the polymer's glass transition temperature, thereby producing plasticized foamed polymer for thermoforming.[72] If no foaming is desired, foaming may be suppressed by thermoforming gas saturated articles under gas pressure.[72]

NaCl is an effective humidity regulator of PP-thermoformed films.[73] Thermoformed cups and trays with dispersed NaCl can be used as packaging material for fresh produce.[73]

9.14 UV LASER

Figure 9.18 shows a schematic diagram of the laser system.[74] Liquid poly(amic acid) mixed with the azodicarbonamide-based blowing agent was spin-coated onto a poly(amic acid) film substrate, and the substrate was then pre-cured in an oven for 10 minutes at 130°C, during which the blowing agent did not decompose, resulting in a slightly cross-linked poly(amic acid) film.[74] Figure 9.19a shows the sparsely-crosslinked polyimide with the blowing agent.[74] Figure 9.19b shows polyimide with closed pores before laser irradiation process.[74] The film was then cured in an oven at the blowing agent decomposition temperature (205°C) for 10 minutes.[74] A highly-crosslinked poly(amic acid) film with micro-closed pores was obtained.[74] The closed-pore thin surface was etched by a UV pulsed laser. Figure 9.19c shows polyimide with open pores after laser irradiation.[74] The nanoporous structure can be applied in lithium-ion batteries, fuel cells, supercapacitors, and sensors because it has a large specific surface area and a large electrical conductivity.[74]

9.15 VACUUM DRYING

A number of methods are used for conversion of organogels into aerogels, including supercritical drying, freeze-drying, and vacuum drying.[75]

A chitosan/wheat gluten biofoam obtained by mixing and phase distribution when a film was cast from an aqueous mixture of chitosan/wheat gluten solution, followed by application of vacuum to remove liquid phase in the absence of any chemical blowing agent.[76] Open pores with sizes of 70-80 μm were observed by microscopy.[76] The density

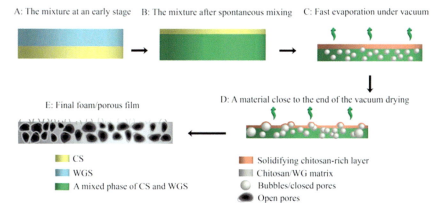

Figure 9.20. The method of the porous structure fabrication using vacuum. [Adapted, by permission, from Chen, F; Gaellstedt, M; Olsson, R T; Gedde, U W; Hedenqvist, M S, *RSC Adv.*, **5**, 94191-200, 2015.]

of foam was 50 kg m⁻³, due to the high porosity (96% air).[76] Figure 9.20 shows the mechanism of formation.[76] Because of the good foaming property of wheat gluten and increasing viscosity of chitosan on evaporation, the viscous/solid chitosan-rich top layer prevents most of the small bubbles formed in the mixed-phase from growing too large, breaking, and collapsing.[76] The chitosan also protects pores from collapsing.[76]

Polystyrene porous microspheres were obtained in a continuous process with a production yield of 91±4 wt%.[77] The foam was collected into the foam collector followed by defoaming, washing, filtration, and vacuum drying.[77]

9.16 WIRE COATING

Compatible LDPE/HDPE blend and LDPE/PP blend with sea-island micro-structure generated uniform cell morphologies suitable for wire coating using supercritical carbon dioxide as a blowing agent.[78]

Arkema developed foam masterbatch for foaming nearly any grade of Kynar PVDF for numerous applications, including wire and cable jacketing market.[79] Foam has the characteristic stability of fluoropolymers when exposed to harsh thermal, chemical, and ultraviolet environments.[79]

Foam insulated wire and cable product was manufactured from one or more fluoropolymers.[80] The base resin comprises tetrafluoroethylene perfluoroalkyl vinyl ether copolymer or/and tetrafluoroethylene-hexafluoropropylene copolymer.[80] Azo, hydrazide, and tetrazole compounds are preferred blowing agents. Boron nitride and talc are used as nucleating agents.[80]

Low loss wire and cable dielectrics are produced from compositions comprised of blowing agent (direct gas or chemical blowing agent), a thermoplastic polymer carrier, and filler.[81] The thermoplastic polymer carrier has a melting point below the decomposition temperature of the chemical blowing agent.[81] The chemical blowing agent can be selected from those based on hydrazine, hydrazide, or azodicarbonamide, or those based on combinations of sodium citrate/citric acid and sodium bicarbonate.[81] The filler can be selected from calcium carbonate, zeolites, and clay.[81] The polymer can be polyethylene, polypropylene, polybutylene, poly(p-phenylene oxide), or polystyrene.[81]

REFERENCES

1 Greene, J D, **US5804112**, *Olympic General Corporation*, Sep. 8, 1998.
2 Lindenfelzer, M E; Anderson, J E; Dix, S E, **US20120228793**, *Mucell Extrusion, LLC,* Sep. 13, 2012.
3 Abbasi, H; Antunes, M; Velasco, J I, *Prog. Mater. Sci.*, **103**, 319-73, 2019.
4 Herner, M; Hatzmann, G; Mueller, G, **DE2145026**, *BASF AG*, Mar. 29, 1973.
5 Grizzle, M L; Preston, L R; Hammel, W J; Evans, P D; Randall, B S, **CA2340555**, *Collins & Aikman Floorcoverings, Inc.*, Jul. 4, 2006.
6 Istrate, O M; Chen, B, *Polym. Int.*, **63**, 2008-16, 2014.
7 Laguna-Gutierrez, E; Escudero, J; Rodriguez-Perez, M A, *Composite Part E: Eng.*, **148**, 156-65, 2018.
8 Li, G; Qi, R; Lu, J; Hu, X; Luo, Y; Jiang, P, *J. Appl. Polym. Sci.*, **127**, 3586-94, 2013.
9 Yao, J; Rodrigue, D, *Cellular Polym.*, **31**, 4, 189-205, 2012.
10 Zhang, B S; Zhang, Z X; Lv, X F; Lu, B X; Xin, Z X, *Polym. Eng. Sci.*, **52**, 218-24, 2012.
11 Saiz-Arroyo, C; Rodriguez-Perez, M A; Tirado, J; Lopez-Gil, A; de Saja, J A, *Polym. Int.*, **62**, 1324-33, 2013.
12 Luo, Y; Zhang, J; Qi, R; Lu, J; Hu, X; Jiang, P, *J. Appl. Sci.*, **130**, 330-7, 2013.
13 Realinho, V; Antunes, M; Velasco, J I, *Polym. Deg. Stab.*, **128**, 260-8, 2016.
14 Rodriguez-Perez, M A; Hidalgo, F; Solórzano, E; de Saja, J A, *Polym. Testing*, **28**, 2, 188-95, 2009.

References

15 Antunes, M; Realinho, V; Velasco, J I; Solórzano, E; Rodríguez-Pérez, M-A; de Saja, J A, *Mater. Chem. Phys.*, **136**, 1, 268-76, 2012.
16 Zander, R J; Ritzema, K; Morse, K; Williams, M P, **US9216526**, *Cadillac Products Automotive Company*, Dec. 22, 2015.
17 Liu, J; Qin, S; Wang, G; Cao, Y, *Polym. Testing*, **93**, 106974, 2021.
18 Yan, Y; Xia, H; Fu, Y; Ni, Q-Q, *Mater Chem. Phys.*, **246**, 122808, 2020.
19 Ma, Z; Zhang, G; Yang, Q; Shi, X; Shi, A, *J. Cellular Plast.*, **50**, 1, 55-79, 2014.
20 Omo, T; Wu, X; Horiuchi, D; Yoda, S, *J. Supercritical Fluids*, **165**, 104963, 2020.
21 Ono, T; Wu, X; Furuya, T; Yoda, S, *J. Supercritical Fluids*, **149**, 26-33, 2019.
22 Álvarez, I; Gutiérrez, C; Rodríguez, J F; de Lucas, A; García, M T, *J. Supercritical Fluids*, **164**, 104886, 2020.
23 Yang, Z; Hu, D; Liu, T; Zhao, L, *J. Supercritical Fluids*, **153**, 104601, 2019.
24 Kohlhoff, D; Nabil, A; Ohshima, M, *Polym. Adv. Technol.*, **23**, 1350-6, 2012.
25 Fan, C; Wan, C; Gao, F; Huang, Y; Xi, Z; Xu, Z; Zhao, L; Liu, T, *J. Cellular Plast.*, **52**, 3, 277-98, 2016.
26 Geissler, B; Feuchter, M; Laske, S; Fasching, M; Holzer, C; Langecker, G R, *J. Cellular Plast.*, **52**, 1, 15-35, 2016.
27 Jing, X; Peng, X-F; Mi, H-Y; Wang, Y-S; Zhang, S; Chen, B-Y; Zhou, H-M; Mou, W-J, *J. Appl. Polym. Sci.*, **133**, 43508, 2016.
28 Gandhi, A; Bhatnagar, N, *Polym. Plast. Technol. Eng.*, **20**, 1812-18, 2015.
29 Hopmann, C; Hendriks, S; Feldges, R, *Polym. Eng. Sci.*, **55**, 2851-8, 2015.
30 Schmitt, H; Creton, N; Prashantha, K; Soulestin, J; Lacrampe, M-F; Krawczak, P, *J. Appl. Polym. Sci.*, **132**, 41341, 2015.
31 Szegda, D; Duangphet, S; Song, J; Tarverdi, K, *J. Cellular Plast.*, **50**, 2, 145-62, 2014.
32 Zepnik, S; Hendriks, S; Kabasci, S; Radusch, H-J, *J. Mater. Res.*, **28**, 17, 2394-2400, 2013.
33 Wang, K; Wu, F; Zhai, W; Zheng, W, *J. Appl. Polym. Sci.*, **129**, 2253-60, 2013.
34 Fox, R; David Frankowski, D; Jeff Alcott, J; Dan Beaudoin, D; Lawrence Hood, L, *J. Cellular Plast.*, **49**, 4, 335-49, 2013.
35 Liao, Q; Tsui, A; Billington, S; Frank, C W, *Polym. Eng. Sci.*, **52**, 1495-508, 2012.
36 Garbacz, T, *Polimery*, **57**, 11-12, 865-68, 2012.
37 Xu, M Z; Bian, J J; Han, C Y; Dong, L S, *Macromol. Mater. Eng.*, **301**, 149-59, 2016.
38 Karlsson, K; Kadar, R; Stading, M; Rigdahl, M, *Cellulose*, **23**, 1675-85, 2016.
39 Peng, X-F; Liu, L-Y; Chen, B-Y; Mi, H-Y; Jing, X, *Polym. Testing*, **52**, 225-33, 2016.
40 Costeux, S; Lantz, D R; Beaudoin, D A; Barger, M A, **WO2013048760**, *Dow Global Technologies LLC*, Apr. 4, 2013.
41 Sameni, J; Jafferb, S A; Tjong, J; Yang, W; Sain, M, *Adv. Ind. Eng. Polym. Res.*, **3**, 1, 36-45, 2020.
42 Long, Y; Sun, F; Liu, C; Xie, X, *RSC Adv.*, **6**, 23726-36, 2016.
43 Lopattananon, N; Julyanon, J; Masa, A; Kaesaman, A; Thongpin, C; Sakai, T, *J. Vinyl Addit. Technol.*, **21**, 134-46, 2015.
44 Aubert, L; Dussault, L, **US20080031843**, *L'Oreal*, Feb. 7, 2008.
45 Lobos, J; Iasella, S; Rodriguez-Perez, M A; Velankar, S S, *Polym. Eng. Sci.*, **56**, 9-17, 2015.
46 Xi, Z; Chen, J; Liu, T; Zhao, L; Turng, L-S, *Chinese J. Chem. Eng.*, **24**, 180-9, 2016.
47 http://www.trexel.com/images/pdfs/TrexelCorporate.pdf
48 Gómez-Monterde, J; Hain, J; Sánchez-Soto, M; Maspoch, M L, *J. Mater. Process. Technol.*, **268**, 162-70, 2019.
49 Jahani, D; Ameli, A; Saniei, M; Ding, W; Park, C B; Naguib, H E, *Macromol. Mater. Eng.*, **300**, 48-56, 2015.
50 Stumpf, M; Spoerrer, A; Schmidt, H-W; Altstaedt, V, *J. Cellular Plast.*, **47**, 6, 519-34, 2011.
51 Lee, R E; Azdast, T; Wang, G; Park, C B, *Int. J. Biol. Molec.*, **155**, 286-92, 2020.
52 Heim, H-P; Tromm, M, *J. Cellular Plast.*, **52**, 3, 299-319, 2016.
53 Wang, W; Ma, Y; Chen, X, **US20150035193**, *Yanfeng Automotive Trim Systems Co., Ltd.*, Feb. 5, 2015.
54 Rohledes, M; Jakob, F, **Foam Injection Molding. Specialized Injection Molding Techniques**. *Elsevier*, 2016, pp. 53-106.
55 Lee, J; Turng, L-S; Dougherty, E; Gorton, P, *Polymer*, **52**, 6, 1436-46, 2011.
56 Zhao, J; Qiao, Y; Wang, G; Park, C B, *Mater. Design*, **195**, 109051, 2020.
57 Sen, I; Dadush, E; Penumadu, D, *J. Cellular Plast.*, **47**, 1, 65-79, 2011.
58 Prociak, A; Michalowski, S; Bak, S, *Polimery*, 57, 11-12, 786-90, 2012.
59 Zauzi, N S A; Ariff, Z M; Khimi, S R, *Mater. Today: Proc.*, **17**, 3, 1001-7, 2019.
60 Lopez-Gil, A; Silva-Bellucci, F; Velasco, D; Ardanuy, M; Rodriguez-Perez, M A, *Ind. Crops Prod.*, **66**, 194-205, 2015.
61 Watkins, R L; Baghdadi, H; Edwards, C; Chang, Y, **EP2970616**, *Nike Innovate CV*, Jan. 20, 2016.

62 Verma, N; Zafar, S; Talha, M, *Manuf. Lett.*, **23**, 9-13, 2020.
63 Raymond, A; Rodrigue, D, *Cellular Polym.*, **32**, 4, 199-212, 2013.
64 Moscoso-Sanchez, F J; Mendizabal, E; Jasso-Gastinel, C F; Ortega-Gudino, P; Robledo-Ortiz, J R; Gonzalez-Nunez, R; Rodrigue, D, *J. Cellular Plast.*, **51**, 5-6, 489-503, 2015.
65 Ramkumar, P L; Kulkarni, D M; Abhijit, V V R; Cherukumudi, A, *Procedia Mater. Sci.*, **6**, 361-7, 2014.
66 Park, C B; Liu, G; Liu, F; Pop-Iliev, R; Zhang, B; D'uva, S, **CA2300776**, Jul. 13, 2004.
67 Aher, B; Olson, N M; Kumar, V, *J. Mater. Res.*, **28**, 17, 2366-73, 2013.
68 Sorrentino, L; Cafiero, L; Iannace, S, *Polym. Eng. Sci.*, **55**, 1281-9, 2015.
69 Kumar, V; Guo, H, **WO2014210523**, *University of Washington*, Dec. 31, 2014.
70 Yeh, S-K; Demewoz, N M; Kurniawan, V, *Polym. Testing,* **93**, 107004, 2021.
71 Sun, X; Turng, L-S, *Polym. Eng. Sci.*, **54**, 899-913, 2014.
72 Branch, G, **US20050203198**, *Microgreen Polymers, Inc.*, Sep. 15, 2005.
73 Sängerlaub, S; Lehmann, E; Müller, K; Wani, A A, *Food Packaging Shelf Life*, **24**, 100482, 2020.
74 Ma, Y-W; Jeong, M Y; Lee, S-M; Shin, B S, *J. Korean Phys. Soc.*, **68**, 5, 668-73, 2016.
75 Chen, F; Gaellstedt, M; Olsson, R T; Gedde, U W; Hedenqvist, M S, *RSC Adv.*, **5**, 94191-200, 2015.
76 Kulkarni, Sadhan C. Jana, S C, *Polymer*, in press, 123125, 2020.
77 Wan, C; Sun, G; Gao, F; Liu, T; Esseghir, M; Zhao, L; Yuan, W, *J. Supercritical Fluids*, 120, 421-31, 2017.
78 Yu, Y; Li, G; Han, W; He, Y, *Chinese J. Chem. Eng.*, in press, 2020.
79 *Addit. Polym.*, **2012**, 4, 2, 2012.
80 Abe, M; Nakayama, A; Nagano, M, **US8901184**, *Hitachi Metals, Ltd.*, Dec. 2, 2014.
81 Reedy, M, **WO2014018768**, *Reedy International Corporation*, Jan. 320, 2014.

10

SELECTION OF FOAMING AND BLOWING AGENTS FOR DIFFERENT POLYMERS

10.1 ACRYLONITRILE-BUTADIENE-STYRENE

The composition includes ABS (100 parts), foam modifier (2-10 parts), cellulosic material (25-50 parts), and blowing agent (0.5-2 parts).[1] Suggested blowing agents include hydrocerol, azodicarbonamide, and sodium bicarbonate.[1] The foam modifier is used to decrease the foam density, increase the foam swell, and improve processing. Acrylic modifier (K415 from Dow) is preferred.[1] Also, high molecular weight SAN can be used.[1] The composition may be formed into a variety of shapes and products using many different processing techniques, e.g., extrusion and molding.[1] One application includes a foam stop located adjacent to the jamb liner of a window.[1]

Cold boxes, refrigerators, freezers may comprise an insulated cooling cabinet having liners made out of ABS in at least one layer.[2] Environmentally acceptable blowing agents, such as 2-fluoro-2,2-dichloroethane (HCFC-141b) and 2,2-dichloro-1,1,1-trifluoroethane (HCFC-123), HFCs, including HFC-245fa, and 1,1,1-trifluoro,3-chloropropene (HFCO-1233zd) can be used in this application.[2]

The invention discloses a preparation method of ABS foam, which comprises the following steps: (1) blending ABS resin, EMA, star SBS, DCP, foaming agent (azodicarbonamide), zinc oxide and stearic acid; (2) compression molding.a ABS foam had uniform cells, low density, good mechanical properties, and good elasticity.[3] It can be used for packaging, automobiles, thermal insulation, buildings, buffer liners, sound absorption, and noise reduction.[3]

The invention relates to a composition comprising a foam blowing component and a polymer component.[4] 1,1,1,4,4,4-Hexafluoro-2-butene and 1,1,1,3,3-pentafluoropropane and their mixtures have been used as blowing agents for polystyrene and acrylonitrile-butadiene-styrene.[4]

REFERENCES

1   Zehner, B E; Ross, S R, **US6784216**, *Crane Plastics Company LLC*, Aug. 31, 2004.
2   Yu, B; Banavali, R, **US20160169575**, *Honeywell International Inc.*, Jun. 16, 2016.
3   CN 109370146B, Nov. 24, 2020.
5   KR20200010206A, Jan 30, 2020.

10.2 ACRYLONITRILE-BUTADIENE-ACRYLATE

Thermal insulation performance in housing and construction industries and high structural strength for automotive, aerospace, and electronic applications are essential features of new lightweight materials.[1] Doubling the value of current thermal insulation materials can save $200 million annually in heating/cooling costs for families in the U.S.[1] In today's average vehicles, as much as 5-10% in fuel savings can be achieved through a 10% weight reduction.[1]

The material, especially for medical instruments or implants, comprises a high-temperature thermoplastic material (PAEK) with a softening temperature above 250°C, which is foamed with titanium hydride and/or zirconium hydride.[2] The implant material has high biocompatibility, and the use of metal-containing blowing agent permits monitoring the implant after implantation into the body with imaging techniques.[2]

The orthopedic implant has a porous, non-metallic, bone interface or outer bone-contacting surface adapted for promoting bone ingrowth into the pores of the surface.[3] The bone-contacting surface is made out of polyaryletherketone.[3] The outer portion has a porosity in the range from 55 to 90 vol%. The skin is removed by machining.[3]

REFERENCES

1 Chiou, N-R; Lee, J L; Yang, J; Yeh, S-K, **WO2009155066**, *The Ohio State University Research Foundation, Nanomaterial Innovation Ltd.*, Dec. 23, 2009.
2 Sinz, I; Zeller, R; Schwiesau, J, **DE10256345**, *Aesculap AG & Co KG*, Dec. 4, 2003.
3 Wallick, M, **US8998987**, *Zimmer, Inc.*, Apr. 7, 2015.

10.3 BISMALEIMIDE RESIN

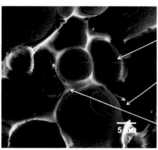

Figure 10.3.1. Three-phase syntactic foams comprising carbon microballoons, bismaleimide, and unreinforced or interstitial void. [Adapted, by permission, from Gladysz, G M; Chawla, K K, **Voids in Materials. From Unavoidable Defects to Designed Cellular Materials. Cellular Materials**, *Elsevier*, 2015, pp. 103-30.]

The high-performance foams with excellent dielectric property can be based on diallyl bisphenol A modified bismaleimide.[1] Foaming agent (AC135) is based on azodicarbonamide.[1] The optimum conditions of prepolymerization are 140°C for 60 min are followed by foaming at 160°C for 35 min.[1] The dielectric constant of the foam decreases with an increase in the blowing agent content.[1]

The syntactic foams have been obtained from carbon microballoons dispersed in bismaleimide binder (Figure 10.1).[2] In the three-phase composite foams (Figure 10.3.1), the compressive strength and modulus increase with increasing volume content of microballoons (and decreasing content of interstitial voids).[2] The strength of the microballoon contributes to the syntactic foam strength.[2]

REFERENCES

1 Xie, X; Gu, A; Liu, P; Liang, G; Yuan, L, *Polym. Adv. Technol.*, **22**, 1731-7, 2011.
2 Gladysz, G M; Chawla, K K, **Voids in Materials. From Unavoidable Defects to Designed Cellular Materials. Cellular Materials**, *Elsevier*, 2015, pp. 103-30.

10.4 BROMOBUTYL RUBBER

A pneumatic tire with a double layer inner liner structure bonded to an inner face of the tire carcass was developed to reduce noise generated by air vibrations inside of the tire's air chamber.[1] The double-layer structure includes foamed and non-foamed layers.[1] The foamed layer is produced by curing a rubber composition containing a nitrogen-releasing foaming agent that decomposes at the vulcanization temperature to release nitrogen gas.[1] The foamed layer has an open-cell structure.[1] The cell cavities have a mean diameter of about 130-200 μm.[1] The p-toluenesulfohydrazide (Porofor TSH 75) is the nitrogen-releasing blowing agent.[1] The p-toluenesulfohydrazide, begins to decompose at 120°C.[1] The mixing step should be conducted to obtain a discharge temperature, not exceeding 100-120°C.[1]

Brominated butyl rubber (density 0.93 g/cm^3; 32 MU (Mooney) at 125°C; bromine content 1.8 wt% based on the weight of the rubber) can be effectively used to increase the coefficient of friction of olefin block copolymer/silicone rubber syntactic foams.[2]

REFERENCES
1　Kanz, C; Frank, U, **US7389802**, *The Goodyear Tire & Rubber Co.*, Jun. 24, 2008.
2　**CN111417686A**, Jul. 14, 2020.

10.5 CELLULOSE ACETATE

The thermoplastic cellulose acetate was foamed using *cis*-1,3,3,3-tetrafluoroprop-1-ene (HFO 1234ze) as a low global warming blowing agent and talc as a nucleating agent.[1] Closed-cell structure was obtained.[1] The expansion ratio increases with increasing HFO 1234ze content.[1] A talc content has only a slight influence on the expansion ratio, but coarse and inhomogeneous cell structure is obtained without talc.[1] Talc helps to produce a foam having higher cell density, smaller cell sizes, and smaller cell wall thickness.[1]

In principle, it is possible to foam cellulose acetate butyrate but, there is still no working production process available that is suitable for industrial use to provide stable foams.[2] A suitable foam can be obtained from cellulose acetate butyrate having a number average molecular weight of more than 30,000 g/mol (30,000-70,000, preferred 40,000), an average butyryl content of at least 30 wt%, and the content of hydroxyl groups 0.8-1.8 wt%. Such polymer has a glass transition temperature in the range of 90-140°C.[2] Pentane in the concentration of 3-10 wt% serves as a suitable blowing agent.[2]

REFERENCES

1 Zepnik, S; Hendriks, S; Kabasci, S; Radusch, H-J, *J. Mater. Res.*, **28**, 17, 2394-400, 2013.
2 Eberstaller, R; Hintermeier, G, **US20130040125**, *Sunpor Kunstrsoff Ges. m.b.H.*, Feb. 14, 2013.

10.6 CHITOSAN

The chitosan-hydroxyapatite, super porous hydrogel composite was obtained by microwave-assisted fabrication for application as bone scaffolds.[1] Sodium bicarbonate was used as a blowing agent.[1] Simultaneous gas foaming, and microwave crosslinking were used.[1] The process was very rapid (1 min.), resulting in a material having superior properties for bone tissue engineering.[1]

Chitosan and wheat gluten biofoam was obtained by mixing, and phase distribution without any use of chemical blowing agent.[2] Soft foams were prepared by vacuum drying with stiffnesses from 0.3 to 1.2 MPa and high rebound resilience.[2] Open pores with sizes of 70-80 μm were produced.[2] The density of foam was 50 kg m^{-3}, due to the high porosity (96% air).[2]

n-Pentane was used as a blowing agent for the production of superabsorbent foam from carboxymethyl chitosan.[3] A stable hydrogel foam was produced that prevented the hydrogel from collapsing.[3] The dried product had a porous structure with a high water-binding capacity.[3] The dried foams had water uptake of 37-107 times their weight depending on the duration of exposure (1 min to 1 h).[3]

The fire fighting foaming composition contains chitosan and a surfactant, such as a protein-derived surfactant or a hydrocarbon surfactant, which is either amphoteric or nonionic.[4] The fire fighting foams can be produced without fluorinated surfactants and other fluorinated compounds (fluorine-containing surfactants have a long lifetime in the environment).[4] The foaming composition is diluted with water and then aerated to form a foam.[4] The foam is distributed over the burning liquid to form a barrier that extinguishes the fire by excluding oxygen.[4]

Chitosan is used in a foamed gel that may be layered onto a suitable backing for use as a wound dressing, or the gel may be directly applied to wounds to perform the hemostatic activity as a result of the action of the chitosan.[5] The foaming is enhanced by the addition of lauryl sulfate.[5] The foaming process utilizes the highly acidic nature of an acid that is used to dissolve the chitosan, such as acetic acid.[5] The addition of a salt such as sodium bicarbonate, which is reactive with the acid, is used to produce a foam.[5]

REFERENCES

1 Beskardes, I G; Demirtas, T T; Durukan, M D; Gümüsderelioglu, M, *J. Tissue Eng. Regen. Med.*, **9**, 1233-46, 2015.
2 Chen, F; Gaellstedt, M; Olsson, R T; Gedde, U W; Hedenqvist, M S, *RSC Adv.* **5**, 94191-200, 2015.
3 Zamani, A; Henriksson, D; Taherzadeh, M J, *Carbohydrate Polym.*, **80**, 4, 1091-1101, 2010.
4 Mulligan, D J; Joslin, N F, **US8431036**, *Kidde Ip Holdings Limited*, Apr. 30, 2013.
5 Scherr, G H, **US20070237811**, Oct. 11, 2007.

10.7 CYANOACRYLATE

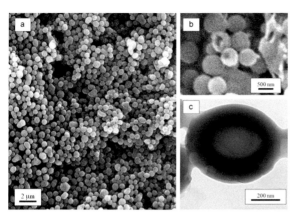

Figure 10.7.1. Submicron spheres. (a) SEM image of cyanoacrylate hollow submicron spheres (CAHSs). (b) SEM image of partially broken CAHSs. (c) High resolution TEM image of a single CAHS. Scale bar shows inset. [Adapted, by permission, from Makuta, T; Yoshihiro, Y; Sutoh, T; Ogawa, K, *Mater. Lett.*, **131**, 310-2, 2014.]

Submicron spheres with polymer shells were obtained by merely pumping the vapor of a commonly used instant adhesive into a citric acid solution to form microbubbles.[1] The main component of the instant adhesive, a cyanoacrylate monomer, is polymerized within seconds in the presence of water, but in the citric acid solution, the polymerization occurs at a much slower rate.[1] The polymer shell is composed of stable biocompatible material used for adhesion to human tissue (skin, blood vessels, and organs).[1] The dimensions of the spheres are less than 1 μm in diameter, making them suitable for use as a hollow nanocarrier in drug delivery applications, including cancer research.[1] The submicron spheres (Figure 10.7.1) were obtained by vaporizing cyanoacrylate, blowing vapor into a microbubble generator equipped with ultrasonic atomizer containing a citric acid solution (0.05 wt%), cooling to allow bubbles to polymerize at the gas-liquid interface.[1]

The addition of 0.4 wt% of MWCNTs and 0.5 wt% graphene into cyanoacrylate showed a thermal conductivity of 0.3195 and 0.3500 W/mK, respectively.[2]

REFERENCES
1 Makuta, T; Yoshihiro, Y; Sutoh, T; Ogawa, K, *Mater. Lett.*, **131**, 310-2, 2014.
2 Anis, B; El Fllah, H; Badr, Y A, *J. Mater. Res. Technol.*, **9**, 3, 2934-45, 2020.

10.8 EPOXY

Epoxy foams were made, by simultaneous crosslinking and foaming, using an amine-amide adduct as a curing agent and polymethyl hydrosiloxane as a blowing agent that can produce hydrogen gas by reacting with a curing agent.[1] The dielectric properties, such as dielectric constant and dissipation factor, increased with the decrease in cell size.[1]

Epoxy/functionalized-MWCNT microcellular foam was obtained using a batch foaming process with supercritical carbon dioxide.[2] The functionalized-MWCNT achieved more uniform dispersion in the epoxy matrix, acted as plasticizing agent to reduce glass transition temperature of composite, and increased CO_2 adsorption, hence greatly enhanced the foamability of the composite.[2] With 20 wt% of functionalized-MWCNT, the average cell size was reduced from 46.32 μm to 0.21 μm, and cell density increased from 2.7×10^7 to 2.74×10^{13} cells/cm^3, respectively.[2] Also, the microcellular foams had enhanced electrical conductivity of 5.2×10^{-4} S/m, electromagnetic interference shielding effectiveness of 22.73 dB, and compressive strength of 26.46 MPa.[2]

Polymer syntactic foams are a class of polymer composites prepared by the incorporation of hollow microballoons in the polymer matrix.[3] The advantage of these foams is their tailor-made properties that can be simply achieved by changing the volume fraction of hollow fillers and matrix.[3] The epoxy-glass microballoon syntactic foams containing 40-60 vol% of glass microballoons were reinforced with halloysite nanotubes.[3] A significant improvement in quasi-static mechanical properties was observed at 0.3 vol% nanotubes.[3]

Syntactic foams are used as lightweight composite materials for spacecraft protection capable of attenuating shock waves (spacecraft shielding).[4] The 0.64 g/cm^3 foam consisted of an epoxy matrix filled with a 55% volume fraction of glass microspheres.[4]

The composite structure comprised a highly-porous open-cell epoxy foam skeleton in which cells were filled with expanded polystyrene.[5] The density of composites varied within 0.22-0.32 g/cm^3 and could be controlled by the compression ratio of the expanded polystyrene granule bed.[5]

Diazirine was used in epoxy resin matrix as the on-demand UV initiated chemical foaming agent at room temperature.[6] Cell dimensions were tunable, ranging from 1 mm to tens nm by varying the matrix curing extent when the UV took effect.[6]

Acetic acid was used to promote the decomposition rate of sodium bicarbonate in the production of epoxy foam.[7] The addition of acetic acid reduced the dielectric constant.[7]

The reaction of the hydroxyalkylamide or hydroxyalkylurea with carboxylic acid releases water that acts as a blowing agent for epoxy resin.[8] The composition does not use chemical blowing agents.[8]

Epoxy formulations derived from vegetable oils have high reactivity.[9] They can be used for the production of lightweight, rigid epoxy foams.[9] Sodium bicarbonate was used as a blowing agent.[9] An exothermicity reducer (mineral compound) is necessary for the reactive formulation to regulate the rate of curing and expansion.[9]

Foamable epoxy-based zonal isolation sealing composition included epoxy resin and a blowing agent.[10] Hydrazide was the blowing agent used in the formulation.[10]

The invention provided a molding produced by a process comprising the following steps: a) introduction of a foamable reactive resin into support material and b) compressive deformation of the support material comprising the foamable reactive resin.[11] The

10.8 Epoxy

invention is related to the production of panels for vehicle construction, including aircraft construction, or as a fire protection layer.[11]

REFERENCES

1. Abraham, A A; Chauhan, R; Srivastava, A K; Katiyar, M; Tripathi. D N, *J. Polym. Mater.*, **28**, 2, 267-74, 2011.
2. Gao, Q; Zhang, G; Fan, X, Xiao, R, *Compos. Part A: Appl. Sci. Manuf.*, **138**, 106060, 2020.
3. Ullas, A V; Jaiswal, B, *Compos. Commun.*, **21**, 100407, 2020.
4. Rostilov, T A; Ziborov, V S, *Acta Astronautica*, **178**, 900-7, 2021.
5. Smorygo, O; Gokhale, A A; Vazhnova, A; Stefan, A, *Compos. Commun.*, **15**, 64-7, 2019.
6. Gao, F; Sun, H; Zhang, H; Cheng, J; Zhang, J, *Mater. Lett.*, **251**, 69-72, 2019.
7. Hamad, W N F W; Teh, P L; Yeoh, C K, *Polym. Plast. Technol. Eng.*, **52**, 754-60, 2013.
8. Finter, J; Jendoubi, E, **US20120055631**, *Sika Technology AG*, Mar. 8, 2012.
9. Mazzon, E; Habas-Ulloa, A; Habas, J-P, *Eur. Polym. J.*, **68**, 546-57, 2015.
10. Zamora, F; Sarkis, R K; Garza, T, **US20150011440**, *Clearwater International LLC*, Jan. 8, 2015.
11. Scherzer, D; Prissok, F; Schutte, M; Alteheld, A; Mertes, J; Hahn,K; Quadbeck-Seeger, H-J, **US20150322230**, *BASF SE*, Nov. 12, 2015.

10.9 ETHYLENE CHLOROTRIFLUOROETHYLENE

A data communication cable comprising a plurality of conductors and the cable separator.[1] The cable separator separates the plurality of conductors. Any chemical or physical blowing agent can be used in this product.[1]

REFERENCES

1   Brown, S M; Thwaites, S A; Siripurapu, S, **EP2973612**, *General Cable Technologies Corporation*, Jan. 20, 2016.

10.10 ETHYLENE-PROPYLENE DIENE RUBBER

Vulcanization of industrial ethylene propylene diene rubber compound was studied with, and without blowing agents.[1] 4,4'-Oxybis benzene sulfonyl hydrazide was used as the blowing agent.[1] The blowing agent induced several differences to the vulcanization reaction:[1]

- decreased a reaction temperature while increasing reaction heat
- eliminated the exothermic peak before vulcanization
- decreased the fully cured resin's glass transition temperature
- reduced operational window

EPDM foams are mostly prepared by thermal decomposition of the chemical blowing agent.[2] Since the cellular structure is formed due to the decomposition of the blowing agent, many parameters, such as its content, dispersion state, and particle size, have a great impact on foam properties.[2] Azodicarbonamide that decomposes between 150 and 210°C, and sulfonyl hydrazide compounds, such as 4,4′-oxybis(benzene sulfonyl hydrazide) that decomposes between 95 and 100°C are typical blowing agents used in EPDM.[2]

Recycled EPDM was mixed with polypropylene and foamed using azodicarbonamide.[3] A foam structure was negatively affected by increasing concentrations of EPDM.[3]

The amount of blowing agent injected had a significant influence on density, foam structure, and surface roughness.[4] The foam structure was also affected by the die pressure.[4] Nitrogen, carbon dioxide, and water are all suitable blowing agents for EPDM.[4]

A stable, reproducible foaming process can be implemented using water as a physical blowing agent for EPDM.[5] The average cell diameters in injection molded elastomer parts blown with water exceed the average cell diameters of chemically foamed parts, and cell size could be easily controlled by the amount of blowing agent.[5] Nucleating agents (e.g., glass fiber, flax fiber, or talc) could be used, to homogenize cellular structures and stabilize the foaming process.[5]

Melt-laminates of wood/natural rubber composite and ethylene-propylene-diene rubber foam were prepared by a compression molding technique.[6] Two different forms of 4,4'-oxybis(benzene sulfonyl hydrazide), OBSH, blowing agent were used; pure OBSH and ethylene-propylene bound OBSH.[6] Ethylene-bound OBSH (3-5 phr) gave EPDM foam with greater number of cells, higher porosity and resistance to water penetration on the foam surface.[6]

EPDM foam is used as a sealing material for various industrial products.[7] Azodicarbonamide, sodium bicarbonate, and N,N'-dinitrosopentamethylenetetramine are preferred blowing agents.[7]

Porous magnetic elastomer, whose mechanical behavior can be controlled by means of a magnetic field, contained ethylene-propylene-diene rubber filled with carbonyl iron particles and foamed with azodicarbonamide.[8]

REFERENCES

1　Restrepo-Zapata, N C; Osswald, T A; Hernandez-Ortiz, J P, *Polym. Eng. Sci.*, **55**, 2073-88, 2014.
2　Plachy, T; Kratina, O; Sedlacik, M, *Compos. Structures*, **192**, 126-30, 2018.
3　Mahallati, P; Rodrigue, D, *Cellular Polym.*, **33**, 5, 233-47, 2014.
4　Michaeli, W; Westermann, K; Sitz, S, *J. Cellular Plast.*, **47**, 5, 483-95, 2011.
5　Hopmann, C; Lemke, F; Binh, Q N, *J. Appl. Polym. Sci.*, **133**, 43613, 2016.
6　Yamsaengsung, W; Sombatsompop, N, *Composites Part B: Eng.*, **40**, 7, 594-600, 2009.

7   Iwase, T; Kawata, J; Kousaka, T; Takahashi, N, **US20130267619**, *Nitto Denko Corporation*, Oct. 10, 2013.
8   Bizhani, H; Katbab, A A; Verdejo, R, **High-Performance Elastomeric Materials Reinforced by Nano-Carbons. Multifunctional Properties and Industrial Applications**, *Elsevier* 2020, pp. 133-147.

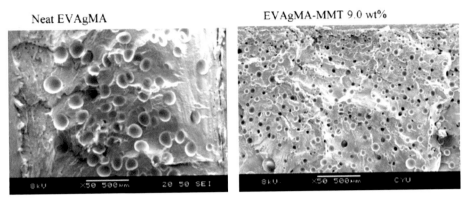

Figure 10.11.1. Cell structure of EVA containing intercalated (left) and exfoliated (right) montmorillonite. [Adapted, by permission, from Hwang, S-s; Liu, S-p; Hsu, P P; Yeh, J-m; Chen, C-l, *Int. Commun. Heat Mass Transfer*, **39**, 383-9, 2012.]

10.11 ETHYLENE-VINYL ACETATE

Ethylene-vinyl acetate copolymer/wood-flour foams having low density (<0.2 g/cm^3) were obtained using chemical blowing agent (activated azodicarbonamide (decomposition point 130-160°C; generates 160-180 cm^3/g of gas, mostly nitrogen)).[1] The morphological form of EVA/wood-flour foams with 20% wood-flour content becomes more uniform with increasing content of stearic acid (best results with 5 wt%).[1] The increase in the content of stearic acid causes that more gas remains in the EVA matrix, and the average cell size increases.[1] The increase in the amount of wood dust generates smaller cells.[5]

Ethylene-vinyl acetate/montmorillonite nanocomposite was manufactured by injection molding, using nitrogen as a blowing agent.[2] Cell sizes decreased with montmorillonite content increasing.[2] In EVA composites, montmorillonite was intercalated but, when EVA grafted with maleic anhydride was used, nanocomposites had an exfoliated layered structure.[2] Figure 10.11.1 shows that the exfoliated montmorillonite is more active in nucleation as it produces smaller cells.[2]

A blowing agent composition with a decomposition temperature of ~100°C has been developed for foaming ethylene-vinyl acetate by Tramaco in Germany.[3]

Good results can be achieved with foaming PE/EVA blends by accurately controlling the blend formulation, mixing, and proper selection of the foaming agent.[4] Azodicarbonamide can be used on existing equipment without modification.[4] The chemical foaming agent is easy to mix and can be directly added through the hopper.[4] Two grades of azodicarbonamide were compared.[4] One gave a better porosity, coloration, and more reduction in part weight; the other, having lower decomposition temperature (melt temperature by DSC of 169.2 vs. 173.4°C), gave a better distribution of pores.[4]

In the production of shoe soles, EVA has the remarkable ability to expand between 30 and 90% compared to the injection mold size, depending on the amount of blowing agent added.[6] Cellogen OT is suitable for this production at a concentration of 6 phr.[6] It releases gas when the temperature of 160°C is achieved.[6] Expansion occurs in sections having a thickness of 10 mm.[6] Surface curvature deformation can be found on components with large section thickness.[6] Thicknesses below 10 mm produce unusual effects.[6]

Ethylene-vinyl acetate, a popular material for the shoe foam material, has a non-uniform porosity distribution, especially near the surface in the low-porosity region called the skin layer.[7] Friction and wear behaviors of crosslinked EVA foams can be adjusted by skin layer and wear surface porosity.[7]

Recycling scrap of ethylene-vinyl acetate foam involves foaming and crosslinking a blend containing virgin EVA resin and scrap EVA foam containing a crosslinking agent (peroxide) and blowing agent (azodicarbonamide).[8] The blend also contains a compatibilizer (1-4 phr), which is a hydrogenated petroleum resin.[8]

Foamed ethylene-vinyl acetate resin contains a biodegradable polyester resin which has a monomer having a double bond.[9] Azodicarbonamide is used as a blowing agent and peroxide as a crosslinker.[9]

Foamed thermoplastic elastomeric polyurethane and ethylene-vinyl acetate copolymer (25-50 wt%) articles for low-density foam, midsole, footwear are made with a combination of a supercritical fluid (carbon dioxide or nitrogen) and a non-supercritical fluid blowing agent (Konz V2894).[10]

Wear-resistant, flame-retardant foam was prepared from ethylene-vinyl acetate copolymer, maleic anhydride-grafted EVA, modified steel slag/hydrotalcite composite material, white corundum, naphthenic oil, foaming agent (azodicarbonamide), accelerator (tetramethyl thiuram disulfide), stearic acid, zinc stearate, zinc oxide, dicumyl peroxide, and antioxidant.[11]

Antistatic composite comprised ethylene-vinyl acetate copolymer, antistatic agent (three-dimensional graphene hollow sphere), foaming agent (azodicarbonamide), stearic acid, zinc stearate, zinc oxide, and dicumyl peroxide.[12]

REFERENCES

1 Kim, J-H; Kim, G-H, *J. Appl. Polym. Sci.*, **131**, 40894, 2014.
2 Hwang, S-s; Liu, S-p; Hsu, P P; Yeh, J-m; Chen, C-l, *Int. Commun. Heat Mass Transfer*, **39**, 383-9, 2012.
3 *Addit. Polym.*, **2001**, 4, 2, 2001.
4 Spina, R, *Mater. Design*, **84**, 64-71, 2015.
5 Zimmermann, M V G; Turella, T C; Santana, R M C; Zattera, A J, *Mater. Design*, **57**, 660-6, 2014.
6 Allen, R D; Newman, S T; Mitchell, S R; Temple, R I; Jones, C L; Boer, C R; Dulio, S, *Robotics Computer-Integrated Manuf.*, **21**, 4-5, 412-20, 2005.
7 Nishi, T; Yamaguchi, T; Hokkirigawa, K, *Biotribology*, **22**, 100128, 2020.
8 Yu, S-C J, **US9074061**, *Nike, Inc.*, Jul. 7, 2015.
9 Chun, J P; Kim, H S; Choi, S Y; Kim, Y J; Yun, K C; Kim, M K; **WO2014021544**, *Samsung Fine Chemicals Co., Ltd.*, Feb. 6, 2014.
10 Baghdadi, H A, **US20150038605**, *Nike, Inc.*, Feb. 5, 2015.
11 **CN110964239A**, Apr. 7, 2020.
12 **CB108752722B**, Nov. 24, 2020.

10.12 FLUORINATED ETHYLENE-PROPYLENE

A foamed article comprises a melt-processible perfluoropolymer (hexafluoropropylene/ tetrafluoroethylene copolymer) and talc (2-15 wt%).[1] The composition is heated to a processing temperature of at least 315°C at which talc functions as a foaming agent.[1] It causes foaming at an expansion rate of 20-50%.[1] Talc is the only foaming agent used in the composition. The hydrogen-containing fluoropolymers are absent from the composition.[1] Use of talc has the benefit that talc is also an acid (HF) scavenger.[1]

The foamable composition for the production of cables included FEP ~96.9 wt%, talc ~1.5 wt%, PTFE ~0.9 wt%, calcium citrate ~0.45 wt%, and Aclyn wax ~0.25 wt%.[2] The foamed insulation had a foaming rate of about 45%.[2]

A foamed molded article is obtained by kneading and foaming a pellet comprising one or more fluorine resins (tetrafluoroethylene perfluoroalkyl vinyl ether copolymer and/ or tetrafluoroethylene-hexafluoropropylene copolymer), a chemical blowing agent (azodicarbonamide, 4,4'-oxybis(benzenesulfonyl hydrazide), or bistetrazole diammonium), and nucleating agent (boron nitride or talc) in an extrusion molding process.[3]

REFERENCES

1　Glew, C A; Boyle, K R; Kent, B L; Hrivnak, J A, **US8877823**, *Cable Components Group, LLC*, Nov. 4, 2014.
2　Glew, C A; Rosa, N M; Glew, C M, **US20190169391A1**, *Cable Components Group LLC*, Oct. 6, 2020.
3　Abe, M; Nakayama, A; Nagano, M, **US8901184**, *Hitachi Metals, Ltd.*, Dec. 2, 2014.

10.13 HYDROXYPROPYL METHYLCELLULOSE

Hydroxypropyl methylcellulose and ethyl hydroxyethyl cellulose are interesting candidates for the production of lightweight, foamed packaging material originating from non-fossil, renewable resources.[1] Using a hot-mold process and water as a blowing agent, nine grades of cellulose derivatives were compared, and six produced stable foam structures.[1] The foamability of these systems depended on viscoelastic properties determined by the water content.[1] There is apparently a limit for chain length suitable for foaming (both high and low molecular weight grades do not perform well).[1]

Hydroxypropyl methylcellulose was successfully foamed during continuous extrusion.[2] A proper selection of the processing parameters was required.[2] Crucial parameters were temperature, pressure, and residence time.[2] The addition of a complementary (to water) blowing agent is needed.[2]

Freeze-drying of polymeric solutions resulted in porous and flexible, but mechanically stable, a soft sponge-like substrate with hydroxypropyl methylcellulose-based solid foams, which were suitable for continuous inkjet printing of pharmaceuticals.[3] The inkjet-printed active pharmaceutical ingredient was released from the dosage forms upon contact with the dissolution medium.[3]

REFERENCES

1 Karlsson, K; Schuster, E; Stading, M; Rigdahl, M, *Cellulose*, **22**, 2651-64, 2015.
2 Karlsson, K; Kadar, R; Stading, M; Rigdahl, M, *Cellulose*, **23**, 1675-85, 2016.
3 Iftimi, L-D; Edinger, M; Bar-Shalom, D; Rantanen, J; Genina, N, *Eur. J. Pharmaceutics Biophamaceutics*, **136**, 38-47, 2019.

10.14 MELAMINE RESIN

Open-cell, melamine resin foams are obtained by heating using hot air, steam, or microwave irradiation.[1] The foams contain up to 50 wt% of organic filler (polymer), which has a melting point above 220°C.[1] The average particle diameter of the filler is similar to the average pore diameter of the foam structure (preferred in the range of 100-500 μm). LDPE wax (Luwax A) is the suggested organic filler.[1] Pentane (20 wt%) is added as a blowing agent.[1] The foam is designed for use as acoustic and thermal insulation in construction, automotive, marine, and rail vehicles, construction of spacecraft, and in the upholstery industry or for insulation of pipes.[1]

A synthetic resin foam of low flammability is composed of resin component (a phenolic resin or a mixture of phenolic resin and melamine resin), hardener component (a reaction product of an organic nitrogen base and a polyol phosphoric acid ester), filler component (ground inorganic fillers; with at least one containing water of crystallization and/or hydroxyl-containing inorganic filler comprises a pH 6-8, e.g., $CaSO_4 \cdot 2H_2O$), and blowing agent (propellant, e.g., pentane and many others).[2]

The melamine foam sponges are marketed under the tradename Mr. Clean Magic Eraser®.[3] The open-cell foam (and melamine foam in particular) shows good soil and/or stain removal performance when used to clean hard surfaces.[3] A cleaning implement comprises hybrid foam (an open-cell melamine-formaldehyde resin).[3] The foam was formed by 20 wt% pentane using microwave radiation.[3]

Carbon foam is a material obtained by heat treating and carbonizing a melamine resin foam in an inert gas atmosphere.[4] It is used in various applications because of its porosity, flexibility, and electrical properties.[4]

A melamine/formaldehyde foam having an open-cell foam structure, with an impregnation applied to the foam structure that comprises at least one particulate phyllosilicate surface-modified with aminosilane has fire-retarding properties.[5]

REFERENCES

1 Steinke, T H; Baumgartl, H; Lenz, W; Hahn, K, **EP2735584**, *BASF SE*, May 28, 2014.
2 Menke, K; Gettwert, V; Fischer, S, **WO2013156562**, *Fraunhofer-Gesellschaft zur Förderung der angewandten Forschung e. V.*, Oct.24, 2013.
3 Gonzales, D A; Deleersnyder, G A; Nessel, P; Steinke, T H, **US20160166128**, *The Procter & Gamble Company*, Jun. 16, 2016.
4 Suzuki, A; Yamashita, J, **US20200243866A1**, *Asahi Kasei Corp*, Jul. 30, 2020.
5 Weiße, S A; Steinke, T H; Vath, B; Lenz, W, **US20200270415A1**, *BASF SE*, Aug. 27, 2020.

10.15 PHENOL FORMALDEHYDE

A halogen-free flame retardant system consisting of ammonium polyphosphate as an acid source, blowing agent (petroleum ether), pentaerythritol as a carbonific agent, and zinc oxide as a synergistic agent was used in this work to enhance flame retardancy of phenolic foams.[1] Heat release rate, total heat release, effective heat of combustion, production, or yield of carbon monoxide, and oxygen consumption were all decreased in the presence of zinc oxide.[1] At the same time, the addition of ZnO caused a decrease in bending and compression strength.[1]

Bark-derived oils produced by hydrothermal liquefaction of outer and inner white birch bark were utilized for the synthesis of biobased phenol-formaldehyde foamable resole resins with different levels of phenol substitution.[2] Hexane was used as a blowing agent.[2] The foams using bio-oil derived from inner bark had lower thermal conductivity and more uniform cell structure than the foams using bio-oil from the outer bark.[2]

The lignin nanoparticle-reinforced phenolic foams show that their properties strongly depend on the cell size distribution of the foams.[3] The amount of blowing agent used for a reinforced foam was 31% less than the amount necessary to produce an unreinforced foam of the same density.[3] n-Pentane was used as a blowing agent.[3]

Lignin-based bio-phenol-formaldehyde foams were produced with pentane and hexane as blowing agents.[4] The lignin foams reduced formaldehyde consumption by up to 50%.[4] The bio-phenol-formaldehyde foams had excellent thermal conductivity (0.03-0.048 W/m·K), combustion properties (LOI: 32-33%), and closed-cell structure, all comparable to commercial phenol-formaldehyde foams.[4]

Reactivity of lignin was improved by depolymerization, demethylation, and phenolation catalyzed with HBr.[5] The obtained product was directly used to prepare phenolic foams without purification.[5] Activation energy of phenolic resin (103.27 kJ/mol) was lower than that of unmodified (113.48 kJ/mol).[5] The modified lignin-based phenolic foam showed had better thermal insulation and mechanical properties than the unmodified lignin-based phenolic foam (Figure 10.15.1).[5]

A phenol-formaldehyde resole was used to produce foam with less than 50% open-cells.[4] 1,1,1,4,4,4-Hexafluoro-2-butene (FEA-1100) was used as a blowing agent.[6]

A phenolic foam was made by foaming and curing a foamable phenolic resin composition that comprised phenolic resin, blowing agent, acid catalyst (benzenesulfonic acid, para-toluenesulfonic acid, xylenesulfonic, naphthalenesulfonic acid, ethylbenzenesulfonic acid, or phenolsulfonic acid), plasticizer (polyester polyol), and inorganic filler (e.g., calcium carbonate).[7] The blowing agent comprised an aliphatic hydrocarbon (75% or more of cyclopentane and 25% or less of isobutene and isopentane).[7] The phenolic foam had a higher pH value (more than 5) compared with the conventional phenolic foams, which helps to reduce corrosion risk when in contact with metallic materials.[7]

"Regular foam" generated from petroleum-derived phenol-formaldehyde plastic and "biofoam" generated from plant-derived phenol-formaldehyde plastic, used globally as floral foam, are eaten by marine and freshwater invertebrates (the freshwater gastropod *Physa acuta*, the marine gastropod *Bembicium nanum*, the marine bivalve *Mytilus galloprovincialis*, adults and neonates of the freshwater crustacean *Daphnia magna*, the marine amphipod *Allorchestes compressa*, and nauplii of the marine crustacean *Artemia* sp.).[8] These microplastics leach phenolic compounds that are toxic to aquatic animals.[8]

10.15 Phenol formaldehyde

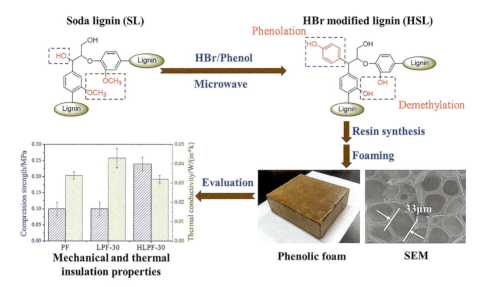

Figure 10.15.1. Lignin modified pheol foam - synthesis and properties. [Adapted, byn permission, from Gao, C; Li, M; Zhu, C; Zhuang, W, *Compos. Part. B: Eng.*, **205**, 108530, 2021.]

REFERENCES

1   Ma, Y; Wang, J; Wang, C; Chu, F, *J. Appl. Polym. Sci.*, **132**, 42730, 2015.
2   Li, B; Feng, S H; Niasar, H S; Zhang, Y S; Yuan, Z S; Schmidt, J; Xu, C, *RSC Adv.*, **6**, 40975-81, 2016.
3   Del Saz-Orozco, B; Oliet, M; Alonso, M V; Rojo, E; Rodríguez, F, *Composites Sci. Technol.*, **72**, 6, 667-74, 2012.
4   Li, B; Yuan, Z; Xu, C, *Eur Polym. J*, **111**, 1-10, 2019.
5   Gao, C; Li, M; Zhu, C; Zhuang, W, *Compos. Part. B: Eng.*, **205**, 108530, 2021.
6   Cobb, M W; Harmer, M A; Kapur, V; Liauw, A Y; Williams, S R, **EP2714786**, *E. I. Du Pont de Nemours and Company*, Apr. 9, 2014.
7   Coppock, V; Zeggelar, R; Takahashi, H; Kato, T, **US8765829**, *Kingspan Holdings (Irl) Limited*, Jul. 1, 2014.
8   Trestrail, C; Walpitagama, M; Nugegoda, D, *Sci. Total Environ.*, **705**, 135826, 2020.

10.16 POLY(3-HYDROXYBUTYRATE-CO-3-HYDROXYVALERATE)

Microbial poly(3-hydroxybutyrate-co-3-hydroxyvalerate) and its blends with cellulose acetate butyrate were extrusion foamed using azodicarbonamide.[1] The concentration of the blowing agent was varied from 0 to 4 phr to achieve maximum density reduction reaching 41%.[1] The average cell diameters were ranging from 58 to 290 μm, and cell densities were ranging from 650 to 180,000 cm^{-3}.[1] A higher melt viscosity and elasticity and reduced gas solubility of the PHBV/CAB blends retard cell coalescence and collapse during foam expansion, producing more uniform cell size distribution and better homogeneity of cellular morphology.[1]

Sodium bicarbonate was used as a blowing agent and citric acid and calcium carbonate as nucleation agents in extrusion foaming of poly(3-hydroxybutyrate-co-3-hydroxyvalerate).[2] The poly(3-hydroxybutyrate-co-3-hydroxyvalerate) is a natural biodegradable polyester having high crystallinity, low melt viscosity, slow crystallization rate, and high sensitivity to the thermal degradation at temperatures above its melting point.[2] This makes it difficult to control the foaming process.[2] The negative gradient temperature profile (100-180-170-160-150-150) was beneficial for minimizing the thermal degradation and to achieve melt strength required to stabilize cell structure.[2] Closed-cell structure with cell sizes between about 50 and 200 μm were obtained.[2]

The poly(3-hydroxybutyrate-co-3-hydroxyvalerate) can be foamed with an organic peroxide.[3] Diperoxyketals, dialkylperoxides, alkylarylperoxides, or their mixtures can be used.[3] The composition can be processed using single-screw extrusion, twin-screw extrusion, buss kneader, two-roll mill, impeller mixing, calendering, blow molding, or thermoforming.[3] Some examples of applications include disposable tableware, a disposable container, packaging material (e.g., food packaging), or insulation.[3]

The closed-cell (more than 80%) foams, for the fabrication of medical products, with densities less than 0.5 g/cm^3, were produced from poly(3-hydroxybutyrate-co-3-hydroxyvalerate).[4] The foam was obtained by heating to temperatures above the melt temperature of the polymer to form a melt polymer system without a substantial loss of the polymer's weight average molecular weight.[4] The blowing agent is one or more of the following: nitrogen, carbon dioxide, air, argon, helium, methane, ethane, propane, butanes, hexanes, or halogenated hydrocarbon.[4]

The application of biobased and biodegradable poly(3-hydroxybutyrate-co-3-hydroxyvalerate) is limited by its high cost and brittleness.[5] Expancel 920 DU 120 microspheres (28-38 μm in diameter) containing low boiling point isopentane as a blowing agent and acrylonitrile/methacrylonitrile/hydroxyethyl methacrylate copolymer as a shell and nanocellulose helped to improve mechanical properties of the foam.[5] The closed-cell foams had density reduced by 2.5-2.7 times.[5] The addition of nanocellulose increased expansion ratio, cell density, and porosity and also led to a more uniform cell size distribution.[5]

REFERENCES

1 Liao, Q; Tsui, A; Billington, S; Frank, C W, *Polym. Eng. Sci.*, **52**, 1495-508, 2012.
2 Szegda, D; Duangphet, S; Song, J; Tarverdi, K, *J. Cellular Plast.*, **50**, 2, 145-62, 2014.
3 Donnelly, Z, **WO2012170215**, *Arkema, Inc.*, Dec. 13, 2012.
4 Connelly, D; Felix, F; Martin, D P; Montcrieff, J; Rizk, S; Williams, S F, **US20150057368**, *Tepha, Inc.*, Feb. 26, 2015.
5 Panaitescu, D M; Trusca, R; Casarica, A, *Int. J. Biol. Macromol.*, **164**, 1867-78, 2020.

10.17 POLY(BUTYLENE SUCCINATE)

Chain extended poly(butylene succinate) foams were produced using supercritical carbon dioxide.[1] The chain extender had multi-epoxy-groups.[1] The incorporation of chain extender caused that the cell size distribution became uniform, the cellular wall became thin, and the cell structure turned oval to pyritohedron.[1] The highest expansion volume ratio was increased by more than 14 times.[1]

A biodegradable poly(butylene succinate) foam was produced by compression molding.[2] The dicumyl peroxide (a crosslinking agent) and trimethylolpropane trimethacrylate (a curing coagent for crosslinking) were used to increase the melt viscosity of PBS.[2] Zinc oxide/zinc stearate were used to reduce the thermal decomposition temperature of azodicarbonamide and to balance the temperatures of vulcanization of PBS and the decomposition of blowing agent.[2] The density of foams decreased with the increase in blowing agent content.[2] The density as low as 0.31 g/cm^3 was obtained.[2]

Carbon nanofiber nanocomposite foams were obtained from biodegradable poly(butylene succinate).[3] Azodicarbonamide, N,N'-dinitroso pentatetramine, and urea activator were used in the formulation.[3] Carbon nanofiber increased the melt viscosity of PBS so that the nanocomposite foams were produced without modifying the chemical structure of PBS.[3] Cell size and blowing ratio increased with the increase in the blowing agent content, blowing temperature, and time (a maximum blowing ratio of 12).[3] The cells were oval-shaped.[3] A small number of carbon nanofibers increased cell density.[3]

The biodegradable poly(butylene succinate)/halloysite nanotube nanocomposite foams were produced using supercritical carbon dioxide as a blowing agent.[4] The cell size decreased, and both cell density and volume expansion ratio increased with the addition of halloysite nanotubes (5 wt% led to the cells of the smallest size).[4] The saturation temperature and saturation pressure were found to significantly influence cell morphology.[4] Higher saturation pressure led to smaller cell size and higher volume expansion ratio.[4]

ZnO, which is frequently added as a catalyst of azodicarbonamide decomposition, was found to give antimicrobial protection to food packaging produced from poly(butylene succinate).[5] Although the minimum inhibition content of ZnO was 6 wt%, which may be too high for catalytic influence.[5]

Poly(lactic acid)/poly(butylene succinate)/cellulose fiber composite for hot cups packaging application was processed by extrusion and injection molding.[6] Presence of cellulose fiber decreased viscosity, increased the number of closed cells, and made cell distribution more uniform.[6] Glass transition temperature and melting temperature were improved by adding cellulose fiber, and decomposition temperature increased when blending with poly(butylene succinate).[6]

Poly(butylene succinate)/cellulose nanocrystal foams, having exceptional thermal insulation, were blown using supercritical carbon dioxide.[7] Branching reaction regulated molecular architecture and increased molecular weight of PBS.[7] Epoxy-based chain extenders were used for branching modification.[7] Thermal conductivity of 0.021 W/m K resulted from modifications.[7] To improve melt viscoelasticity and increase cell nucleation number, a chain extender (epoxy-based) and polyhedral oligomeric silsesquioxane were introduced to PBS through the melt blending method.[8] The cell size and volume expansion ratio reached 6.80±1.21 μm and 7.48±0.03 times, respectively.[8]

Poly(butylene succinate)/carbon nanotubes/polytetrafluoroethylene ternary nanocomposite foams had superior electrical conductivity due to the synergistic effect of "ball milling" (played by PTFE having a diameter of ~5-10 μm) and supercritical fluid-assisted processing.[9] Adding PTFE dramatically improved pore structure, leading to a two-orders-of-magnitude increase in pore density.[9] Conductivity of blend with PTFE (54.05 S/m at 3 wt% CNT) was four times higher than that without it.[9] Figure 10.17.1 illustrates reasons.[9]

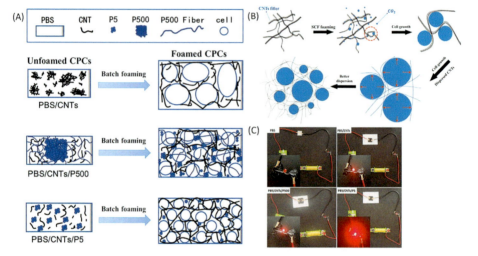

Figure 10.17.1. PBS/CTN composite - formation and properties, [Adapted, by permission, from Huang, A; Lin, J; Chen, S; Peng, X, *Compos. Sci. Technol.*, **201**, 108519, 2021.]

The implants may be made from fibers and meshes of poly(butylene succinate) and its copolymers or by 3D printing, and fibers may be oriented.[10] Coverings and receptacles made from forms of poly(butylene succinate) and its copolymers have also been developed for use with cardiac rhythm management devices and other implantable devices.[10]

REFERENCES

1 Zhou, H; Wang, X; Du, Z; Li, H; Yu, K, *Polym. Eng. Sci.*, **55**, 988-94, 2015.
2 Li, G; Qi, R; Lu, J; Hu, X; Luo, Y; Jiang, P, *J. Appl. Polym. Sci.*, **127**, 3586-94, 2013.
3 Lim, S K; Lee, S I; Jang, S G; Lee, K H; Choi, H J; Chin, I-J, *J. Macromol. Sci., Part B: Phys.*, **30**, 100-10, 2011.
4 Wu, W; Cao, X; Lin, H; He, G; Wang, M, *J. Polym. Res.*, **22**, 177, 2015.
5 Petchwattana, N; Covavisaruch, S; Wibooranawong, S; Naknaen, P, *Measurement*, **93**, 442-8, 2016.
6 Vorawongsagul, S; Pratumpong, P; Pechyen, C, *Food Packaging Shelf Life,* **27**, 100608, 2021.
7 Yin, D; Mi, J; Zhou, H; Tion, H, *Carbohydrate Polym.*, **247**, 116708, 2020.
8 Yin, D; Mi, J; Zhou, H; Wang, X; Fu, X, *Polym. Deg. Stab.*, **167**, 228-40, 2019.
9 Huang, A; Lin, J; Chen, S; Peng, X, *Compos. Sci. Technol.*, **201**, 108519, 2021.
10 Williams, S F; Rizk, S; Martin, D P, **US20190269817A1**, *Tepha Inc*, Oct. 23, 2020.

10.18 POLY(ε-CAPROLACTONE)

Biodegradable foams for packaging applications can be produced from poly(ε-caprolactone)/starch using carbon dioxide as a blowing agent.[1] The cell size of the foam was reduced with the increase in starch concentration.[1] The crystallinity was reduced after the material was foamed.[1]

Bimodal, porous scaffolds for bone regeneration were prepared *via* supercritical carbon dioxide foaming.[2] Saturation and foaming temperatures and pressures, as well as depressurization time, were selected to optimize the pore structure of the foams and to induce the formation of a macro-porosity suitable for bone cell adhesion and colonization.[2] The double scale pore size distribution with a macro-porosity of mean pore size of 134±80 µm and 121±49 µm, respectively, and a micro-porosity in the 1-10 µm range were obtained.[2]

PCL foams were produced by thermally-induced phase separation.[3] Tetrahydrofuran/methanol (solvent/non-solvent) mixture was used for induction of liquid-liquid phase separation of PCL solutions.[3] Lower PCL concentration, lower THF content, and higher quench temperature lead to larger pore sizes.[3] By selectively tuning the process parameters, foams with controlled pore sizes (10-450 µm), porosity (83-91%), and morphology (cellular, bead-like, microspherical) were obtained.[3] Crystallinity increased with decreasing polymer concentration and increasing solvent quality due to the enhanced chain mobility.[3]

Water-soluble drugs have shown burst release, and insoluble drug slower release from supercritical carbon dioxide foamed PCL samples.[4]

REFERENCES
1 Ogunsona, E; D'Souza, N A, *J. Cellular Plast.*, **51**, 3, 245-68, 2015.
2 Salerno, A; Zeppetelli, S; Di Maio, E; Iannace, S; Netti, P A, *J. Supercritical Fluids*, **67**, 114-22, 2012.
3 Onder, O C; Yilgor, E; Yilgor, I, *Polymer*, **136**, 166-78, 2018.
4 Asikainen, S; Paakinaho, K; Seppälä, J, *Eur. Polym. J.*, **113**, 165-75, 2019.

10.19 POLYACRYLONITRILE

A super abrasive resin product includes a super abrasive grain component, an oxide component, and a continuous phase defining a network of interconnected pores.[1] The oxide component consists of an oxide of a lanthanoid, and the continuous phase includes a thermoplastic polymer component (polyacrylonitrile).[1] The blowing agent of the super abrasive product precursor includes discrete particles of a shell that includes copolymer polyacrylonitrile and polyvinylidenechloride.[1]

Thermally expandable microspheres are special physical foaming agents with a core-shell structure formed by encapsulating blowing agents (low-boiling point alkanes) with thermoplastic polymers, such as polyacrylonitrile, polymethylmethacrylate, and polyurethane.[2] When the heating temperature exceeds the glass transition temperature of the thermoplastic shell, the shell softens, and the alkanes rapidly vaporize to form a hollow structure.[2]

REFERENCES

1   Upadhyay, R; Vedantham, R, **US20150360346**, *Saint-Gobain Abrasives, Inc.*, Dec. 17, 2015.
2   Hu, L; Wang, J; Qin, L; Xu, H;, Yang, Z, *Int. J. Adhesion Adhesives*, **105**, 102783, 2021.

10.20 POLYAMIDE

A polyamide is prepared by polymerizing one anionic lactam (caprolactam and/or lauryl lactam), and incorporation of one chemical blowing agent.[1] Temperatures of polymerization and foaming are the same (140°C).[1] Sodium bicarbonate and citric acid, as marketed by Clariant under the trade name Hydrocerol® are used as blowing agents.[1]

A foamable polyamide composition contains a polyamide comprising carboxylic group(s) (PA-6 or PA-66), thermoplastic rubber (EPDM or EPM), and a compound having an isocyanate group (oligomeric polyisocyanate, such as hexamethylene diisocyanate trimer or isophorone diisocyanate trimer).[2]

The polyamide foam includes polyamide resin compounded with composite chain extender including epoxy chain extender, and maleic anhydride grafted polypropylene wax.[3] The polyamide foams have improved properties, including a smooth surface, low density, and small cell size.[3]

REFERENCES

1 Scherzer, D; Desbois, P, **EP2511076**, *BASF SE*, Oct. 17, 2012.
2 Oark, J H; Yu, Y C, **WO2015062868**, *Rhodia Operations*, May 7, 2015.
3 Zha, M; Wang, M, **US20200317878A1**, *Advansix Resins and Chemicals LLC*, Oct. 8, 2020.

10.21 POLYCARBONATE

Unimodal or bimodal cell-size distributions can be obtained using supercritical carbon dioxide as a blowing agent in the production of polycarbonate foams.[1] The glass transition temperature of the polymer/gas mixture is found to present a good linear relationship with the absorbed carbon dioxide concentration (foaming takes place above T_g of the polymer/gas mixture) (Figure 10.21.1).[1] A two-step batch depressurization is applied to produce bimodal foams containing both small and large cells.[1] The bimodal foams exhibit significantly improved tensile properties compared to the unimodal foams.[1] The extensional stress-induced nucleation promotes the generation of nanocellular structures around the expanding larger cells.[1]

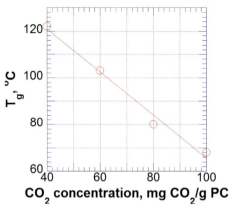

Figure 10.21.1. Glass transition temperature of PC as a function of CO_2 concentration. [Data from Ma, Z; Zhang, G; Yang, Q; Shi, X; Shi, A, *J. Cellular Plast.*, **50**, 1, 55-79, 2014.]

The microcellular injection molding process is used for the production of polycarbonate foam with water/salt solution as the physical blowing agent.[2] Salt crystals of 10-20 μm recrystallized during molding acted as nucleating agents in the PC foamed parts (foamed parts molded without nucleating agent exhibit much larger and fewer bubbles).[2] Parts blown with water have a stronger specific Young's modulus and specific ultimate strength than the commercial microcellular injection molded parts with nitrogen used as a blowing agent.[2]

The microcellular foams produced by injection molding with a physical blowing agent (MuCell™ Technology by Trexel) show a significantly higher notched impact strength than the compact polycarbonate if the compact material is brittle under the same testing parameters.[3] If the compact polycarbonate breaks toughly, the notched impact strength of the foamed material is always lower.[3] The gas counter pressure and precision mold opening were used to obtain a microcellular foam with cell diameters around 10 μm.[3] A significantly higher notched impact strength was obtained than for the unfoamed material under the same testing conditions.[3] It is possible to improve the notched impact strength by 400% as compared to the resin material.[3]

Highly expanded open-cell polycarbonate foams were obtained with nitrogen in MuCell™ technology.[4] Expansion ratios ranging from 1 to 8 and open-cell content up to 85% were obtained for acoustic insulation foams.[4] The sound wave transmission loss was increased by up to 2.5-fold while thermal conductivity of foams was decreased up to fivefold as compared with the unfoamed polymer.[4]

A polycarbonate resin, polyester resin, and a blowing agent were kneaded and extruded to produce foam with an apparent density of 40-400 kg/m^3.[5] More than 80% were closed-cells.[5] Cyclopentane and n-butane are used as blowing agents.[5]

Polycarbonate microcellular foam with novel "coral reef-like" structures was developed to improve mechanical properties of the foam by incorporating green asphalt-based microporous organic polymer into polycarbonate *via* melt mixing.[6] Good compatibility

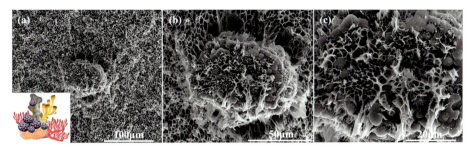

Figure 10.21.2. SEM micrographs of "coral reef-like" structure at different magnifications. [Adapted, by permission, from Zhu, J; Tan, D; Li, L; Zhang, S; Chen, Y, *Eur. Polym. J.*, **134**, 109780, 2020.]

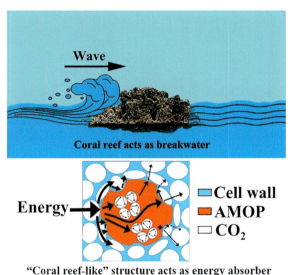

Figure 10.21.3. Schematic description of toughening mechanism of "coral reef-like" structures. [Adapted, by permission, from Zhu, J; Tan, D; Li, L; Zhang, S; Chen, Y, *Eur. Polym. J.*, **134**, 109780, 2020.]

between asphalt-based microporous organic polymer and polycarbonate facilitated the formation of "coral reef-like" structures during the foaming process, ascribed to a strong interfacial bond between asphalt-based microporous organic polymer and polycarbonate matrix.[6] Due to the toughening and strengthening effects of "coral reef-like" structures, the impact and compressive strength of asphalt-based microporous organic polymer/polycarbonate with 1.5 wt% asphalt-based microporous organic polymer achieved 3.2 and 1.2 times increases when polycarbonate was used as a control.[6] Figure 10.21.2 shows elements of its morphological structures.[6] The mechanism of the effect of "coral reef-like" structures on the impact strength of materials is presented in Figure 10.21.3.[6] It can be utilized to fabricate polymeric materials of high toughness.[6] During impact, foam is subjected to external forces, and cracks are generated.[6] When the cracks encounter the "coral reef-like" structures during crack propagation, cracks are terminated or deflected instead of damaging "coral reef-like" structures.[6] A large amount of energy is consumed by the deflection or termination.[6] The composite is suitable as thermal insulation material because of its lightweight, high mechanical performance, and low thermal conductivity.[6]

REFERENCES

1. Ma, Z; Zhang, G; Yang, Q; Shi, X; Shi, A, *J. Cellular Plast.*, **50**, 1, 55-79, 2014.
2. Peng, J; Turng, L-S; Peng, X-F, *Polym. Eng. Sci.*, **52**, 1464-73, 2012.
3. Bledzki, A K; Rohleder, M; Kirschling, H; Chate, A, J. Cellular Plast., 46, 415-40, 2010.
4. Jahani, D; Ameli, A; Saniei, M; Ding, W; Park, C B; Naguib, H E, *Macromol. Mater. Eng.*, **300**, 48-56, 2015.
5. Ishikawa, T; Kogure, N; Okuda, A, **EP2423250**, *JSP Corporation*, Dec. 17, 2014.
6. Zhu, J; Tan, D; Li, L; Zhang, S; Chen, Y, *Eur. Polym. J.*, **134**, 109780, 2020.

10.22 POLYCARBONATE/ABS

Injection-molded components with functionally graded foam structures were produced from polycarbonate/ABS blend.[1] The production of foam injection molded components with different foaming ratios was possible by changing core movements in the closed injection mold after the cavity has been filled volumetrically with plastic melt containing blowing agent.[1] Density, morphology, mechanical properties can be tailored to the application.[1]

REFERENCES

1   Heim, H-P; Tromm, M, *J. Cellular Plast.*, **52**, 3, 299-319, 2016.

10.23 POLYCHLOROPRENE

The fire retardant material contains polychloroprene and chloroparaffin that are expanded to a final density of less than 200 kg/m^3.[1] The product of foaming is useful for thermal insulation, acoustic insulation, acoustic damping, and vibration damping.[1] Azodicarbonamide was the foaming agent used together with either zinc borate or zinc oxide as the nucleating agent.[1] Thermal conductivity of 0.034-0.038 W/mK, limiting oxygen index of 53.7-59.3, and smoke growth rate of 142-172 m^2/s was achieved.[1]

Polymeric gloves having varied thicknesses can be made out of polychloroprene.[2] Sulfosuccinamates are effective in providing a low viscosity at low shear rates during foaming as well as providing a stable foam.[2]

Chloroprene foam containing carbon fiber and having a thickness of 1.7 mm had an EMI-SE of 60 dB.[3]

REFERENCES

1 Quante, H; Weidinger, J; Zauner, C, **US20100311855**, *Armacell Enterprise GmbH*, Dec. 9, 2010.
2 Khor, A; Pham, T H; Lucas, D; Baki, K A M; Abd Latif, M S B, **WO2015143476**, *Ansell Limited*, Oct. 1, 2015.
3 Singh, A K; Shishkin, A; Koppel, T; Gupta, N, **Porous materials for EMI shielding. Materials for Potential EMI Shielding Applications**, *Elsevier*, 2020, pp.287-314.

10.24 POLYDIMETHYLSILOXANE

Polysiloxane foams (precursors to silicon oxycarbide foams) were prepared *via* simultaneous cure and foaming of liquid methylsiloxane resins.[1] The relatively environment-friendly ethanol was used as a blowing additive and concentrated aqueous ammonia as a catalyst.[1] The uniform foam was obtained by using siloxane-based surfactants with bubble nucleation by mechanical stirring.[1] The precursor foams were then pyrolyzed in nitrogen at 1000°C to produce SiOC foams with densities of 0.17-0.42 g/cm^3.[1]

A microporous, open-cell foam of a hierarchical structure from a composition comprising foamable liquid polymer (polydimethylsiloxane), blowing agent (isopropanol), and viscosity modifier (powdered sucrose) for the foam-forming composition.[2] The open-cell polymer foam, due to its hydrophobic character, is capable of absorbing relatively low surface tension liquids (e.g., hydrocarbon oils) and reject high surface tension liquids (e.g., water).[2] These foams are applicable in maritime oil spill clean-up operations.[2]

Polydimethylsiloxane foam with high compressive resistance and motion-energy absorption efficiency was made using thermo-expandable hollow microspheres (hydrocarbons in the thermoplastic shell).[3] PDMS/microsphere=2:1 foams had a compressive modulus of 10.2 MPa and compression strength of 0.8 MPa (neat PDMS 4.3 MPa in modulus and 0.5 MPa in strength at 10% strain).[3] After the large compression of 70%, the modulus decreased to 2.5 MPa (only 25% of the initial value).[3] High ideal efficiency of motion-energy absorption of 38.3% was obtained in the case of foam but only 15.8% at 70% strain for neat PDMS.[3] Figure 10.24.1 shows morphological changes in cells after compression.[3] The softening phenomenon and high motion-energy absorption were ascribed to the breaking of microspheres.[3]

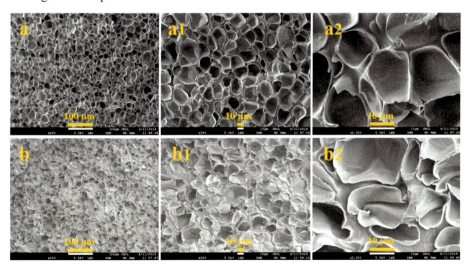

Figure 10.24.1. SEM images of PDMS/micrsophere=2:1 foams without compression (a), with 50 repeated compression up to 40% strain (b), the magnification images of a and b were shown in a1,a2 and b1, b2, respectively. [Adapted, by permission, from Cai, J-H; Huang, M-L; Chen, X-D; Wang, M, *Appl. Surf. Sci.*, **540**, Past 1, 148364, 2021.]

MXene-decorated polymer foam beads were encapsulated in polydimethylsiloxane that resulted in superb EMI-SE of 23.5-39.8 dB with only 0.0225-0.0449 vol% MXene (plus 0.02 vol% PANI) and low R coefficient of 0.20-0.31.[4]

Flexible and superhydrophobic PDMS/CNF/PDMS foam composites were prepared.[5] Carbon nanofibers with hollow structures were decorated onto the skeleton of polydimethylsiloxane foam with strong interfacial adhesion.[5] CNFs enhanced surface roughness and hydrophobicity and served as capillary tubes, improving the oil adsorption and oil/water separation performance.[5]

Graphene oxide sheets or nanoribbons were assembled onto the silicone rubber foams foam surface by bonding onto PDMS foam an ultralow loading (≤0.10 wt%) of GO derivatives (Figure 10.24.2).[6] Significant improvements in thermal stability and flame retardancy of the PDMS foam without affecting its density and elasticity (31-40% reduction in peak heat release rate and 80-95% improvement in total smoke release).[6]

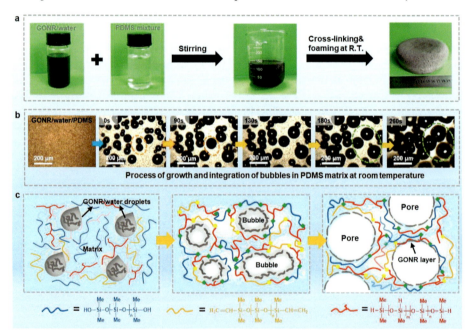

Figure 10.24.2. Fabrication process and schematic illustration of silicone rubber foam (SiRF) nanocomposites. (a) Digital images of the fabricating process of graphene oxide nanoribbons (GONR) coated SiRF at ambient temperature. (b) Optical microscopy image of GONR/water/PDMS mixture before and during the foaming process (transparent PDMS phase and black GONR-based suspension). (c) Schematic illustration of GONR/water foaming PDMS processes, showing the formation of GONR layer on the SiRF skeleton. [Adapted by permission, from Cao, C-F; Wang, P-H; Zhang, J-W; Tang, L-C, *Chem. Eng. J.*, **393**, 124724, 2020.]

Lightweight polydimethylsiloxane foam materials with outstanding mechanical flexibility and high-temperature stability as well as excellent flame resistance are attractive for various potential applications.[6]

Multifunctional elastomer foam with robust mechanical properties, highly piezoresistive sensitivity, special temperature-sensitive characteristics, and excellent EMI shielding was fabricated using thermo-expandable microspheres in polydimethylsiloxane/multiwalled carbon nanotubes.[7] Addition of expandable microspheres (051DU40 from Akzo-Nobel) enhanced piezoresistivity, facilitating quick response (60 ms), excellent reliability, and extraordinary stability.[7]

The above are just a few examples of the vibrant fields of applications of polydimethylsiloxane foams.

REFERENCES

1　Strachota, A; Cerný, M; Chlup, Z; Depa, K; Šlouf, M; Sucharda, Z, *Ceramics Int.*, **41A**, 10, 13561-71, 2015.
2　Shojaei-Zadeh, S; Fechtmann, M, **US20140246374**, *Rutgers, The State University of New Jersey*, Sep. 4, 2014.
3　Cai, J-H; Huang, M-L; Chen, X-D; Wang, M, *Appl. Surf. Sci.*, **540**, Past 1, 148364, 2021.
4　Jia, X; Shen, B; Zhang, L; Zheng, W, *Carbon*, **173**, 932-40, 2021.
5　Duo, Z; Long, B; Gao, S; Gao, J, *J. Hazardous Mater.*, **402**, 123838, 2021.
6　Cao, C-F; Wang, P-H; Zhang, J-W; Tang, L-C, *Chem. Eng. J.*, **393**, 124724, 2020.
7　Cai, J-H; Li, J; Chen, X-D; Wang, M, *Chem. Eng. J.*, **393**, 124805, 2020.

10.25 POLYETHERKETONE

A foamed material is made by heating a mixture that includes a polymer, especially a polyaryletherketone polymer (polyetherketone and polyetheretherketone), and a decomposable material, such as magnesium hydroxide or aluminum hydroxide/hydrated alumina, to a foaming temperature at or above the decomposition temperature of the decomposable material.[1] In the method, the decomposable material decomposes to produce water, which produces foaming within the polymeric material.[1] The average particle size of foaming filler was from 0.9-3 µm.[1] Magnesium hydroxide (Hydrofy GS 2.5) was the most efficient blowing agent (10% addition brought extrudate to the density of 0.5 g cm^{-3}).[1]

High-performance polymers such as polyimide, polyethersulfone, and polyetheretherketone have high glass transition temperature and small free volume.[2] CO_2 presents the low gas diffusivity in these polymers, and thus the CO_2 saturated polymers exhibit stable gas content for a time after being desorbed.[2] Once they possess excellent microcellular foaming behavior, the novel *in situ* foaming fused deposition modeling method could be suitable for them to manufacture the hierarchical porous parts.[2]

Polyetheretherketone microporous foams with multilayer structure were obtained by supercritical carbon dioxide foaming method.[3] The cell structure of the PEEK foams was multi-layer arranged in the polymer crystalline zone.[3]

Porous polymeric films find a wide range of applications as a permselective medium for use in energy harvesting and storage, filtration, separation, and purification of gases and fluids, CO_2 and volatile capture, electronics, devices, structural supports, packaging, labeling, printing, clothing, drug delivery systems, bioreactor, etc.[4] These films are made by dry and/or wet methods with its multilayer structure constructed by coextrusion, lamination, and coating from polyarylketones.[4]

REFERENCES

1 Seargeant, K M, **EP1373381**, *Victrex Manufacturing Limited*, Jan. 2, 2004.
2 Li, M; Jiang, J; Hu, B; Zhai, W, *Compos. Sci. Technol.*, **200**, 108454, 2020.
3 Zhou, D; Xiong, Y; Yuan, H; Zhang, L, *Compos. Part B: Eng.*, **165**, 272-8, 2019.
4 Song, K; Song, J M; Song, J M, **US20180043656A1**, *Liso Plastics LLC*, Feb. 15, 2018.

10.26 POLYETHERIMIDE

Supercritical carbon dioxide at 20 MPa was used to produce polyetherimide nanofoams in a hot press.[1] PEI can absorb 10 wt% carbon dioxide and can reach an equilibrium concentration in 100 h at 45°C.[1] When the clamping force was increased from 1 to 10 tons, the average cell size was increased from 40 to 4000 nm or by a factor of 100.[1] The nanofoams had a bulk porosity (void fraction) in the range of 25-64%, and the average cell sizes in the range of 40-100 nm.[1] A transition from microscale cells to nanoscale cells was observed at gas concentrations in the range of 94-110 mg CO_2/g PEI.[3]

The porosity, pore size distributions, and the density of the foams obtained by carbon dioxide blowing were affected by pressure drop, pressure loss rate, and temperature at the die.[2] Nanosilica particles 0.08-0.6 wt% increased only the foam density but did not control nucleation rate and pore size distribution, most likely, because of their poor dispersibility and agglomerated state in a single screw extruder.[2]

A separator spline formed of a foamed, halogen-free polymeric material for separating twisted pair of conductors is described.[4] The halogen-free polymeric material can be a polyetherimide (PEI) or polyetherimide/polysiloxane copolymer.[4] Dihydrooxadiazinone was used as a blowing agent and talc as a nucleating agent.[4]

A thermoformed polyetherimide/poly(biphenyl ether sulfone) article was produced in three steps, including preparation of a polyetherimide/poly(biphenyl ether sulfone) blend, foaming the composition to yield a foamed polyetherimide/poly(biphenyl ether sulfone) material, and molding the foam under the effect of heat and pressure.[5]

Microcellular polyetherimide/graphene composite foams with a density of 0.3 g/cm^3 and EMI-SE of 5-13 dB were fabricated by water vapor-induced phase separation method.[6]

Polyetherimide bead foam products were fabricated *via* microwave-assisted foaming and selective sintering.[7] During this process, the pre-expanded PEI beads were coated with carbon nanotubes and were then saturated with compressed carbon dioxide/tetrahydrofuran.[7] Upon microwave irradiation, the surface localized CNTs were intensively heated and served as hotspots, which increased the expansion ratio of PEI beads to 8.2 fold.[7] The foam having density of 0.28 g/cm^3 had a strong inter-bead bonding strength (up to 7 MPa).[7]

MWNTs-COOH@Fe_3O_4 hybrid with ferromagnetism was prepared, and polyetherimide (PEI)/MWNTs-COOH@Fe_3O_4 nanocomposite foams with continuous gradient structure were fabricated *via* water vapor-induced phase separation method under a magnetic field.[8] Along the magnetic field direction, a continuous gradient distribution of MWNTs-COOH@Fe_3O_4 formed in PEI matrix, while the average cell size gradually decreased, the cell wall thickness and apparent density increased, forming gradient cell structure.[8]

Microporous polyetherimide/carbon nanotube foams with high strength and segregated structure have been developed by sinter molding and supercritical CO_2 foaming (Figure 10.26.1).[9] The segregated composite foams exhibited excellent conductivity and EMI shielding effectiveness of 7.98 S/m and 32.3 dB, respectively, at a low percolation threshold of 0.06 vol%.[9]

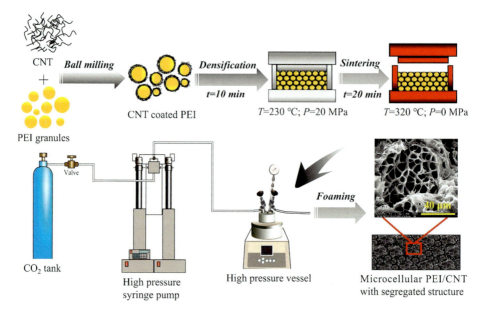

Figure 10.26.1. Illustration showing a fabrication process of the PEI/CNT composite foams with segregated structure. [Adapted, by permission, from Feng, D; Liu, P; Wang, Q, *Composites Part A: Appl. Sci. Manuf.*, **124**, 105463, 2019.]

Polymer foams based on polyetherimides fulfill the legal specifications demanded by the aviation industry for aircraft interiors.[10] Specifically, the demands on fire characteristics, stability to media, and mechanical properties constitute a great challenge.[10] Process for producing the PEI particle foam comprised extruding a composition consisting 80 to 99.5 wt% of PEI, 0.5 to 10 wt% of blowing agent (volatile liquid or chemical blowing agents).[10]

REFERENCES

1   Aher, B; Olson, N M; Kumar, V, J. Mater. Res., 28, 17, 2366-73, 2013.
2   Aktas, S; Gevgilili, H; Kucuk, I; Sunol, A; Kalyon, D M, *Polym. Eng. Sci.*, **54**, 2064-74, 2014.
3   Miller, D; Chatchaisucha, P; Kumar, V, *Polymer*, **50**, 23, 5576-84, 2009.
4   Tryson, G R; Rice, B, **EP2896054**, Sabic Global Technologies BV, Jul. 22, 2015.
5   Kenkare, N; Kwan, K S; El-Hibri, M J, **WO2014170255**, Solvay Specialty Polymers USA, LLC, Oct. 23, 2014.
6   Zhu, S; Zhou, Q; Wang, M, Ye, C, *Compos. Part B: Eng.*, **204**, 108497, 2021.
7   Feng, D; Liu, P; Wang, Q, *Mater. Sci. Eng: B*, **262**, 114727, 2020.
8   Sun, X; Zhao, X; Ye, L, *Composites Part A: Appl. Sci. Manuf.*, **126**, 105579, 2019.
9   Feng, D; Liu, P; Wang, Q, *Composites Part A: Appl. Sci. Manuf.*, **124**, 105463, 2019.
10 Traßi, C; Holleyn, D; Bernhard, K, **US20200207939A1**, Evonik Operations GmbH, Jul. 2, 2020.

10.27 POLYETHERSULFONE

Microcellular foaming of thin (~100 μm) polyethersulfone films containing varying trace concentrations of tetrahydrofuran using carbon dioxide as a physical blowing agent was conducted to obtain a membrane.[1] Tetrahydrofuran present in the polymer at concentrations above 0.04 wt% led to a transition from the closed cellular structure into open-cellular morphologies.[1] The open structure includes small spot-like openings (diameters between 10 and 100 nm) in the cell walls.[1]

High-temperature resistant polysulfone insulation for the pipe is a syntactic foam having a degree of foaming of up to about 50%.[2] The thermal insulation layer comprises polysulfone having Vicat softening point greater than 200°C and thermal conductivity of less than 0.40 W/mK.[2] The thermal conductivity can be further decreased through foaming.[2] Foaming with the use of chemical or physical blowing agents or hollow microspheres is appropriate.[2]

Particle foam based on polyethersulfone has high-temperature resistance, flame retardancy and is lightweight with improved stiffness and strength, according to BASF.[3] It is suitable for complex-shaped components in cars, airplanes, and trains.[3] The expandable PESU granulate is pre-foamed into beads with low densities between 40 and 120 g/L and can be processed into molded parts with complex 3D geometries.[3]

The microcellular foaming of PSU by supercritical CO_2 was assisted by a co-blowing agent (ethanol).[4] The interaction energy of the blowing agent with polymer chains increased with solvent content.[4] The expansion ratio of foams increased when adding co-blowing agents, and the foaming temperature window expanded upon the addition of co-blowing agents.[4]

Poly(biphenyl ether sulfone)/polyethersulfone foam had improved compressive strength and impact performance.[5] Foam is used in lightweight applications such as transport and building materials.[5]

REFERENCES

1 Krause, B; Docrrigter, M F; van der Vegt, N F A; Strathmann, H; Wessling, M, *J. Membrane Sci.*, **187**, 1-2, 181-92, 2001.
2 Edmondson, S J; Wong, D; Mockel, M, **WO2014131127**, *Shawcor, Ltd.*, Sep. 4, 2014.
3 Anon. *Reinforced Plastics*, **63**, 1, 13-14, 2019
4 Hu, D-d; Gu, Y; Liu, T; Zhao, L, *J. Supercritical Fluids,* **140**, 21-31, 2018.
5 Kenkare, N; Kwan, K S; Looney, W W; El-Hibri, M J; Gopalakrishnan, V; Rich, J, **US20190390058A1**, *Solvay Specialty Polymers USA LLC*, Dec. 26, 2019.

10.28 PES/PEN BLENDS

High-performance thermoplastic blends based on polyethersulfone (PES) and poly(ethylene 2,6-naphthalate) (PEN) were foamed with supercritical carbon dioxide to obtain foams with improved heat deflection temperature, extended processing range, and controlled cellular morphology.[1] The cellular morphology was strongly influenced by the initial morphology of the blend (PES as a dispersed phase in PEN-based blends acted as blowing agent reservoir).[1]

REFERENCES

1 Sorrentino, L; Cafiero, L; Iannace, S, *Polym. Eng. Sci.*, **55**, 1281-9, 2015.

10.29 POLYETHYLENE

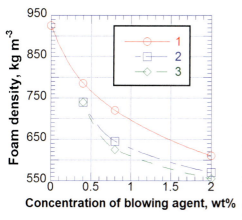

Figure 10.29.1. Foam density vs. concentration of blowing agent. 1 – Hostatron P1941, 2 – Adcol blow X1020, 3 – Hydrocerol PLC 751. [Data from Garbacz, T; Palutkiewicz, P, *Cellular Polym.*, **34**, 4, 189-214, 2015.]

Sodium bicarbonate+citric acid (Hostatron P 1941), azodicarbonamide (Adcol-blow X1020 and Hydrocerol PLC 751) were compared in the foaming of several polymers, including polyethylene.[1] Figure 10.29.1 shows that there is little difference between the densities of foams obtained with different blowing agents.[1] The concentration of blowing agent has a predominant influence.[1]

The cell size and cell density of ultra-high molecular weight polyethylene were affected by the combined effects of crystal, temperature, and pressure (higher foaming temperatures led to larger cells and lower cell densities).[2] Higher pressure led to lower cell density and larger average cell diameter when foaming during the heating stage due to the reduction of crystals and melt strength.[2] While in foaming during the cooling stage, higher saturation pressure led to a higher cell density due to the increase of solubility of CO_2, and the cell density decreased as the pressure further increased due to the cell coalescence.[2]

High-density polyethylene, flax fiber, and azodicarbonamide were used in injection molding formulation.[3] Cell size, cell density, and skin and core thicknesses were affected by blowing agent, natural fiber contents, and mold temperature.[3] A better microcellular asymmetric structure was obtained with higher fiber and blowing agent contents and higher average mold temperature.[3]

Low-density polyethylene foams having high rebound resilience were obtained using a blend with polyethylene-octylene elastomer.[4] Azodicarbonamide was used as a blowing agent. The rebound resilience of LDPE was improved by increasing the flexibility of cell walls (addition of polyethylene-octylene elastomer) and the cell density and decreasing the foam density.[4] The rebound resilience was affected by crystallinity, crosslinking degree, foam density, and cell density.[4] The rebound resilience increased from 33% to 47% as the polyethylene-octylene elastomer content increased to 30 phr (cell density increased from 191 to 2217 cell/cm^3).[4]

The effect of injection molding parameters and blowing agent addition on selected properties, surface state, and structure of HDPE parts was studied.[5] The mold temperature has the main influence on the properties and surface of molded parts from solid and foamed HDPE.[5] The weight, density, mechanical properties, and gloss increased with the increase in the mold temperature.[5] The number and size of pores were also influenced by the mold temperature.[5]

The blowing agent injection location in the extrusion barrel affected the residence time inside the barrel, and, consequently, the foam microstructure.[6] The higher gas residence time resulted in foams with lower cell size, higher expansion ratio, and enhanced cell density.[6]

A peroxide-modified linear low-density polyethylene, comprising the reaction product of a peroxide and a linear low-density polyethylene was used for the production of foams.[7] The foamed LLDPE is suitable for use in the wire and cable industry.[7] A foaming level is less than 20%.[7] The blowing agent is a masterbatch that contains 50 wt% azodicarbonamide mixed with LLDPE base resin.[7]

Foamed linear medium-density polyethylene parts were prepared by rotational molding in biaxial mode, using azodicarbonamide.[8] An exothermic chemical blowing agent increased the peak internal air temperature.[8] The part cooling was slower because the gas bubbles acted as thermal insulating material.[8] Addition of 20 wt% of wood particles and 0.4 wt% azodicarbonamide gave the best balance of foam properties.[9]

The temperature gradient between the top and bottom plates in compression molding of LMDPE containing azodicarbonamide was used to produce symmetric and asymmetric structural foams.[10] Material with asymmetric foam density had different flexural strength depending on the side on which force was applied.[10] The apparent modulus was always higher when the load was applied to the side having the highest density.[10]

HDPE/PA6 foam blends were obtained using azodicarbonamide as a blowing agent.[11] Because of the very different thermal and rheological properties of the polymers, a good cellular structure can only be obtained by a careful selection of temperature profile and feeding strategy.[11] A coupling agent was found necessary to stabilize the blend before foaming.[11]

Efficient dissolution and modification strategy were proposed for the fabrication of superhydrophobic polyethylene foam with 98.6% porosity using sodium chloride as a sacrificial template and superhydrophobic nano-silica particles as a surface chemical modifier.[12] A microscale porous and interconnected 3D framework was formed when NaCl was dissolved and the nano-silica particles adsorbed on the surface of interconnected pores to form nanoscale structures.[12] The foam had a water contact angle of 158±2° and a sliding angle of 4±2°.[12] The foam maintained its superhydrophobicity under 980 Pa pressure of water droplet impact and 8.1 kPa pressure of water flow impact.[12] The foam exhibited exceptional mechanical durability against knife scratching, tape-peeling, bending-twisting, and ultrasonication in ethanol because of unique hierarchical micro-nanostructure and the stable adsorption of nano-silica particles.[12]

Window and door frames are often a source of heat transfer through the window or door.[13] In the winter, heat may escape an interior space through the frames.[13] In the summer, heat may enter the interior space.[13] Crosslinked polyethylene foam was used to cover the outer frame members extending between the structure and the one or more transparent panes of the multiple transparent panes to improve insulation properties of frames.[13]

A method of making foam using renewable resources was patented.[14] The foam was made using green polyethylene polymers made from renewable sugarcane ethanol.[14] The use of these polymers to make foam has the potential to reduce carbon dioxide gas emissions by more than half.[14] The foam can be used in a variety of applications and can also be made with blends of renewable LDPE and non-renewable LDPE.[14]

Low-density polyethylene resins are widely used in various foam applications such as cushion/protective packaging, construction, insulation, sporting goods, medical applications, etc.[15] It is desirable to develop modified high-density polyethylene that is suitable for foam applications in terms of good processability and wide foam density range but with a higher stiffness than that achieved with LDPE.[15] HDPE was treated with 275 ppm of 2,5-dimethyl-2,5-di (tert-butylperoxy) hexane peroxide (Triganox 101 E-10) at 200°C in twin-screw extruder.[15] 3% isobutane, 0.5% talc, and 1% Hydrocerol CF-40E were used in formulation.[15]

REFERENCES

1 Garbacz, T; Palutkiewicz, P, Cellular Polym., 34, 4, 189-214, 2015.
2 Liu, J; Qin, S; Wang, G; Zhang, H; Zhou, H; Gao, Y, *Polym. Testing*, **93**, 106974, 2021.
3 Tissandier, C; Gonzalez-Nunez, R; Rodrigue, D, *J. Cellular Plast.*, **50**, 5, 449-73, 2014.
4 Wang, W; Gong, W; Zheng, B, *Polym. Eng. Sci.*, **53**, 2527-34, 2013.
5 Bociaga, E; Palutkiewicz, P, *Polym. Eng. Sci.*, **53**, 780-91, 2013.
6 Gandhi, A; Bhatnagar, N, *Polym. Plast. Technol. Eng.*, **54**, 1812-8, 2015.
7 Pujari, S; Kmiec, C J; **WO2016032715**, *Dow Global Technologies LLC*, Mar. 3, 2016.
8 Moscoso-Sanchez, F J; Mendizabal, E; Jasso-Gastinel, C F; Ortega-Gudino, P; Robledo-Ortiz, J R; Gonzalez-Nunez, R; Rodrigue, D, *J. Cellular Plast.*, **51**, 5-6, 489-503, 2015.
9 Raymond, A; Rodrigue, D, *Cellular Polym.*, **32**, 4, 199-212, 2013.
10 Yao, J; Rodrigue, D, *Cellular Polym.*, **31**, 4, 189-205, 2012.
11 Reyes-Lozano, C A; Ortega Gudino, P O; Gonzalez-Nunez, R; Rodrigue, D, *J. Cellular Plast.*, **47**, 2, 153-72, 2010.
12 Wu, T; Xu, W-H; Guo, K; Xie, H; Qu, J-p, *Chem. Eng. J.*, in press, 127100, 2021.
13 Holtby, Q A J, **US20200308898A1**, *Katch Kan Holdings Ltd*, Oct. 1, 2020.
14 Ramesh, N; Yap, C; Smith, L, **US20200062915A1**, *Sealed Air Corp*, Feb. 27, 2020.
15 Doufas, A K; Li, L; Luce, J T; **US20200087477A1**, *ExxonMobil Chemical Patents Inc*, Mar. 19, 2020.

10.30 POLY(ETHYLENE-CO-OCTENE)

Poly(ethylene-co-octene)/multiwall carbon nanotube microcellular foams were prepared by melt blending process and expansion using supercritical carbon dioxide.[1] In neat poly(ethylene-co-octene) foam, the cell opening resulting from cell coalescence was observed in the cell wall.[1] When multiwalled carbon nanotubes, MWCNT, were incorporated, they oriented themselves in the cell walls.[1] This caused strain hardening and reinforcement of the cell walls.[1] As a result, a dramatic decrease in the open-cell content was observed.[1]

Poly(ethylene-co-octene) foam material was prepared using supercritical carbon dioxide foaming.[2] The foam prepared by this process was different from general microcellular thermoplastic because it had resilience.[2]

REFERENCES
1 Zhai, W; Wang, J; Chen, N; Naguib, H E; Park, C B, *Polym. Eng. Sci.*, **52**, 2078-89, 2012.
2 Yang, C; Wu, G, **Radiation Cross-Linking for Conventional and Supercritical CO$_2$ Foaming of Polymer. Radiation Technology for Advanced Materials**, *Academic Press*, 2019, pp 115-39.

10.31 POLY(ETHYLENE TEREPHTHALATE)

The melt strength of four types of poly(ethylene terephthalate) grades was compared.[1] The two long-chain branched poly(ethylene terephthalate)s had higher melt viscosity and greater elasticity than the other two linear ones.[1] Only one long-chain branched poly(ethylene terephthalate) showed strain hardening, and this grade was successfully extrusion foamed with carbon dioxide.[1] The extruded poly(ethylene terephthalate) foams were produced with a volume expansion ratio of 9.5, the average cell size of 265 μm, cell density of 4.5$\times$10^5 cells/cm^3 and a narrow cell size distribution.[1]

Reactive extrusion with pyromellitic dianhydride was used to modify poly(ethylene terephthalate) for foaming.[2] Broad foaming temperature window using carbon dioxide was obtained for PET modified with 0.8 and 0.5 wt% PMDA.[2] PET foams with the expansion ratio between 10 and 50 times, the cell diameter between 15 and 37 μm, and the cell density between 6.2$\times$10^8 and 1.6$\times$10^9 cells/cm^3 were produced.[2]

The microcellular injection molding (MuCell) of *in situ* polymerization-modified PET was performed using supercritical nitrogen.[3] Pentaerythritol and pyromellitic dianhydride were used as modifying monomers.[3] A uniform cell structure with an average cell size of around 35 μm was obtained.[3] The PET modified with pyromellitic dianhydride generated a slightly finer cell structure and a higher cell density than the pentaerythritol-modified PET.[3]

A commercial bottle grade polyester containing modified organoclay exhibited enhanced thermal and dynamic-mechanical properties and improved foaming behavior.[4] The organoclay was obtained *via* intercalation of bis-(hydroxyethyl terephthalate) into sodium montmorillonite layers.[4] Addition of clay increased decomposition temperature but also decreased melt viscosity, which suppressed foaming.[4] To compensate for viscosity reduction, pyromellitic dianhydride was used in the reactive extrusion process of poly(ethylene terephthalate) nanocomposites.[4]

A foam sheet having a mean bubble diameter of approximately 10 μm constituted a problem that transmission of light increased when the sheet thickness decreased.[5] Therefore, the desired reflectivity could not be maintained.[5] A thermoplastic resin foam having a high reflectivity even when formed into a thin film can be produced by adding a specific melt-type crystallization nucleating agent to a thermoplastic resin.[5] The poly(ethylene terephthalate) is a preferred resin for this solution because of its crystallinity and good foaming properties.[5] In the foaming step, the melt-type crystallization nucleating agent is the starting point of the formation of a bubble nucleus.[5] As a result, a foam having fine pores with a mean bubble diameter of less than 1 μm can be obtained.[5] Such a foam is molded, and a reflecting material having a high reflectivity can thus be obtained.[5] Many melt-type nucleating agents can be used, but N,N'-dicyclohexyl-2,6-naphthalenedicarboxamide is preferable because of its transparency and particle diameter of generated crystal.[5] The amount used is in the range of 0.25 to 1.5 wt%.[5]

The strain hardening and fast relaxation were beneficial for maintaining a uniform bubble wall.[6] The initial bubble wall geometry had a strong impact on bubble wall uniformity.[6] Open-cell foam was produced by aggravating the initial non-uniformity because of the addition of polytetrafluoroethylene particles.[6]

PTFE was *in situ* fibrillated and entangled into a network of poly(ethylene terephthalate).[7] PTFE fibrils network significantly improved crystallization and rheological behaviors of PET.[7] PTFE fibrils network affected cell nucleation and growth of PET microcellular foams at rheological percolation of 1 wt% at which the smallest cell diameter (2 μm) and the highest cell density (10^{11} cells/cm^3) were obtained.[7]

Recycled poly(ethylene terephthalate) was processed with additives, such as chain extender, nucleating agent, antioxidant, blowing agent (Hydrocerol-2224, azodicarbonamide), antistatic agent, processing aid, and impact modifier.[8] The resultant composite foaming sheet had high crystallization temperature (217°C), good flexural strength (623 kg/cm^2), great tensile strength (506 kg/cm^2), high melting temperature (245°C), good flexural modulus (22,881 kg/cm^2), moderate surface electric resistance (8.6×10^{11} Ω/sq.), low density (0.85 g/cm^3), small average cell size (160 μm), and high notched Izod impact strength (155 J/m).[8] The composite was prepared by thermo-chemical extrusion foaming process to be used for electronic packages.[8]

Supercritical CO_2 extrusion foaming was used to prepare PET foam, which was subjected to isothermal treatment in the post-process to improve the crystallization.[9] Crystal perfection was caused by migration and rejection of the structural defects at the crystallites induced by slow crystallization, and the crystallinity rapidly increased with the rise of isothermal temperature, especially above the glass transition temperature.[9]

REFERENCES

1 Fan, C; Wan, C; Gao, F; Huang, C; Xi, Z; Xu, Z; Zhao, L; Liu, T, *J. Cellular Plast.*, **52**, 3, 277-98, 2016.
2 Xia, T; Xi, Z; Liu, T; Pan, X; Fan, C; Zhao, L, *Polym. Eng. Sci.*, **55**, 1528-35, 2015.
3 Xi, Z; Zhang, F; Zhong, H; Liu, T; Zhao, L; Turng, L-S, *Polym. Eng. Sci.*, **54**, 2739-45, 2013.
4 Scamardella, A M; Vietri, U; Sorrentino, L; Lavorgna, M; Amendola, E, *J. Cellular Plast.*, **48**, 6, 557-76, 2012.
5 Ikeda, H; Saito, M; Inamori, K, **EP2610288**, *Furukawa Electric Co., Ltd.*, Jul. 3, 2013.
6 Ge, Y; Liu, T, *Chem. Eng. Sci.*, **230**, 116213, 2021.
7 Jiang, C; Han, S; Chen, S; Wang, X, *Polymer*, in press, 123171, 2021.
8 Lai, C-C; Yu, C-T, Wang, F-M; Chen, C-M, *Polym. Testing*, **74**, 1-6, 2019.
9 Yao, S; Hu, D; Xi, Z; Zhao, L, *Polym. Testing*, **90**, 106649, 2020.

10.32 POLYIMIDE

A polyimide film with closed and open pores was formed by using azodicarbonamide. Samples were precured at 130°C and cured at 205°C in sequence so that the closed pores might not coalesce in the film.

A polyimide foam does not have a property of preventing moisture permeation; a problem of water condensation may occur when it is used as a thermal insulation material.[2] To remediate this problem, a surface layer including a flame-retardant resin composition containing a chlorine-containing elastomer (chlorosulfonated polyethylene) and a flame retardant is laminated with a polyimide foam.[2] The polyimide foam that is used in this invention is formed of polyimide, which has a glass transition temperature exceeding 300°C (preferably aromatic polyimide formed of an aromatic tetracarboxylic acid component and an aromatic diamine compound).[2]

Flame resistance and smoke releasing behavior of isocyanate-based polyimide foam produced by free-foaming require improvement that can be accomplished by incorporation of silica aerogel layers into cells of isocyanate-based polyimide foam by *in situ* growth process of silica sol.[3] The silica aerogel layers were firmly attached to the pores and surfaces of cells, providing exceptional thermal insulation, flame protection, and preserving the original cellular structure.[3] The silica aerogel layers improved limiting oxygen index (from 22 to 33%), decreased heat release rate, the peak of HRR (from 174 to 72 kW/m^2), and specific optical density of smoke.[3]

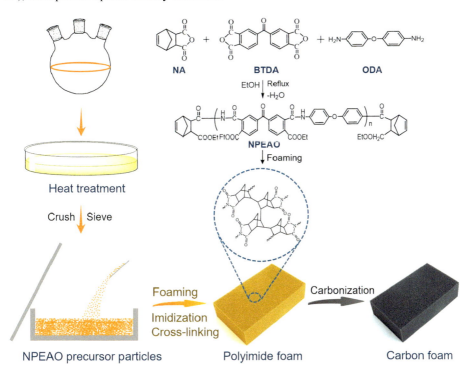

Figure 10.32.1. Schematic illustration of fabrication of carbon foams derived from thermosetting polyimide foam. [Adapted, by permission, from Li, J; Ding, Y; Yu, N, He, X, *Carbon*, **158**, 45-54, 2020.]

Polyimide foam with excellent acoustic behavior was prepared by regulating cellular structure on a large scale and introducing a sharp hole structure.[4] The average cellular window size decreased by 48% and pore distance increased by 76%.[4]

Rigid carbon foams were fabricated by carbonization of thermosetting polyimide foam.[5] Benefiting from the crosslinked networks, the prepared carbon foams had excellent thermal and dimensional stability with low shrinkage of 47% after they were carbonized at 1500°C (Figure 10.32.1).[5] The foams had excellent electromagnetic interference shielding effectiveness and specific EMI shielding effectiveness of 54 dB and 593.4 dB cm^3/g respectively at 10 GHz with the thickness of 2 mm.[5]

REFERENCES

1 Ma, Y-W; Jeong, M Y; Lee, S-M; Shin, B S, *J. Korean Phys. Soc.*, **68**, 5, 668-73, 2016.
2 Hosoma, T; Yamamoto, S, **EP2738000**, *UBE Industries, Ltd.*, Jun. 22, 2016.
3 Sun, G; Duan, T; Liu, C; Han, S, *Polym. Testing*, **91**, 106738, 2020.
4 Ren, X; Wang, J; Sun, G; Han, S, *Polym. Testing*, **84**, 106393, 2020.
5 Li, J; Ding, Y; Yu, N, He, X, *Carbon*, **158**, 45-54, 2020.

10.33 POLY(LACTIC ACID)

Poly(lactic acid) has poor melt strength, and for this reason, it is difficult to foam due to cell coalescence during foaming.[1] A few percent of polytetrafluoroethylene particles stabilized PLA foams against bubble coalescence and collapse.[1] The PTFE-containing foams, blown with azodicarbonamide, were stable even if held under molten conditions for an extended period of time.[1] Because of their low surface energy, the PTFE particles adsorbed on the inner surface of the foam bubbles and formed an interfacial "shell" that prevents coalescence.[1]

The polymer crystallinity and viscoelastic properties are critical during the foaming process.[2] The addition of chain extender (Joncryl 436) increased viscosity and cold crystallization temperature of polylactide.[2] The addition of 1.0-1.5 wt% chain extender was instrumental in obtaining fine cellular structure and low density (0.7 g/dm^3) of foamed PLA.[2]

The high melt viscosity polylactide was obtained using *in situ* reaction of carboxyl-ended polyester and solid epoxy.[3] PLA foams were prepared by a chemical compression molding.[3] Azodicarbonamide was used as a blowing agent, and its decomposition temperature was modulated by the addition of zinc oxide.[3] The foam having a density as low as 0.16 g/cm^3 was made.[3]

The microcellular poly(lactic acid) foams with various crystallinities, cell morphologies, and densities were prepared using carbon dioxide.[4] Crystallization of about 20% was reached in poly(lactic acid) samples after carbon dioxide saturation.[4] The highest crystallinity of 38.2% was achieved for the foamed poly(lactic acid).[4] The open-cell structure tended to decrease the tensile properties of PLA foams.[4]

High-density polylactic acid sheets were foamed with carbon dioxide and nitrogen.[5] Branched and linear chain extenders were used to improve the elongational properties of the polymer.[5] Nitrogen gave smaller cell sizes compared to supercritical carbon dioxide at lower nucleating agent content.[5] When the nucleating agent amount was increased, the quality of foam sheets got worse with nitrogen.[5] This was because of the higher degassing pressure of nitrogen compared to carbon dioxide.[5] The higher degassing pressure causes an earlier phase separation inside the die.[5] The linear chain extender led to an increase in cell size.[5]

Poly(lactic acid)/maleic-anhydride-grafted-polypropylene blend was prepared by melt blending method.[6] The blend was foamed using supercritical carbon dioxide.[6] The maleic anhydride reacted with poly(lactic acid), and a small amount of branched polymer was formed.[6] Homogeneous and finer cellular morphology of poly(lactic acid)/maleic-anhydride-grafted-polypropylene foams with high expansion ratio could be achieved with a proper content of maleic-anhydride-grafted-polypropylene in the blends.[6]

Foam injection molding was conducted by using a blowing agent under high pressure and temperature to produce parts having a cellular core and compact solid skin (the so-called "structural foam").[7] The 3 wt% talc improved the morphology of foams.[7]

A review includes a discussion of fundamental studies of PLA/gas mixtures, crystallization behavior, foaming mechanisms, the effect of molecular architecture and configuration of PLA blends and composites, and more.[8]

A foamed thermoplastic material contains polyhydroxyalkanoate, poly(lactic acids), or their mixtures.[9] The polylactide was foamed using organic peroxide.[9] Peroxyesters such as peroxycarbonates can lower the density of compositions containing poly(lactic acid).[9]

The transesterification reaction between poly(lactic acid) and polycarbonate was promoted by using tetrabutyl titanate as a catalyst.[10] The blends were foamed by using a batch-foaming process with carbon dioxide as a blowing agent.[10] Cell density of poly(lactic acid)/polycarbonate blend increased at low-catalyst content while decreased at high catalyst content.[10] The change was caused by the changes of interfacial properties of blend phases through transesterification reaction, which reduced crystallinity of polycarbonate component.[10]

In situ microfibrillated poly(ε-caprolactone)/poly(lactic acid) composites had enhanced rheological properties, crystallization kinetics, and foaming ability (Figure 10.33.1).[11] The composites demonstrated spherical PLA domains; however, after stretching, the spherical particles transformed to fibrilar structures and developed a 3D network of entangled PLA fibers.[11] The fibrillated samples induced open-cell foams, unlike pure PCL and undrawn PCL/PLA samples.[11]

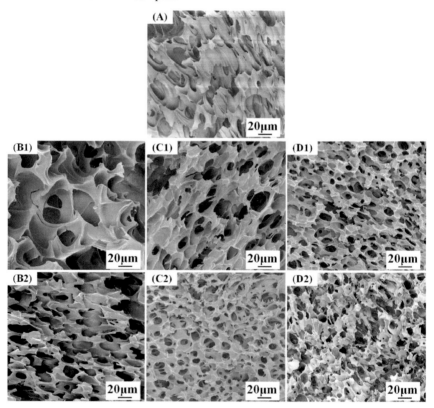

Figure 10.33.1. SEM images of the cryo-fractured surfaces of foamed samples of neat PCL (A), PCL/PLA (95/5 wt%) with spherical (B1) and nanofibrillar (B2) PLA domains, PCL/PLA (90/10 wt%) with spherical (C1) and nanofibrillar (C2) PLA domains, PCL/PLA (80/20 wt%) with spherical (D1) and nanofibrillar (D2) PLA domains. [Adapted, by permission, from Qiao, Y; Li, Q; Jalali, A; Jiang, J, *Compos. Part B: Eng.*, in press, 108594, 2021.]

Ultra-low density poly(lactic acid)/carbon nanotube nanocomposite foam (volume expansion ratio of 49.6 times) was fabricated by a simple, effective, environmentally friendly, and CO_2-based foaming methodology.[12]

Poly(lactic acid) crystallization was induced by saturating process of CO_2 blown PLA foam, which resulted in microcellular foam having a cellular size of 2 μm and cellular density of 10^{13} cell/cm^3 at 130°C and 10 MPa.[13]

REFERENCES

1   Lobos, J; Lasella, S; Rodriguez-Perez, M A; Velankar, S S, *Polym. Eng. Sci.*, **56**, 9-17, 2016.
2   Ludwiczak, J; Kozlowski, M, *J. Polym. Environ.*, **23**, 137-42, 2015.
3   Luo, Y; Zhang, J; Qi, R; Lu, J; Hu, X; Jiang, P, *J. Appl. Polym. Sci.*, **130**, 330-7, 2013.
4   Ji, G; Wang, J; Zhai, W; Lin, D; Zheng, W, *J. Cellular Plast.*, **49**, 2, 101-17, 2013.
5   Geissler, B; Feuchter, M; Laske, S; Fasching, M; Holzer, C; Langecker, G R, *J. Cellular Plast.*, **52**, 1, 15-35, 2016.
6   Wang, X; Liu, W; Li, H; Du, Z; Zhang, C, *J. Cellular Plast.*, **52**, 1, 37-56, 2016.
7   Pantani, R; Volpe, V; Titomanlio, G; *J. Mater. Process. Technol.*, **214**, 12, 3098-107, 2014.
8   Nofar, M; Park, C B, Prog. Polym. Sci., 39, 10, 1721-41, 2014.
9   Donnelly, Z, **US20140107240**, *Arkema, Inc.*, Apr. 17, 2014.
10 Bao, D; Liao, X; He, G; Huang, E; Yang, Q; Li, G, *J. Cellular Plast.,* **51**, 4, 349-72, 2015.
11 Qiao, Y; Li, Q; Jalali, A; Jiang, J, *Compos. Part B: Eng.*, in press, 108594, 2021.
12 Li, Y; Yin, D; Liu, W; Wang, X, *Int. J. Biol. Macromol.*, **163**, 1175-86, 2020.
13 Yu, K; Ni, J; Zhou, H, Mi, J, *Polymer*, **200**, 122539, 2020.

10.34 POLYMETHYLMETHACRYLATE

Microcellular foaming of reinforced core/shell polymethylmethacrylate was done using supercritical carbon dioxide in a single-step process.[1] The key parameter was the depressurization rate.[1] The larger depressurization rates generated larger cell sizes.[1]

Nanocellular foams with an open porous structure were produced from polymethylmethacrylate when the polymer was saturated with carbon dioxide at 5 MPa at -20 to -30°C and foamed at 50°C.[2] Higher foaming temperatures (e.g., 70°C) contributed to the increase in the open-cell structures.[2] Temperatures higher than 70°C resulted in the worm-like structures.[2] The transition from microcellular to nanocellular foams begun at a carbon dioxide concentration from 27.5 to 31 wt%.[2] In addition to the saturation temperatures for carbon dioxide, a temperature range of -56.6 to 31.1°C and a pressure range of 0.518 to 30 MPa were suitable to achieve liquid carbon dioxide.[2]

Polymethylmethacrylate microspheres were used in the implant having controlled porosity.[3] The implant material was used for filling bone defects, bone regeneration, and tissue engineering of bone.[3]

In the continuous process of extrusion of nanoporous foam, polymethylmethacrylate (dried before feeding) was fed into a single-screw extruder. Aerosil 300 (nucleating agent at a concentration of 0.25 wt%) was fed together with PMMA, and polymer melt was saturated with carbon dioxide. The polymer/carbon dioxide mixture was cooled over a period of 30 minutes to a dissolution temperature, and foam was extruded.[4]

Two-step foaming at high pressure (>50 MPa) and rapid decompression (>1.25 GP/s) resulted in high cell nucleation (10^{14} -10^{15} nuclei cm^{-3}), the cell size of 42-58 nm, and porosity of 27-58%.[5] A fine cell structure (30.5±10.4 nm) was observed with scanning transmission electron microscopy.[5] Almost no cells were observed at a saturation temperature of 0°C by SEM observation, but finer cells (about 30 nm) existing locally in the formed polymer were observed by STEM observations (Figure 10.34.1).[5] The use of high saturation pressure provided a sufficient driving force for nucleation even at low foaming temperature and high uniformity in cell size even at high foaming temperature.[5]

Polymethylmethacrylate-block-polybutylacrylate-block-polymethylmethacrylate (MAM) in PMMA acts as a CO_2 reservoir, promoting heterogeneous nucleation to obtain nanofoam.[6] Styrene-acrylonitrile (SAN) was added to PMMA to adjust the solubility of CO_2 in PMMA.[6]

The addition of 1 and 2 wt% of carbon nanotubes to PMMA foam increased tensile strength by 8% and 22%, respectively, as compared to pure foam.[7] The optimum percentage of carbon nanotubes was 3.3 wt%, which gave optimum modulus and tensile strength of 87 MPa and 4.02 GPa, respectively.[7] Further increase in nanotube concentration reduced tensile strength due to nanotube agglomeration.[7]

Multilayer hot melt pressing and supercritical carbon dioxide foaming were used to produce microporous PMMA foams.[8] Nucleation and directional growth of the cells were promoted by multilayer introduction into the polymer matrix.[8] When the distance of the multilayer interface was smaller than the cell's critical nucleation size, the microcellular foams had uniform, multilayer cell structure, and the unit cell shape was well controlled.[8]

10.34 Polymethylmethacrylate   183

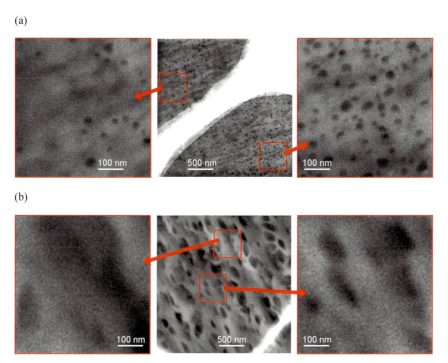

Figure 10.34.1. High annular dark field-scanning transmission electron microscopy image of PMMA foams obtained by two-step foaming process with saturation temperature of 0 (a) and 20°C (b). Saturation pressure and foaming temperature were 100 MPa and 50°C. [Adapted, by permission, from Ono, T; Wu, X; Horiuchi, S, Furuya, T; Yoda, S, *J. Supercritical Fluids*, **165**, 104963, 2020.]

REFERENCES

1 Reglero Ruiz, J A; Viot, P; Dumon, M, *J. Appl. Polym. Sci.*, **118**, 320-31, 2010.
2 Kumar, V; Guo, H, **WO2014210523**, *University of Washington through Its Center for Commercialization*, Dec. 31, 2014.
3 Lao, J C A; Lacroix, J; Jallot, E D A; Dieudonne, X, **CA2914130**, *Universite Blaise Pascal-Clermont-Ferrand II*, Dec. 11, 2014.
4 Costeux, S; Lantz, D R; Beaudoin, D A; Barger, M A, **WO2013048760**, *Dow Global Technologies LLC*, Apr. 4, 2013.
5 Ono, T; Wu, X; Horiuchi, S, Furuya, T; Yoda, S, *J. Supercritical Fluids*, **165**, 104963, 2020.
6 Yeh, S-K; Liao, Z-E; Wang, K-C; Tseng, T-W, *Polymer*, **191**, 122275, 2020.
7 Pour, M H N; Payganeh, G; Tajdari, M, *Eur. J. Mech. - A/Solids*, **83**, 104019, 2020.
8 Zhou, D; Xiong, Y; Yuan, H; Zhang, L, *Compos. Part B: Eng.*, **165**, 272-8, 2019.

10.35 POLYOXYMETHYLENE

Polyoxymethylene foam was produced using azodicarbonamide.[1] Melt flow index controlled the spherical cell size and its uniform distribution, including cell density.[1] By varying compression pressure, it was possible to fine-tune the microstructure and control the impact strength, toughness, and tensile modulus of the POM foam.[1]

The impact strength of polyoxymethylene foams increased with increasing cell density because cell walls absorbed impact energy and bubbles suppressed crack propagation.[2]

REFERENCES
1 Mantaranon, N; Chirachanchai, S, *Polymer*, **96**, 54-62, 2016.
2 Ma, H; Gong, P; Zhai, S; Li, G, *Chem. Eng. Sci.*, **207**, 892-902, 2019.

10.36 POLYPROPYLENE

Sodium bicarbonate+citric acid (Hostatron P 1941), azodicarbonamide (Adcol-blow X1020 and Hydrocerol PLC 751) were compared in foaming of several polymers, including polyethylene.[1] There was little difference between the densities of foams obtained with different blowing agents.[1] The concentration of blowing agent had a predominant influence.[1]

Microcellular injection molding of polypropylene and glass fiber composites was performed using supercritical nitrogen.[2] The cell morphology and glass fiber orientation were influenced by cooling and shear effects.[2] The optimal conditions for injection molding were supercritical nitrogen content of 0.4 wt%, injection speed of 60%, melt temperature of 190°C, and a mold temperature of 70°C.[2] Microfoamed parts had an average cell size of less than 30 μm.[2]

In cellulose fiber reinforced polypropylene foamed with supercritical carbon dioxide, the cell density increased with the increase of cellulose content.[3] The void fraction decreased with the addition of cellulose.[3] The type of polypropylene had a minimal influence on the outcome of blowing.[3]

Polypropylene medium density foams (0.3-0.6 kg m^{-3}) were obtained by compression molding using azodicarbonamide.[4] The density, open-cell content, and blowing agent concentration had a significant influence on the mechanical performance of medium-density PP foams.[4]

Polypropylene foam was produced by continuous extrusion using carbon dioxide.[5] A poor cell morphology and narrow foaming window were observed because of low melt strength.[5] PTFE addition has improved process outcome.[5] The PTFE particles were deformed into fine fibers under shear or extensional flow.[5] This significantly increased the melt strength of polypropylene from 0.005 N to 0.03 N for 4.0 wt% PTFE at 230°C. PTFE also improved cell morphology and broadened foaming window.[5]

Recyclable, high mechanical strength and thermal insulation polypropylene foams were manufactured by supercritical CO_2 foaming in the presence of β nucleating agent (N, N'-dicyclohexyl-2, 6-naphthalenedicarboxamide).[6] The PP foam had a variety of advantages, such as high expansion ratio, continuous honeycomb polygonal cells, the very low thermal conductivity of 26.4 mW/m•K, and strong tensile and compressive strength.[6]

Graphene nanoplatelets (surface area of about 400 g/cm^2) were finely dispersed within the PP matrix.[7] The presence of

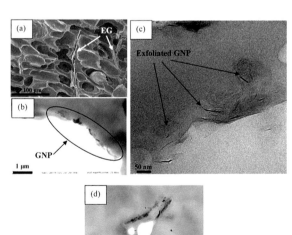

Figure 10.36.1. (a) Expanded graphite, EG, located between adjacent cells; (b), (c) Graphene, GNP, adsorbed on a cell wall; (d) a bubble formed next to a GNP flake. [Adapted, by permission, from Ho, Q B; Kontopoulou, M, *Polymer*, **198**, 122506, 2020.]

expanded graphite and graphene nanoplatelets suppressed cell coalescence and promoted heterogeneous nucleation, resulting in cell densities of the order of 10^7 cells/cm^3 and a fine microcellular structure with an average cell diameter in the range of 20-60 μm.[7] Figure 10.36.1 shows the distribution of expanded graphite and graphene nanoplatelets in the foam.[7] Graphene nanoplatelets with high surface area induced secondary cell nucleation, resulting in a secondary population of smaller cells in the range of 1-10 μm.[7]

Polypropylene/carbon black composite beads were expanded using supercritical CO_2 as a blowing agent *via* the bath-foaming process, and high expansion ratios of 11–25 were achieved even when the content of carbon black reached 30 wt%.[8] The beads were further processed by steam-chest molding to foamed sheets of low density, excellent compressive strength, and tensile toughness.[8] The PP composite foams containing 20 and 25 wt% carbon black exhibited ultralow reflection loss of -39.3 and -60 dB over the frequency range of 1.7-18 GHz, as well as two effective bandwidths (RL <-10 dB) of 11 and 15 GHz, respectively.[8]

Crystal melting behavior of polypropylene under different CO_2 pressures was *in situ* studied.[9] The semi-molten state of linear polypropylene was suitable for batch foaming.[9] Linear polypropylene foam with an ultra-high expansion ratio of 45 was fabricated.[9] The fabricated high expansion foam had a low thermal conductivity of 37.2 mW m^{-1} K^{-1}.[9]

Benzene triamide-based nucleating agent was used to control cell nucleation during foaming of isotactic polypropylene.[10] It forms supramolecular nanostructures in the polymer melt, acting first as nucleating sites for foam formation and second as nuclei for the polymer crystallization.[10] The foam, having a 50% reduction of density, was made by injection molding with nitrogen.[10] The addition of only 0.02 wt% of an additive was sufficient to obtain a remarkable reduction of the cell sizes.[10]

The core movement in the closed injection mold after the cavity has been filled volumetrically with plastic melt containing blowing agent has been applied to foam injection molding of polypropylene to modulate density to produce components with different foaming ratios.[11] The pull and foam process permitted an increase in stiffness, created spacers, or joining surfaces.[11]

Polypropylene/nano-crystalline cellulose foams were produced through extrusion compounding combined with injection molding.[12] Structural foams having a foamed core surrounded by two unfoamed skins were produced, and a closed-cell structure was obtained.[12] The nano-crystalline cellulose had a nucleating effect increasing the number of cells and decreasing their size.[12] Increased mold temperature caused a decrease in skin thickness because there was more time for cell nucleation and growth before the part solidified.[12]

The supercritical carbon dioxide was used for the production of foam by injection molding of isotactic polypropylene containing nano-calcium carbonate.[13] The nano-calcium carbonate was found to be a very efficient nucleating agent.[13] A high-pressure drop release rate controlled by backpressure resulted in the formation of small cell sizes.[13]

Four PP samples with different molecular architectures, namely isotactic polypropylene, ethylene-propylene block copolymer, ethylene-propylene random copolymer, and branched PP, were used to observe the relationship between the molecular structure and polymer foamability.[14] The foaming temperature window of branched PP was more than

34°C, whereas that of linear PPs was as narrow as 2-4°C.[14] Unlike in the case of linear PPs, the foam density of branched PP was insensitive to the foaming temperature.[14]

Using visualization mold, it was observed that in the high-pressure foam injection molding process, the overall cell density did not change with the injection speed, but the cell density increased significantly with the blowing agent's concentration and with a nucleating agent.[15]

Polypropylene composition for the production of foamed molded articles for finished parts for the automotive industry contains 10-20 wt% talc.[12] The blowing agent is nitrogen.[16] The foamed articles are used as dashboards, instrument panels, or other interior trim components for a car.[16]

Hydrocerol® CF40 (Clariant; sodium bicarbonate-citric acid nucleating agents (40 percent concentration of sodium bicarbonate and citric acid in a wax and polyethylene base in a pellet form)) has been used for the production of automotive parts.[17] On the polymers side, the composition of heterophasic polypropylene, propylene homopolymer, a copolymer of propylene and ethylene, high melt strength polypropylene, and a high-density polyethylene have been used.[17]

The melt grafting reaction using macromonomer vinyl polydimethylsiloxane and comonomer styrene was used to prepare a high melt strength polypropylene.[18] The melt strength of the grafted polymer increased by more than 12 times compared to that of the virgin isotactic polypropylene.[18] This resulted in smaller cell size, higher cell density, and higher expansion ratio as compared to the virgin iPP when supercritical carbon dioxide was used as a blowing agent.[18]

Because of the low water solubility in hydrophobic polymers such as polypropylene, it is difficult to produce low-density foams using water as a blowing agent.[19] A thermoplastic starch was used as an effective water carrier.[19] An open-cell PP/TPS blend foam with an apparent density as low as 19.5 kg m^{-3} with the open-cell content of 95.6% was prepared.[19]

Polypropylene/polydimethylsiloxane blends were batch-foamed at different saturation pressures of carbon dioxide.[20] The better solubility of carbon dioxide in PDMS made it as a carbon dioxide reservoir to induce more nucleation in blends than in the virgin polymer.[20] An excessively high PDMS concentration, however, had a negative effect on cell density.[20]

The distribution of nanoclay influenced foam morphology of PP/EPDM/organoclay nanocomposites.[21] The nanocomposite foams were produced *via* a batch process using supercritical nitrogen.[21] The foaming occurred mostly in the PP phase.[21] The organoclays had more affinity to PP and were primarily distributed in the PP phase.[21] Their nucleating abilities improved foaming.[21]

REFERENCES

1 Garbacz, T; Palutkiewicz, P, *Cellular Polym.*, **34**, 4, 189-214, 2015.
2 Xi, Z; Sha, X; Liu, T; Zhao, L, *J. Cellular Plast.*, **50**, 5, 489-505, 2014.
3 Kuboki, T, *J. Cellular Plast.*, **50**, 2, 113-28, 2014.
4 Saiz-Arroyo, C; Rodriguez-Perez, M A; Tirado, J; Lopez-Gil, A; Jose A de Saja, J A, *Polym. Int.*, **62**, 1324-33, 2013.
5 Wang, K; Wu, F; Zhai, W; Zheng, W, *J. Appl. Polym. Sci.*, **129**, 2253-60, 2013.
6 Yang, C; Zhang, Q; Zhang, W; Wu, G, *Polym. Deg. Stab.*, in press, 109406, 2021.
7 Ho, Q B; Kontopoulou, M, *Polymer*, **198**, 122506, 2020.
8 Wu, F; Li, Y; Lan, X; Zheng, W, *Compos. Commun.*, **20**, 100358, 2020.

9 Hou, J; Zhao, G; Wang, G; Zhang, L; Dong, G; Li, B, *J. Supercritical Fluids*, **145**, 140-50, 2019.
10 Stumpf, M; Spoerrer, A; Schmidt, H-W; Altstaedt, V, *J. Cellular Plast.*, **47**, 6, 519-34, 2011.
11 Heim, H-P; Tromm, M, *J. Cellular Plast.*, **52**, 3, 299-319, 2016.
12 Yousefian, H; Rodrigue, D, *J. Appl. Polym. Sci.*, **132**, 42845, 2015.
13 Xi, Z; Chen, J; Liu, T; Zhao, L; Turng, L-S, *Chinese J. Chem. Eng.*, **24**, 180-9, 2016.
14 Yu, C; Wang, Y; Wu, B; Xie, Y; Yu, C; Chen S; Li, W, *Polym. Testing*, **30**, 8, 887-92, 2011.
15 Shaayegan, V; Wang, G; Park, C B, *Chem. Eng. Sci.*, **155**, 27-37, 2016.
16 Langenfelder, D; Rohrmann, J, **WO2013178509**, *Basell Poliolefine Italia S.R.L.*, Dec. 5, 2013.
17 Kastner, E; Tranninger, M; Primat, C, **WO2016005301**, *Borealis AG*, Jan. 14, 2016.
18 Zhou, S; Zhao, S; Xin Z, *Polym. Eng. Sci.*, **55**, 251-9, 2015.
19 Xu, M Z; Bian, J J; Han, C Y; Dong, L S, *Macromol. Mater. Eng.*, **302**, 149-59, 2016.
20 Li, B; Wu, Q; Zhou, N; Shi, B, *Int. J. Polym. Mater.*, **60**, 51-61, 2011.
21 Keramati, M; Ghasemi, I; Karrabi, M; Azizi, H, *Polym. J.*, **44**, 433-8, 2012.

10.37 POLYSTYRENE

Unsaturated fluoropropenes having greenhouse warming potential values <15 (100 times less than HFC-134a) have been assessed as hydrofluorocarbon (HFC-134a) replacement for styrenic extrusion foaming because it has high greenhouse warming potential.[1] HFO-1261zf (3-fluoropropene) has good solubility and sufficiently high diffusion coefficient in polystyrene, and it offers excellent foaming capabilities.[1] Regular large cell size and low density can be produced, even with high concentrations of HFO-1261zf (3-fluoropropene). HFO-1243zf (1,1,1-trifluoropropene) and HFO-1234ze (1,1,1,3-tetrafluoropropene) are also good alternatives to HFC-134a, if co-blowing agents are used to minimize nucleation, which helps in controlling bubble formation.[1]

Large-cell extruded foams with zero-ozone depletion potential were produced for insulating billet applications.[2] Polystyrene copolymer containing acrylonitrile/1,1,1,2-tetrafluoroethane (HFC-134a)/1,1-difluoroethane (HFC-152a)/water system was used in this application.[2] Because of higher molecular weight, HFC-134a has lower diffusivity in styrenic polymers, which makes it a good choice for insulating applications but, at the same time, it has high nucleation potential, which is moderated by the presence of the other two blowing agents.[2]

Building on the best thermal performance of HFC-134a, carbon dioxide was used as a co-blowing agent to produce styrofoam with good insulating properties.[3] The addition of IR attenuators, such as graphite or carbon black, was used to further improve the thermal insulating properties of foams.[3]

Forming an expanded polystyrene foam was performed with a combination of *trans*HFO-1234ze (3-8 wt%) and *trans*HCFO-1233zd (2-10 wt%).[4] The resultant foam had a low density (lower than 40 kg/m^3) and excellent thermal conductivity (not greater than 20 mW/mK).[4]

Polar polystyrene copolymers for enhanced foaming were developed based on copolymerization using fluorinated(meth)acrylate, glycidyl(meth)acrylate, hexafluorobutyl acrylate as co-monomers having a polar function.[5] This helps to increase the solubility of carbon dioxide in the copolymer.[5]

The cell nucleation in nanosilica/polystyrene composites was investigated.[6] The silica surface was modified with vinyltriethoxysilane.[6] Foams were prepared using supercritical carbon dioxide.[6] The cell density increased with the decrease in nanosilica size and the increase of silica loading.[6] Further increase in cell density was observed when the surface of silica was treated with silane.[6]

Spherical polystyrene beads containing 2-propanol as an expansion agent were prepared, which were then expanded using microwave energy.[7] The expansion of 2.5 times of the initial volume was achieved.[7]

Styrene-butadiene-styrene and calcium carbonate nanoparticles were melt blended with polystyrene and extrusion-foamed using supercritical carbon dioxide as a blowing agent to toughen and reinforce polystyrene foams simultaneously.[8] The SBS and calcium carbonate had a significant influence on cell structure and the compressive properties of the composite foams.[8]

The multiwalled carbon nanotube/polystyrene nanocomposites were expanded using azodicarbonamide.[9] Foaming increased the percolation threshold, reduced DC and AC conductivities, widened the insulator–conductor transition window, and reduced the dissipation factor of composites.[9]

A superhydrophobic film was obtained by the microcellular plastic foaming method.[10] Polystyrene film was foamed using azodicarbonamide.[10]

Methyl methacrylate was dissolved in polystyrene with supercritical carbon dioxide at a temperature of 60°C and a pressure of 8 MPa. Polymerization of MMA was conducted at 100°C and 8 MPa in the presence of a crosslinking agent (diurethane dimethacrylate).[11] Carbon dioxide was initially playing the role of plasticizer and then a blowing agent.[11] The crosslinking agent was used to control the elasticity of PMMA domains.[11] The difference in elasticity delayed the bubble nucleation in the PMMA domains and caused the formation of a bimodal distribution of cell sizes.[11] The smaller cells were ranging from 10 to 30 μm in diameter and were located in the walls of the large cells having 200-400 μm in diameter.[11]

Polystyrene foams having bimodal cell morphology can be produced in the extrusion foaming process using carbon dioxide and water as co-blowing agents and two particulate additives as nucleation agents.[12] One particulate (nanoclay 30B) decreased water foaming time, so both carbon dioxide and water-induced foaming simultaneously, while the other (activated carbon from wood particles) increased the carbon dioxide nucleation rate with little effect on the carbon dioxide foaming time.[12]

Expanded off-grade polystyrene is commonly known as waste in the manufacturing process of expanded polystyrene.[13] Impact modifier such as styrene-ethylene-butylene-styrene helps to adjust properties that off-grades can be returned to use.[13]

Fabrication of aerogel foams from syndiotactic polystyrene was done using the silica-based Pickering emulsion method.[14] A large fraction of micrometer size voids (macrovoids) was created inside porous thermo-reversible sPS gels by dispersing water droplets without the use of surfactants before gelation of sPS solution.[14] Hydrophobic silica particles having a diameter of 50-70 nm were used to stabilize Pickering emulsion of micrometer size water droplets in emulsions of sPS solution in toluene at 110°C.[14]

A new composite electrode was developed from graphite powder and expanded polystyrene.[15] 75 wt% graphite content was appropriate for preparing the composite electrode with working potential similar to that of carbon paste electrodes, superior mechanical stability, and much faster response to ferrocyanide, close to reversible and similar to that of the much more expensive glassy carbon electrode.[15]

N-doped porous carbons prepared by carbonization of foam polymer materials and urea are promising anode materials of lithium-ion batteries for their low cost and high performance.[16] Polystyrene was mixed with melamine cyanurate (Melanic Mc; 5 wt%) and azobisformamide (3 wt%) to prepare precursors of porous carbon.[16]

Polystyrene foam exhibits a ductile to brittle transition when the loading mode changes from compression to tension.[17] PS foam can be compressed to a strain of 70% without failure.[17] However, it exhibits a brittle failure at around a strain of 2% when loaded in tension.[17] The elastic modulus in tension is nearly one order larger than that measured in compression.[17] A homogeneous deformation can be observed in uniaxial deformation tests, where PS foam exhibits a near-zero Poisson's ratio.[17]

REFERENCES

1. Vo, C V; Fox, R T, *J. Cellular Plast.*, **49**, 5, 423-38, 2013.
2. Fox, R; Frankowski, D; Alcott, J; Beaudoin, D; Hood, L, *J. Cellular Plast.*, **49**, 4, 335-49, 2013.
3. Vo, C V; Bunge, F; Duffy, J; Hood, L, *Cellular Polym.*, **30**, 3, 137-55, 2011.
4. Bowman, J M; Williams, D J, **WO2014015315**, *Honeywell International Inc.*, Jan. 23, 2014.
5. Wang, W; Knoeppel, D W; Greenberg, M, **US20150057383**, *Fina Technology, Inc.*, Feb. 26, 2015.
6. Zakiyan, S E; Famili, M H N; Ako, M, *J. Mater. Sci.*, **49**, 6225-39, 2014.
7. Sen, I; Dadush, E; Penumadu, D, *J. Cellular Plast.*, **47**, 1, 65-79, 2011.
8. Jing, X; Peng, X-F; Mi, H-Y; Wang, Y-S; Zhang, S; Chen, B-Y; Zhou, H-M; Mou, W-J, *J. Appl. Polym. Sci.,* **135**, 43508, 2016.
9. Arjmand, M; Mahmoodi, M; Park, S; Sundararaj, U, *J. Cellular Plast.*, **50**, 6, 551-62, 2014.
10. Zhang, Z X; Li, Y N; Xia, L; Ma, Z G; Xin, Z X; Kim, J K, *Appl. Phys. A*, **117**, 755-9, 2014.
11. Kohlhoff, D; Nabil, A; Ohshima, M, *Polym. Adv. Technol.*, **23**, 1350-6, 2012.
12. Zhang, C; Zhu, B; Li, D; Lee, L J, *Polymer*, **53**, 12, 2435-42, 2012.
13. Direksilp, C; Threepopnatkul, P, *Energy Procedia*, **56**, 135-41, 2014.
14. Kulkarni, A; Jana, S C, *Polymer*, **212**, 123125, 2021.
15. Surkov, A M; Queiroz; R G; Rinco, R S;Baccaro, A L B, *Talanta*, **223**, Part 2, 121780, 2021.
16. Huang, J; Lin, Y; Ji, M; Xu, J, *Appl. Surf. Sci.*, **504**, 144398, 2020.
17. Tang, N; Lei, D; Huang, D; Xiao, R, *Polym. Testing*, **73**, 359-65, 2019.

10.38 POLYURETHANE

In injection molding, the mold is partially filled with reactants, such as isocyanate, polyol, catalysts, blowing, and chemical agents.[1] The reaction and crosslinking begin.[1] Chemical reactions are exothermic, and they increase the foam temperature.[1] Expansion of the foam is due to carbon dioxide production and volume expansion.[1] Polyurethane foam formation includes several complex phenomena such as chemical reactions, heat generation, and blowing agent formation.[1]

The pour foam formulations have two components. On one side, they contain polyols, chain extenders, and water.[2] In the second component, they contain isocyanate, blowing agent, catalyst, and surfactant.[2] Mixing both components initiates a polymerization reaction that is accelerated by a catalyst, the formation of carbon dioxide, evaporation of blowing agent, and cell nucleation.[2] The amount and the type of catalyst has to be chosen in such a way that the end of polyurethane formation occurs when the right size of bubbles is achieved.[2] Sprayed foams have a density in the range of 50-70 kg/m^3.[2]

Three environmentally-friendly blowing agents have been used to fabricate rigid polyurethane foams from MDI and polypropylene glycols.[3] Water induced the earliest bubble; temperature and pressure rised, whereas HFC 365mfc gave the slowest reactivity.[3] Water blown foams had the largest cell size and the lowest compression strength, the highest glass transition temperature, and the greatest dimensional stability at high temperature.[3] HFC 365mfc gave the smallest cell size and the greatest compression strength.[3] HFC 245fa gave the intermediate values, except for the lowest thermal conductivity of the foam.[3]

Liquid nucleating agent improved thermal insulating properties of rigid polyurethane foams obtained using HFC-365mfc.[4] The cell diameter decreased from 228 to 155 μm, and the thermal conductivity decreased from 0.0227 to 0.0196 kcal/mh°C when the perfluoroalkane (PF 5056 from 3M) content was 0.0 to 2.0 parts per hundred polyol by weight.[4] The perfluoroalkane likely acted as a nucleating agent.[4]

The molded PU foam was produced using gas counter-pressure, which enabled the use of high amounts of carbon dioxide.[5] Rigid foams were produced with a density of 116 kg/m^3 using only carbon dioxide.[5] Flexible foams had a density of 61 kg/m^3.[5]

The polyol premix compositions comprise a combination of a hydrohaloolefin blowing agent (1,3,3,3-tetrafluoropropene (J234ze); 1-chloro-3,3,3-trifluoropropene (1233zd), 1,1,1,4,4,4-hexafluorobut-2-ene (1336mzzm)), a polyol, a silicone surfactant, and a catalyst system that includes a bismuth-based metal catalyst (BiR$_3$).[6] The polyol premix has a cream time of less than about 6-10 seconds.[6]

The production of polyurethane foam had a polyol component, which contained a physical blowing agent in the form of an emulsion and the isocyanate-reactive component.[7] Both components were mixed and injected into the mold cavity, followed by a reduction of pressure below ambient pressure.[7] The small droplets of blowing agent formed nucleation sites.[7] The physical blowing agent was selected from the group of hydrocarbons, halogenated ethers, and/or perfluorinated hydrocarbons with 1 to 6 carbon atoms.[7] The vacuum-assisted foaming process led to molded polyurethane foams possessing more even density distribution.[7] The mixture had a gel time of 35 s.[7] The polyurethane foam had density in the range of >28 kg/m^3 and <45 kg/m^3.[7]

10.38 Polyurethane

Rigid polyurethane blowing agent enhancers are low molecular weight alcohols (methanol, ethanol, and any isomer of propanol, butanol, pentanol, hexanol, heptanol, octanol, nonanol, and decanol) and/or ethers (dipropylene glycol dimethyl ether and diethylene glycol monomethyl ether, propylene glycol monopropyl ether) that assist the action of blowing agents comprising a hydrohalocarbon compound.[8] The production of foams with enhancer permits the use of a reduced amount of blowing agent to obtain the same effect as with the normal concentration of blowing agent without enhancer.[8]

The 2,4,4,4-tetrafluorobutene-1 is used alone or in combination with a hydrofluoroolefin, hydrofluorocarbon, hydrochlorofluoroolefin, or a hydrocarbon.[9] The blowing agent is effective in the manufacture of thermosetting foams.[9]

Bio-sourced non-isocyanate polyurethane microcellular foams were prepared using supercritical carbon dioxide.[10] These low-density foams offer low thermal conductivity.[10] A poly(methylhydrogenosiloxane) was used as a blowing agent to foam the non-isocyanate polyurethane by reaction with diamines.[11] The foams exhibited a glass transition temperature between -18°C and 19°C and thermal stability above 300°C.[11]

Polypropylene glycol-grafted polyethyleneimines reversibly absorb and release carbon dioxide to blow polyurethanes.[12] These adducts are thermally unstable and can release carbon dioxide to blow polyurethanes, whose polymerization is exothermic.[12] The thermal conductivity of these foams is higher than that of traditional polyurethane foams (0.040 vs. 0.020–0.027 W/mK), but they have no VOC emission.[12]

By-products obtained from wood treatment were used as raw materials for producing flexible polyurethane foams.[13] Lignin was used to replace polyol.[13] Water was used as a blowing agent, and an isocyanate index (NCO/OH) was less than 100 to improve the flexibility of the foam.[13] The quality of foam was comparable to the presently used by the furniture industry.[13]

Rapeseed oil-based polyurethane foams were modified with glycerol and cellulose micro/nanocrystals.[14] Rapeseed oil-based polyol was synthesized by chemical modification by epoxidation followed by oxirane ring-opening with diethylene glycol.[14] The use of glycerol as a hydroxyl component, water as a reactive blowing agent, and micro/nanocellulose gives green polyurethane foam.[14]

The biodegradable polyurethane foams were synthesized from a starch/petroleum-based polyol, 2,4/2,6-toluene diisocyanate, using a one-shot method with water as the blowing agent.[15] Chain extenders such as starch, diethanolamine, and glycerol were studied.[15] Starch increased biodegradability.[15]

The cream time, rise time, tensile, and compressive strengths at room temperature, glassy and rubbery state moduli, glass transition temperature, and shape fixity and recovery increased with the addition and an increasing amount of multiwalled carbon nanotubes in polyurethane foam with shape memory effects.[16]

The isocyanate-functionalized silica nanoparticles were chemically incorporated into polyurethane during the synthesis of flexible PU foam from polypropylene glycol and toluene diisocyanate following the one-shot method with water as the blowing agent.[17] The shape fixity, shape recovery, and strain energy storage significantly increased with reduced hysteresis loss in shape memory foam.[17]

A computer-based simulation data for rigid polyurethane foam-forming reactions was compared with experimental data for six blowing agents, including methyl formate

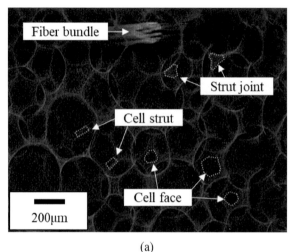

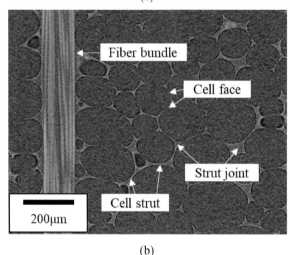

Figure 10.38.1. (a) 3D X-ray image (upward: foaming direction) and (b) 2D X-ray image (x-y plane) of glass fiber-reinforced polyurethane foam. Cells are organically connected to each other by strut and strut joint. The cell face is covered by a thin membrane so that it is difficult to observe in Figure 10.38.1 (b) (2D x-ray image). However, cell face can be easily observed in Figure 10.38.1 (a). Several strands of fibers form groups and penetrate the foam. [Adapted, by permission, from Kim, M-S; Kim, J-D; Kim, H-H; Lee, J-M, *Int. J. Mech. Sci.*, **194**, 106188, 2021.]

and C5-C6 hydrocarbons.[18] Modeling density data agreed with experimental measurements.[18]

The morphometric characterization revealed different morphometry evolution across the sample thickness according to density.[19] The inhomogeneities were higher for the high-density foams, which concentrate more material near the top and bottom surfaces.[19] The volume fraction varied from 40% near the surfaces to about 20% in the middle.[19] Medium density foam was more homogeneous than the high-density foam.[19] The greatest morphometric inhomogeneities in the high-density foams influenced the failure pattern.[19] The highest porosity and the lowest volume fraction in the central region promoted failure initiation.[19]

Glass fiber-reinforced polyurethane foam subjected to a low-energy repetitive impact shows performance degradation.[20] The critical impact energy causing performance degradation was 1,092-1,228 J/m².[20] In the impact performance, the peak stress, impact duration, and peak strain decreased by 13.5%, increased by 28.5%, and increased by 35.2%, respectively, after 80 repetitions of 1,500 J/m² impact.[20] Under the same impact condition, the elastic modulus decreased by 44.5% compared to that of the intact specimen.[20] Figure 10.38.1 examines the morphology of glass-fiber-reinforced polyurethane foam.[20]

Strong interfacial interaction facilitated organic montmorillonite nanoscale dispersion in thermoplastic polyurethane.[21] The thermoplastic polyurethane/organic montmorillonite topological network regulated cell growth.[21] Hydrogen bonding between thermoplastic polyurethane and organic montmorillonite was beneficial for cell stabiliza-

10.38 Polyurethane

tion.[21] The organic montmorillonite regulated cell growth/stabilization without affecting cell nucleation (Figure 10.38.2).[21] Optimal montmorillonite loading was 1 wt%.[21]

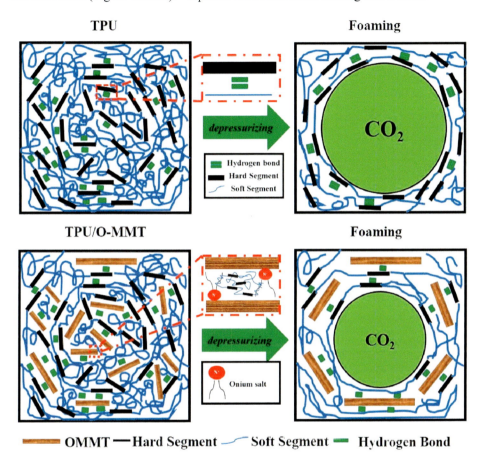

Figure 10.38.2. Schematic diagram of cell growth in the foaming process. [Adapted, by permission, from Lan, B; Li, P; Luo, X; Gong, P, *Polymer*, **212**, 123159, 2021.]

Rigid polyurethane foam panels contained phase change materials.[22] Inclusion of phase change material enhanced the thermal regulation of indoor space.[22] The performance of the rigid polyurethane panel containing phase change material was at the maximum when the external and internal mean temperatures were closer to the melting peak temperature of the phase change material.[22] The efficiency of the rigid polyurethane panels incorporating phase change material was not exclusively dependent on the presence of phase change material but was also affected by the imposed temperature profile and the thermal properties of the phase change material.[22]

Artificial leather was composed of fabric layer, binder layer laminated to the upper portion of the fabric layer, a water-based polyurethane foam layer laminated to the upper portion of the fabric layer, skin layer laminated to the upper portion of the water-based polyurethane foam layer; and a surface treatment layer laminated to the upper portion of the skin layer.[23] Open cells were formed in the water-based polyurethane foam layer through mechanical foaming.[23] The artificial leather was designed for automobile seats.[23]

Soft polyurethane foam contains a polyol component, isocyanate component, foaming agent (water), cell opening agent, foam stabilizer (high-activity silicone), and catalyst.[24] The catalyst is C11, or higher amine, and the cell opening agent is a polyol in which the content of ethylene oxide is more than 50 mol%.[24] The foam was developed for vehicle seat pad.[24]

REFERENCES

1. Samkhaniani, N; Gharehbaghi, A; Ahmadi, Z, *J. Cellular Plast.*, **49**, 5, 405-21, 2013.
2. Stirna, U; I Beverte, I; V Yakushin, V; Cabulis, U, *J. Cellular Plast.*, **47**, 4, 337-55, 2011.
3. Lim, H; Kim, E Y; Kim, B K, *Plast. Rub. Compos.*, **39**, 8, 364-9, 2010.
4. Lee, Y; Jang, M G; Choi, K H; Han, C; Kim, W N, *J. Appl. Polym. Sci.*, **133**, 43557, 2016.
5. Hopmann, C; Latz, S, *Polymer*, **56**, 29-36, 2015.
6. Yu, B; Williams, D J, **WO2014133986**, *Honeywell International Inc.*, Sep. 4, 2014.
7. Albers, R; Heinemann, T; Vogel, S; Loof, M; Gu, Z; Atsushi, U; Sangjo, S; Shihu, Z; **WO2014019962**, *Bayer Materialscience AG*, Feb. 6, 2014.
8. Miller, J W; **US9321892**, *Air Products and Chemicals, Inc.*, Apr. 26, 2016.
9. Chen, B B; Costa, J S; Bonnet, P, **WO2011050017**, *Arkema, Inc.*, Apr. 28, 2011.
10. Grignard, B; Thomassin, J-M; Gennen, S; Poussard, L; Bonnaud, L; Raquez, J-M; Dubois, P; Tran, M-P; Park, C B; Jerome, C; Detrembleur, C, *Green Chem.*, **18**, 2206-15, 2016.
11. Cornille, A; Dworakowska, S; Bogdal, D; Boutevin, B; Caillol, S, *Eur. Polym. J.*, **66**, 129-38, 2015.
12. Long, Y; Sun, F; Liu, C; Xie, X, *RSC Adv.*, **6**, 23726, 2016.
13. Bernardini, J; Anguillesi, I; Coltelli, M-B; Cinelli, P; Lazzeri, A, *Polym. Int.*, **64**, 1235-44, 2015.
14. Mosiewicki, M A; Rojek, P; Michalowski, S; Aranguren, M I; Prociak, A, *J. Appl. Polym. Sci.*, **132**, 41602, 2015.
15. Kang, S M; Kang, M S; Kwon, S H; Park, H; Kim, B K, *Polym. Eng.*, **34**, 6, 555-9, 2014.
16. Kang, S M; Kwon, S H; Park, J H; Kim, B K, *Polym. Bull.*, **70**, 885-93, 2013.
17. Kang, S M; Kim, M J; Kwon, S H; Park, H; Jeong, H M; Kim, B K, *J. Mater. Res.*, **27**, 22, 2837-43, 2012.
18. Al-Moameri, H; Zhao, Y; Ghoreishi, R; Suppes, G J, *J. Appl. Polym. Sci.*, **132**, 42454, 2015.
19. Belda, R; Palomar, M; Giner, E, *Mater. Sci. Eng.:C*, **120**, 111754. 2021.
20. Kim, M-S; Kim, J-D; Kim, H-H; Lee, J-M, *Int. J. Mech. Sci.*, **194**, 106188, 2021.
21. Lan, B; Li, P; Luo, X; Gong, P, *Polymer*, **212**, 123159, 2021.
22. Amaral, C; Silva, T; Vicente, R, *Energy*, **216**, 119213, 2021.
23. Son, S H; Jung, Y B, **US20200040522A1**, *LG Hausys Ltd*, Feb. 6, 2020.
24. Sanefuji, K; Watanabe, M; Seguchi, H, **US20200140642A1**, *Bridgestone Corp*, May 7, 2020.

10.39 POLYVINYLALCOHOL

Flame-retarded polyvinylalcohol foam with intrinsic flame retardant characteristics was prepared through continuous extrusion.[1] Water was used as a plasticizer and blowing agent.[1] The foam had apparent density 0.25 g/cm^3, average cell diameter 450 µm, cell density 10^3 cells/cm^3, UL94 V-0, LOI 35%, tensile strength 1.8 MPa, and elongation at break 57.1%.[1]

A porous PVAl scaffold contains a pore-forming agent.[2] Formation includes heating to melt the PVAl, cooling the melted PVAl and mixing the PVAl with a heat-decomposable pore-forming agent, repeating the freezing/thawing of the mixed PVAl to cure the PVAl mixture, and stirring the cured PVAl mixture with a hydrochloric acid solution at a high temperature of 65°C or more to produce foam.[2] The pore-forming agent is sodium bicarbonate (NaHCO$_3$) or ammonium hydrogen carbonate (NH$_4$HCO$_3$).[2]

Porous foams were prepared from cellulose nanofibers and polyvinylalcohol by unidirectional freezing, resulting in a homogeneous pore structure.[3] The cellulose nanofibers were derived from paper mill sludge, a by-product of paper manufacturing wastewater treatment.[3] Sodium tetraborate decahydrate (borax) was used as a crosslinking agent.[3] The density of the cellulose nanofibers/PVAl foams were 0.03 g cm^{-3} with a compressive strength of 116 kPa at 20% strain.[3] The foams were competitive to commercial expanded polystyrene foam.[3]

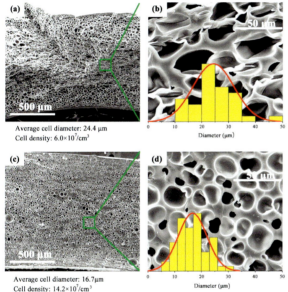

Figure 10.39.1. SEM micrographs and cell size distributions of (a, b) PVA and (c, d) PVA/0.1 wt% graphene oxide nanocomposite foams obtained at 80°C and 15 MPa with the depressurization rate of 5 MPa/s. [Adapted, by permission, from Liu, P, Chen, W; Bai, S; Wang, Q; Duan, W, *Compos. Part A: Appl. Sci. Manuf.*, **107**, 675-84, 2018.]

Nanocomposite foam based on polyvinylalcohol and graphene oxide was prepared by the supercritical fluid foaming.[4] Water was used as a benign solvent, and graphene oxide sheets were fully exfoliated and homogeneously dispersed in PVAl matrix with assistance of mild sonication.[4] Figure 10.39.1 shows the effect of graphene oxide on PVAl foam morphology.[4]

Carbon nanotube hybrid aerogels with high mechanical properties were fabricated by interconnecting structural networks of CNTs using polyvinylalcohol through amine-functionalized iron oxide.[5]

Polyvinylalcohol foam had a pore size ranging from 0.3 to 1 mm.[6] PVA foam had interconnected hollow cells and above 90% of volume was air.[6] Open pores were connected three dimensionally.[6] This physical structure could be used to deliver various functions, such as system for infusing medications, e.g., insulin.[6] It absorbs an aqueous solution so as to saturate the foam material by at least 95% in a time between 0.1 and 1 minutes.[6]

REFERENCES

1 Guo, D; Bai, S; Wang, Q, *J. Cellular Plast.*, **51**, 2, 145-63, 2015.
2 Shim, Y-B; Jung, H-H; Choi, Y-R; Jang, J-W, **US20120108689**, *Korea Bone Bank Co., Ltd.*, May 3, 2012.
3 Adu, C; Rahatekar, S; Jolly, M, *Mater. Lett.*, **253**, 242-5, 2019.
4 Liu, P, Chen, W; Bai, S; Wang, Q; Duan, W, *Compos. Part A: Appl. Sci. Manuf.*, **107**, 675-84, 2018.
5 Park, O-K; Lee, J H, *Compos. Part B: Eng.*, **144**, 229-36, 2018.
6 Dang, K H; Chattaraj, S; Fusselman, H S; Hoffman, L P; Zhang, G, **US20180200412A1**, *Medtronic Minimed Inc*, Jul. 19, 2018.

10.40 POLYVINYLCHLORIDE

Sodium bicarbonate+citric acid (Hostatron P 1941), azodicarbonamide (Adcol-blow X1020 and Hydrocerol PLC 751) were compared in foaming of several polymers, including polyethylene and polypropylene.[1] Similar to these polymers, PVC shows a little difference between densities of foams obtained with different blowing agents.[1] The concentration of blowing agent has a predominant influence.[1]

Azodicarbonamide was used in the foaming of PVC plastisols for the production of vinyl flooring.[2] Similar to the above observation, the concentration of blowing agent had the principal influence on the expansion ratio.[2] The addition of calcium carbonate had only a little effect on foaming.[2]

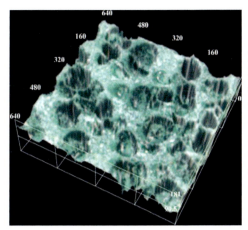

Figure 10.40.1. Confocal micrograph of cross-section of a single layer of cable coating using cellular PVC containing 0.4 wt% azodicarbonamide-based Adcol-blow UP-0Xb+X1020. [Adapted, by permission, from Garbacz, T, Polimery, 57, 11-12, 865-8, 2012.]

Cable coating was done by extrusion.[3] Figure 10.40.1 shows that pores of uniform size are evenly distributed in the cross-section of the material.[3] On other photographic material, solid outer skin is also visible.[3]

Flexible polyvinylchloride foams have excellent mechanical properties.[4] Clear correlations have been obtained between the molecular weight and structure of plasticizer and plastisol's rheological behavior.[4] The linear phthalates are more compatible with the resin yielding higher plastisol viscosity than branched phthalates.[4] Behavior of PVC plastisol and its foaming is different than the foaming of any other polymer.[4] At an elevated temperature, the plasticizer begins to swell the polymer particles, and the viscosity of plastisol rises.[4] On further heating, the swollen particles interact with each other, and a process known as gelation occurs in which borders between grains begin to disappear, and free plasticizer becomes exhausted.[4] Viscosity markedly increases.[4] At even higher temperatures, the PVC crystallites and grains fuse, and the system forms a homogeneous material.[4] This process is known as a fusion.[4] At a certain point of these changes, the chemical blowing agent begins to decompose and produce gas, forming bubbles. If the viscosity of plastisol is not sufficient at this point, the rising pressure of the gas will break the skin, escape, and form series of small craters on the surface (some of which might eventually heal). If viscosity is too high and the pressure buildup in the bubble is not sufficient to expand it, the material will not foam, but gas will diffuse out of the material. This shows how important it is to synchronize both processes of viscosity increase and production of gas within the sample. Azodicarbonamide is the major blowing agent used in PVC foaming, and its temperature of decomposition, which is normally above 200°C, can be lowered by the addition of costabilizer, the so-called kicker. The amount and the type of kicker control the extent of the reduction of temperature of decomposition. The temperature of decomposition has to fit the kinetics of viscosity development.

A linear relationship exists between the molecular weight of plasticizers and the corresponding plastisols' glass transition temperatures containing these plasticizers.[4] Plasticizers having better gelation properties are more suitable for foaming.[4] Usually, benzyl butyl phthalate with very high gelation power is used to regulate the gelation kinetics of PVC plastisol.

To achieve the highest level of fusion (i.e., the maximum fusion factor as well as the minimum fusion time), different compounds were studied.[5] The increase in nanoclay content reduces the fusion time.[5]

Fine particles of azodicarbonamide having average diameters of 5, 8, 11, and 22 μm were used together with 20 wt% ZnO at doses of 0-3.0 wt% PVC.[6] The maximum reduction of density by 46% was achieved when particles of 22 μm were used at 2 wt%.[6]

Ca stearate and Zn stearate heat stabilizers used instead of organotin stabilizers increased foam formation and decreased pore sizes and regularity in pore size distribution.[7] Samples containing zeolite, $CaCO_3$, cellulose, or luffa flour had lower pore volume than unfilled samples.[7]

A composition comprising 4,4'-oxybis(benzenesulfonylhydrazide) and sodium bicarbonate is an efficient blowing agent for plasticized PVC.[8] Good foam structure and smooth surface were obtained for the following composition: PVC (K 67) 100, DINP 35, Mesamoll 35, $CaCO_3$ 5, 4,4'-oxybis(benzenesulfonylhydrazide) 0.875, and sodium bicarbonate 2.625 (expansion rate 1.9) (also good foam was obtained when only 3.5 phr of sodium bicarbonate was used in the same composition; expansion rate 1.2).[8]

The invention reported on an acrylic process aid useful in vinyl foam extrusion.[9] The process aid was an acrylic copolymer containing from 50 to 79 wt% of methyl methacrylate monomer units and had a glass transition temperature of less than 65°C.[9] PVC and CPVC foams containing the acrylic process aid fused faster at the same temperature or fused at the same time at lower temperatures.[9]

The PVC composite material included 40-60 parts by weight of PVC, 40-60 parts by weight of calcium carbonate, 0.2-0.6 parts by weight of composite foaming agent (0.15 wt% of sodium bicarbonate and 0.35 wt% azodicarbonamide), 3-5 parts by weight of foam regulator (ZD530), 2-4 parts by weight of toughener, 0.8-1.2 parts by weight of lubricant (stearic acid+PE wax), and 2-3 parts by weight of stabilizer (Ca/Zn stearate=1:1).[10] The formulation was used for the production of foam board and flooring.[10]

The expandable composition contained a polymer selected from a group consisting of polyvinylchloride, polyvinylidenechloride, polyvinylbutyrate, polyalkyl(meth)acrylate, and their copolymers, an expanding agent (azodicarbonamide, Unifoam AZ Ultra 1035), foam stabilizer (ZnO), and di-2-ethylhexyl terephthalate as a plasticizer.[11] The formulation was suitable for the production of expandable composition for floor coverings, wallpapers, or synthetic leather.[11]

The thermal radiation propagating through the foam contributes to the total thermal transfer.[12] Both flake aluminum powder that mainly reflects infrared and the flake graphite powder that mainly absorbs infrared were added to the rigid crosslinked PVC foam.[12] The results demonstrated that the reduction of radiative thermal conductivity of the crosslinked-PVC foam with flake aluminum powder was more significant than the use of the crosslinked-PVC foam with flake graphite powder.[12] The experimental results were in

10.40 Polyvinylchloride

good agreement with the theoretical prediction.[12] Figure 10.40.2 illustrates the effect of opacifier (aluminum flakes) on the thermal transfer of energy.[12]

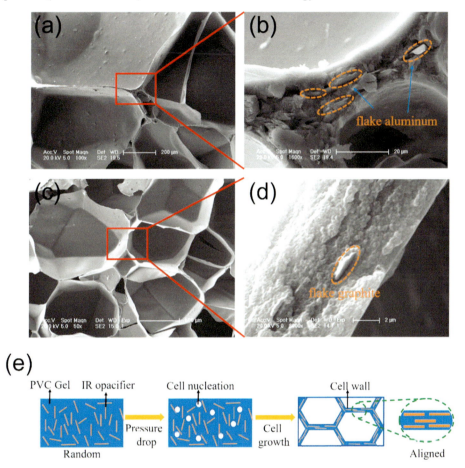

Figure 10.40.2. SEM images of the cellular morphologies of crosslinked-PVC/opacifier composite foams ((a) and (c)) and dispersed state of fillers in the cell walls ((b) and (d)): (a) and (b) FPVC/aluminum-2.0; (c) and (d) FPVC/graphite-2.0; (e) Schematic drawing for IR opacifier dispersion in crosslinked-PVC/opacifier composite foams. [Adapted, by permission, from You, J; Xing, H; Xue, J; Jiang, Z; Tang, T, *Compos. Sci. Technol.*, in press, 108566, 2021.]

Thermal insulation and excellent weatherability of microcellular foams based on lightweight polymer were obtained using chlorinated polyethylene rubber/polyvinylchloride compounds that were electron beam radiation crosslinked and chemically foamed (azodicarbonamide).[13] Thermal conductivity of these foams was as low 0.0374 W/mK.[13]

A multi-purpose tile system was developed for flooring.[14] The rigid closed-cell foam used in the base layer provided the tile with a desired rigidity and robustness, preventing damaging.[14] The rigid base layer was made from closed-cell PVC foam free of plasticizer.[14] The foam also contained 3-9% of toughening agent.[14]

Expandable polyvinylchloride paste contained polyvinylchloride resin, and tris(2-ethylhexyl) trimellitate plasticizer, zinc-based stabilizer; and azodicarbonamide.[15]

Nanoparticles of multiwalled carbon nanotubes were functionalized with a sodium hypochlorite solution.[16] The foam cell sizes decreased with the addition of MWCNTs.[16] The dispersion of nanoparticles in the PVC was increased by functionalization.[16]

Method of measuring polymer foam and apparatus were developed.[17]

REFERENCES

1   Garbacz, T; Palutkiewicz, P, *Cellular Polym.*, **34**, 4, 189-214, 2015.
2   Radovanovic, R; Jaso, V; Pilic, B; Stoiljkovic, D, *Hem. ind.*, **58**, 6, 701-7, 2014.
3   Garbacz, T, *Polimery*, **57**, 11-12, 865-8, 2012.
4   Zoller, A; Marcilla, A, *J. Appl. Polym. Sci.*, **121**, 1495-505, 2011.
5   Moghri, M; Khakpour, M; Akbarian, M; Saeb, M R, *J. Vinyl Addit. Technol.*, **21**, 51-9, 2015.
6   Petchwattana, N; Covavisaruch, S, *Mater. Design,* **32**, 5, 2844-50, 2011.
7   Demir, H; Sipahioglu, M; Balköse, D; Ülkü, S, *J. Mater. Process. Technol.*, **195**, 1-3, 144-53, 2008.
8   Stevens, R, **US20150267024**, *Lanxess Limited*, Sep. 24, 2015.
9   Lavallee, P R, **WO2011041195**, *Arkema Inc.*, Apr. 7, 2011.
10 Fang, Q, **US20160177579**, *Zhejiang Tianzhen Bamboo & Wood Development Co., Ltd.*, Jun. 23, 2016.
11 Becker, H G; Grass, M; Huber, A, **CA2817803**, *Evonik Oxeno GmbH*, May 31, 2012.
12 You, J; Xing, H; Xue, J; Jiang, Z; Tang, T, *Compos. Sci. Technol.*, in press, 108566, 2021.
13 Zhang, Z X; Wang, C; Wang, S; Wen, S; Phule, A D, *Radiat. Phys. Chem.*, **173**, 108890, 2020.
14 Boucke, E A; Song, J, **US20200063443A1**, *I4F Licensing NV Tower IPCO Co Ltd*, Feb. 27, 2020.
15 Hurley, J M, **US20190177502A1**, *Manduka LLC*, Jun. 13, 2019.
16 Ghasemi, I; Farsheh, A T; Masoomi, Z, *J. Vinyl Addit. Technol.*, **18**, 161-7, 2012.
17 Steward, S T; Van Rheenen, P R, **WO2013095876**, *Rohm & Haas Company*, Jun. 27, 2013.

10.41 POLY(VINYL CHLORIDE-CO-VINYL ACETATE)

Poly(vinyl chloride-co-vinyl acetate) plastisols were foamed with different commercial plasticizers.[1] Effect of chemical nature of plasticizers and their molecular weight have been observed, regarding the interaction (swelling and early stages of gelation) between the resin and plasticizer.[1] The heat evolved in the swelling process was a nearly linear decreasing function of the molecular weight of a plasticizer.[1] The maximum temperature of the process corresponding to the azodicarbonamide decomposition was an increasing function of the plasticizer's molecular weight.[1]

The extensional viscosity was responsible for the behavior of the system when undergoing the extensional stress produced by the released gases, such as in foaming.[2] Depending on the structure and thus on the compatibility of the plasticizer used, each plastisol developed its properties and structure accordingly.[2] Diisononyl cyclohexane-1,2-dicarboxylate (DINCH) having alicyclic ring was less compatible compared with the two other plasticizers (bis(2-ethylhexyl)-1,4-benzenedicarboxylate and diisononyl phthalate).[2] They were not able to withstand pressure from gases released during the foaming process, yielding foams of poorer quality, unless the temperature of the blowing agent decomposition was modulated with a kicker.[2]

REFERENCES

1 Zoller, A; Marcilla, A, *J. Appl. Polym. Sci.*, **121**, 3314-21, 2011.
2 Zoller, A; Marcilla, A, *J. Appl. Polym. Sci.*, **128**, 354-62, 2013.

10.42 POLYVINYLIDENEFLUORIDE

Foams composed of neat PVDF and immiscible blends of PVDF with polystyrene exhibit poor cell characteristics.[1] Miscible blends of PVDF with polymethylmethacrylate yield foams with improved morphologies.[1] PVDF/PMMA melt viscosity decreases with increasing PMMA content and supercritical carbon dioxide concentration.[1] The cell density increases as the PMMA fraction is increased and the foaming temperature is decreased.[1]

The foamed PVDF is manufactured in a process using a foaming agent (azodicarbonamide or p-toluenesulfonylsemicarbazide; 0.5 phr), and nucleating agent.[2] Foamed tubes, pipes, rods, sheets, and conduit are especially useful.[2] Foaming increases impact resistance, improves insulation properties, and reduces dielectric constant.[2] A preferred nucleating agent is calcium carbonate.[2] Nucleating agents, having smaller particle size and have rougher surfaces, are preferred.[2]

Developing thermal management materials with high thermal conductivity is critical to microelectronics.[3] High thermally conductive polyvinylidenefluoride composites were fabricated by using graphene foam and dopamine-modified graphene nanoplatelets as fillers.[3] Modified graphene nanoplatelets welded onto the graphene foam skeleton to repair the structural defects of graphene foam, which was conducive to improving the composite's thermal conductivity by forming efficient thermal conductive pathways in the matrix (Figure 10.42.1).[3] The obtained graphene foam/modified graphene nanoplatelets/PVDF composite (graphene foam: 0.27 vol%, modified graphene nanoplatelets: 8.82 vol%) exhibited an in-plane thermal conductivity of 6.32 W m^{-1} K^{-1}, which was much higher than that of graphene foam/graphene nanoplatelets/PVDF composite (4.26 W m^{-1} K^{-1}) at the same graphene foam and graphene nanoplatelets contents.[3]

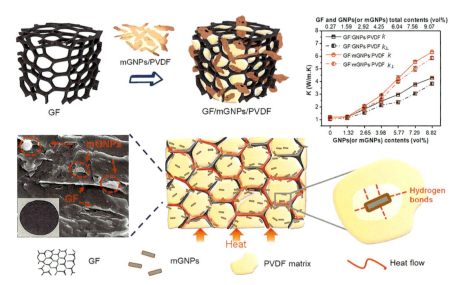

Figure 10.42.1. Schematic diagram of the fabrication process for the modified graphene nanosheets and graphene foam, GF/modified graphene nanosheets, mGNPs/PVDF composites, their distribution morphology, and heat flow transport mechanism. [Adapted, by permission, from Yang, T; Jiang, Z; Han, H; Hu, J, *Compos. Part B: Eng.*, **205**, 108509, 2021.]

10.42 Polyvinylidenefluoride

The PMMA/(PVDF-microcrystals) system was formed by self-assembly through CO_2 induced crystallization.[4] PVDF-microcrystals presence increased cell nucleation density and helped to refine cellular structures.[4] The porosity and cellular structure of the foams was regulated in a wide range using blend composition and foaming process.[4] The homogeneous PMMA/PVDF blends had a single glass transition temperature, which decreased linearly with the increase in PVDF content.[4] The increase in PVDF content reduced CO_2 requirement.[4] PVDF-microcrystals were uniformly separated from the blends due to CO_2 induced crystallization.[4] Because of self-assembly of PVDF-microcrystals, the foam had higher nucleation density and finer cell size than that of neat PMMA, and the rupture and coalescence problems of cells were effectively avoided.[4] With saturation pressure at 13.8 MPa and PVDF content of 20%, the nanocellular polymer foam had higher than 70% porosity, the cell size of 287 nm, and a cell nucleation density of more than 10^{14} cells/cm^3.[4]

Lightweight PVDF/Ni-chains composite foams with uniform morphology were fabricated by batch-foaming process.[5] Remarkable electrical conductivity (0.01 S/m) and magnetic properties were achieved by the condensed Ni-chains networks.[5] The composite foams had high shielding effectiveness (26.8 dB) dominated by absorption and excellent thermal insulation performance (0.075 W/m·K).[5] Figure 10.42.2 illustrates the effect of Ni-chains.[5]

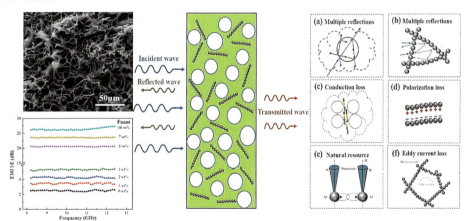

Figure 10.42.2. Schematic illustration of microwave transmitting across the PVDF/Ni-chains composite foams and related dissipation mechanism: (a) and (b) Multiple reflections by microcellular structure and Ni-chains; (c) Conduction loss by conductive Ni-chains pathway; (d) Interfacial polarization loss between insulator-conductor interfaces; (e) and (f) Magnetic loss including natural resonance and eddy current loss. [Adapted, by permission, from Zhang, H; Zhang, G; Bao, Q; Kim, J-K, *Chem. Eng. J.*, **379**, 122304, 2020.]

Polyvinylidenefluoride microfoam was produced using a chemical blowing agent in a continuous process.[6] With the use of 10% masterbatch formulation (Hydrocerol; azodicarbonamide, paraffin, and ZnO) (contains 2% chemical blowing agent in the final product), the cell density was increased while the cell size and foam density were decreased; the average cell size, cell density, and void fraction were 50 μm, 7.7×10^6 cells/cm^3, and 33%, respectively.[6] Decrease of die temperature from 135 to 130°C caused cell density to increase and cell size to decrease, while void fraction decreased from 58% to 39%.[6] This was due to the loss of melt strength upon increasing the temperature of the melt PVDF as it exited the die.[6]

REFERENCES

1. Siripurapu, S; Gay, Y J; Royer, J R; DeSimone, J M; Spontak, R J; Khan, S A, *Polymer*, **43**, 20, 5511-20, 2002.
2. Zerafati, S; Stabler, S M, **EP2449012**, *Arkema, Inc.*, May 9, 2012.
3. Yang, T; Jiang, Z; Han, H; Hu, J, *Compos. Part B: Eng.*, **205**, 108509, 2021.
4. Shi, Z; Ma, X; Zhao, G; Wang, G; Zhang, L; Li, B, *Mater. Design*, **195**, 109003, 2020.
5. Zhang, H; Zhang, G; Bao, Q; Kim, J-K, *Chem. Eng. J.*, **379**, 122304, 2020.
6. Sameni, J; Jaffer, S A; Tjong, J; Yang, W; Sain, M, *Adv. Ind. Eng. Polym. Res.*, **3**, 1, 36-45, 2020.

10.43 NATURAL RUBBER

Potassium oleate was used as a blowing agent to create open-cell natural rubber foam.[1] The addition of treated natural fibers (bagasse and palm oil fibers) resulted in a decrease in the average cell size and an increase in the number of foam cells but only up to 5 wt%.[1]

Benzenesulfonyl hydrazide was used as a blowing agent for natural rubber designated for thermal insulation for refrigeration and air conditioning systems.[2] The lowest value of thermal conductivity of 0.040 W/mK was obtained with 6 wt% of blowing agent.[2]

Acoustic absorbing foam materials are produced from dry natural rubber (NR) with the addition of sodium bicarbonate as a blowing agent.[3] The viscoelastic and damping properties are governed by the average cell size, relative density, crosslink density and number of cells per unit volume.[3] The lowest foaming temperature of 140°C yielded the natural rubber foam with the highest relative density, crosslink density, smallest average cell size, and greatest number of cells per unit volume, and, thus, having a superior sound absorption coefficient.[3]

Nanocomposite foams were fabricated from 60/40 wt% ethylene vinyl acetate/natural rubber blends by using azodicarbonamide as a blowing agent.[4] A nanofiller acted as a blend compatibilizer.[4] Sodium montmorillonite was more effective in compatibilization.[4] Nanofiller content of 5 phr was recommended for the production of EVA/NR nanocomposite foams.[4]

Microcellular natural rubber was prepared using supercritical CO_2 technology.[5] Cell size and size distribution decreased with increasing saturation pressure and time.[5] Crosslinking level of rubber reduced cell size and size distribution.[5]

REFERENCES

1 Tomyangkul, S; Pongmuksuwan, P; Harnnarongchai, W; Chaochanchaikul, K, *J. Reinf. Plast Compos.*, **35**, 8, 672-81, 2016.
2 Padakan, R; Radagan, S, *Agric. Natural Resources*, in press, 2016.
3 Najib, N N; Ariff, Z M; Bakar, A A; Sipaut, C S, *Mater. Design*, **32**, 2, 505-11, 2011.
4 Lopattananon, N; Julyanon, J; Masa, A; Kaesaman, A; Thongpin, C; Sakai, S, *J. Vinyl Addit. Technol.*, **21**, 134-46, 2013.
5 Tessanan, W; Phinyocheep, P; Gibaud, A, *J. Supercritical Fluids*, **149**, 70-8, 2019.

10.44 STARCH

Porous bionanocomposites based on halloysite nanotubes as nanofillers and plasticized starch as polymeric matrix were prepared by melt-extrusion.[1] Water was used as a natural blowing agent.[1] The expansion ratio and the porosity increased with increasing the die temperature.[1] Addition of more water permits reduction of the foaming temperature.[1] Halloysite gives double benefits: it acts as a nucleating agent, increasing the porosity and as a barrier agent increasing the proportion of small cells.[1] Nanocomposites were suitable for biomedical applications.[1]

Starch was converted into polyetherols by reactions with alkylene carbonates and used for the production of polyurethane foam.[2] The polyurethane foam had apparent density, water uptake, and polymerization shrinkage similar to the common polyurethane foams.[2]

Three-dimensional carbon/carbonaceous foams have attracted great research interest in recent years and have been widely used in many military and civilian fields.[3] Replacing traditional raw materials (e.g., pitches and petrochemical raw material derived resins) with renewable biomass resources is becoming one of the trends in research of carbonaceous foams.[3] Carbonaceous foam was prepared from starch.[3]

Starch foams with silanized starch addition showed lower water absorption, lower moisture absorption, higher density, and higher impact resistance.[4] Except for the increased density, all property changes were desirable in packaging.[4] Methyltrimethoxysilane presented higher surface hydrophobicity and lower water absorption.[4]

Starch was oxidized with hydrogen peroxide to different degrees of oxidation.[5] Oxidation caused chain scission that decreased paste viscosity.[5] Oxidation produced carboxyl groups that counteracted chain scission at low oxidation degree.[5] Oxidation improved starch foam expansion due to the decreased paste viscosity.[5] For the highest oxidant concentration (70%), the paste viscosity was too low to allow gas bubble entrapment; thus, the foam density increased.[5]

REFERENCES

1 Schmitt, H; Creton, N; Prashantha, K; Soulestin, J; Lacrampe, M-F; Krawczak, P, *J. Appl. Polym. Sci.*, **132**, 41341, 2015.
2 Lubczak, R; Szczęch, D; Lubczak, J, *Polym.Testing*, **93**, 106884, 2021.
3 Fu, C; Guo, S; Lei, H; Huo, J, *J. Anal. Appl. Pyrolysis*, **149**, 104858, 2020.
4 Bergel, B F; Araujo, L L; Campomanes Santana, R M, *Carbohydrate Polym.*, **241**, 116274, 2020.
5 Barbosa, J V; Martins, J; Magalhães, F D, *Ind. Crops Products*, **137**, 428-35, 2019.

11

ADDITIVES

Many groups of additives are included in the foaming formulations. They can be roughly divided into additives that affect the decomposition of a foaming agent, affect the chemical properties of the polymer, and participate in the physical modification of plastic material that is processed by foaming. The last group of additives is the largest. The additives involved in foaming formulations are discussed below in alphabetical order.

11.1 ACTIVATORS, ACCELERATORS, AND KICKERS

The above three names are given to additives that perform the same function; namely, they affect the decomposition temperature of the solid blowing agents. In some cases, they are called activators since they initiate the production of gas (although at a lower temperature than the decomposition temperature of the foaming agent alone). They are also called accelerators because at any given temperature above the decomposition temperature, they accelerate the production of gas. And finally, in PVC processing, some PVC thermal stabilizers are called kickers because they "kick" the decomposition of foaming additive (usually azodicarbonamide) lowering it, relative to the type of kicker and its amount.

Chapter 2 showed that the following activators were used in practical formulations: carbamide (urea compound), zinc 2-ethylhexanoate, zinc benzenesulfinate, zinc carbonate, zinc ditolyl sulfinate, zinc oxide, and zinc stearate. Out of this range, zinc oxide is the most common.

In the production of medium density polyethylene parts by rotational molding, azodicarbonamide was used as a foaming agent and zinc oxide to decrease its decomposition temperature.[1] Zinc oxide was also used for the same purpose in PVC formulation.[2]

Azodicarbonamide decomposes on heating to give a high volume of gas, which consists mainly of nitrogen and carbon monoxide.[3] These decomposition products are suitable to create a fine and uniform cell structure of foam with a little tendency to shrink, a property that is fundamental in the production of soft foams, such as plasticized PVC or rubber foams.[3] The decomposition temperature of azodicarbonamide can be reduced from 200-220°C to as low as 125°C by the addition of suitable activators (kickers), but useful decomposition rates are usually only achieved at 140°C and above.[3]

In polyurethanes, the activator is usually added to lower temperature of decomposition of blowing agent that gas production will take place before crosslinking.[4]

In foaming EVA formulations, blowing accelerators may be used if a selected blowing agent reacts or decomposes at temperatures higher than 170°C, such as 220°C or more, at which the surrounding polymer would be degraded if heated to this high activation temperature.[5] Blowing accelerators may include cadmium salts, cadmium-zinc salts, lead salts, lead-zinc salts, barium salts, barium-zinc salts, zinc oxide, titanium dioxide, triethanolamine, diphenylamine, or sulfonated aromatic acids and their salts.[5]

Activators and accelerators are also used in epoxy resins, EPDM, NBR, PVC/NBR, rubber, and SBR. Acetic acid was used as an activator in the production of hard epoxy foam using sodium bicarbonate as a blowing agent.[6] It was added in proportions from 1:1 to 1:4 of activator:blowing agent.[6] Acetic acid had promoted the production of CO_2 from the decomposition of sodium bicarbonate.[6] The density, mechanical properties, and crosslink density decreased with the incorporation of acetic acid.[6]

Introduction of relatively new plasticizer (DINCH) showed that its lower compatibility with poly(vinyl chloride-co-vinyl acetate) creates a problem in foaming because the plastisols prepared with DINCH were not able to withstand the pressure evolved by the released gases during the foaming process, which resulted in a poor quality foam as compared to other plasticizers.[7] However, the selection of the proper amount of kicker (ZnO) helped to delay the decomposition process of the foaming agent (azodicarbonamide) to allow this plastisol to develop the required melt strength and consequently lead to foams of good quality.[7] Zinc oxide was used as a kicker to decrease the decomposition temperature of the blowing agent to 170°C in the production of starch foam by extrusion.[8]

Barium/zinc carboxylate is known to be a fast kicker and thermal stabilizer of PVC formulations.[9] Metals are used in this kicker in the following proportions: Ba:Zn=3:1 to 3:5.[9] Typical application of this kicker is in the production of artificial leather and wallcoverings where the high speed of the process is important.[9] Ca/Zn stabilizers are very common, non-toxic stabilizers, and they also reduce the temperature of decomposition of foaming agent.[7] They are slower than Ba/Zn kicker but act at lower temperatures.[9] K/Zn carboxylates are another group of common kickers.[9]

In PVC plastisols processed by extrusion, the azodicarbonamide product used was Genitron SCE, which consisted of 75 wt% azodicarbonamide and 25 wt% kicker (zinc oxide/zinc stearate).[10]

In foaming natural rubber, water incorporated into the rubber compound by direct mixing prior to pre-mixing with tetramethylthiuram disulfide was likely responsible for the formation of thick skin during foaming and curing because it decreased the efficacy of the accelerator functions of the tetramethylthiuram disulfide, thereby delaying the curing process in the outer region of the specimen.[11]

11.2 CATALYSTS

Unlike the previously discussed additives, catalysts act on the polymer to develop proper rheological properties sufficient to hold foaming gas within the polymer matrix and thus help to regulate the extent of foaming and the size of pores.

Pentamethyldiethyltriamine and dimethylcyclohexylamine were used in the reactive system as catalysts to control the rate of crosslinking and related size of foam cells.[12] Stannous octoate and triethylenediamine were also used for the same purpose in another polyurethane formulation when chlorofluorocarbons and hydrochlorofluorocarbons were used as blowing agents.[13] Bismuth-based catalysts were used in polyurethane systems blown with hydrohaloolefins.[14] Amine catalyst system was developed to improve stability of polyurethane systems using halogen-containing blowing agents.[15] Water-blown, sprayable PU foam was catalyzed with a catalyst comprising a tertiary amine, a carboxylic acid salt of an alkali metal, a salt of an N-[(2-hydroxy-5-alkylphenyl)alkyl]-N-alkylamino carboxylic acid and a bis(dialkylamino)alkyl ether.[16]

Resole can be cured at 150-250°C without a cure accelerator, but a long time was needed to complete the cure and the slow reaction permitted air bubbles to escape from the resin.[10] p-Toluenesulfonic acid and nitric acid were used to increase the cure rate.[17]

Polysiloxane foams as precursors to silicon oxycarbide foams were prepared *via* simultaneous cure and foaming of liquid methylsiloxane resins, using the environment-friendly ethanol as a liquid blowing agent and concentrated aqueous ammonia as a catalyst.[18]

In the production of polyurethane foam from polyetherols from starch, triethylamine was used as a catalyst.[19] Increase in catalyst percentage resulted in fast growth time, pore elongation, and decrease in compression strength, while at lower catalyst percentage, the viscous surface of foaming mixture and under-crosslinking were obtained.[19]

11.3 CROSSLINKING AGENTS

The final structure and performance of porous materials greatly depend on crosslinking methods during the gelation process.[20] Physical crosslinking is generally based on weaker forces, such as hydrogen bonding and electrostatic interaction.[20] While chemical crosslinking usually produces stronger interaction and network, such as covalent bonding and polymerization.[20] Hydrogen bonding between nanocellulose molecules or between nanocellulose and other additives is the most commonly used gelation method.[20] The electrostatic interaction usually occurs between the negatively-charged nanocellulose and positively-charged crosslinking agents.[20] Divalent cations are widely used to form ionic crosslinking between anionic nanocellulose *via* electrostatic action.[20] Covalent crosslinking is formed through the action of covalent bonds to achieve crosslinking between nanocellulose and reactive coupling agents.[20] Both covalent and polymerization crosslinking achieve gelation by the formation of new covalent bonds, which can significantly enhance the mechanical strength of the network.[20]

A method of recycling scrap of ethylene-vinyl acetate foam involves foaming and crosslinking a blend containing virgin EVA resin and scrap EVA foam, as well as a crosslinking agent and blowing agent for the EVA resin.[21] Dicumyl peroxide is a suitable peroxide crosslinking agent.[21] Azodicarbonamide in the form of a masterbatch containing 50% EVA was used as a blowing agent.[21] Similar formulation was used to manufacture the spongy soles, in which EVA was a biodegradable resin, azodicarbonamide the foaming agent, dicumyl peroxide a crosslinking agent, and ZnO an accelerator.[22]

Low-density shoe soles can also be obtained from polyurethane containing crosslinking and chain extending agent, the proportions of which regulate the compression set.[23] Water was a blowing agent in this technology.[23]

Hybrid spray foams utilize a urethane reactant, a crosslinker, and an (optional) epoxy and/or acrylic resin along with a blowing agent and rheology modifier to produce a quick-setting foam that remains in place until the foam forms and cures.[24] The blowing agent is a low-boiling point fluorocarbon.[24] The polyfunctional amine crosslinking agent is a part of this two-component system.[24]

Radiation crosslinking improved the foaming of atactic polypropylene by the supercritical carbon dioxide and led to a wider temperature window for foaming.[25] The cell structure of polypropylene foam was more uniform, and the foaming temperature window was increased by 10°C.[25] The best cell morphology was observed at a dose of 30 kGy, giving an average diameter and cell density of 16.4 μm and 5.7×10^7 cells/cm^3, respectively.[25]

A biodegradable poly(butylene succinate) foam was produced by compression molding using dicumyl peroxide as a crosslinking agent and trimethylolpropane trimethacrylate as a curing coagent.[26] Zinc oxide/zinc stearate were used to reduce the thermal decomposition temperature of azodicarbonamide to balance vulcanization of resin and the decomposition of blowing agent.[26]

Silane crosslinked polypropylene was prepared by grafting of silane onto the polymer backbone in a melt process using a twin-screw extruder and then crosslinking in warm water.[21] Benzoyl peroxide was used as the initiator, silane as a monomer, styrene as assisted crosslinking agent.[27] Azodicarbonamide was used as the blowing agent and talcum powder as the nucleation agent.[27] Crosslinking improved foamability of polypropylene and the quality of foam.[27]

Peroxide crosslinking of PVC foam was used to improve the dimensional stability at higher temperatures and to increase the resistance to mechanical stresses that produce deformation.[28] Trimethylolpropane-trimethacrylate, 2,5-dimethyl-2,5-bis(t-butyl peroxy) hexane, and 6-dibutylamine-1,3,5-triazine-2,4-dithiol were used as crosslinking agents and azodicarbonamide as blowing agent.[28]

Crosslinking of the conjugated polymer chains by phytic acid has been widely used to make porous and stable networks.[29] The introduction of phytic acid can efficiently prevent aggregation of particles, the formation of the impenetrable film, and to provide interconnected polypyrrole conductive network with porous morphology.[29] Hierarchical structure in the interconnected porous network improved electron transfer properties by facilitating rapid ionic and electronic mobility and increasing charge storage ability.[29]

Multilayer foam structure comprising at least one foamed polypropylene/polyethylene layer with TPU cap layer has been produced by coextrusion combined with irradiation.[30] Ionizing radiation is often unable to produce a sufficient degree of crosslinking in polypropylene and polyethylene-based materials.[30] Thus, crosslinking promoters can be added to the compositions that are fed into the extruders to promote crosslinking.[30] Chemical crosslinking typically involves peroxides, silanes, or vinylsilanes.[30] In peroxide crosslinking processes, the crosslinking typically occurs in the extrusion die.[30] For silane and vinylsilane crosslinking processes, the crosslinking typically occurs during post extrusion in a secondary operation where the crosslinking of the extruded material is accelerated with heat and moisture.[30]

Foamed sealing compound (applied to vehicle tires for the reduction in weight and improvement of driving dynamics) was obtainable by emulsion polymerization of the conjugated diene in the presence of crosslinker I (trimethylolpropane trimethacrylate) and crosslinker II (divinylbenzene).[31]

The examples of crosslinking during foaming are just a few of many existing contributions. They show that crosslinking is an essential part of the foaming process, which is credited with the improvement of the physical-mechanical properties of foams.

11.4 CURING AGENTS

Curing agents and crosslinking agents are frequently just a name difference, as the above information about poly(butylene succinate) shows.[26] Similarly, in polyurethane systems,

11.4 Curing agents

the crosslinking reactions are part of the curing system.[23-24] There are still many curing reactions, which have not been mentioned in the previous section, and they are included under this heading.

Sulfur is the most common crosslinking agent used in the vulcanization of various kinds of rubber. It links one chain to another through unsaturations in the rubber chains. In the vulcanization of EPDM rubber compounds, the blowing agent presence seems to make a difference.[32] The blowing agent induces several differences to the vulcanization reaction: it decreases reaction temperature while increasing reaction heats, it eliminates the exothermic peak before vulcanization and decreases the fully cured resin's glass transition temperature, and it shortens the operational window.[32]

A composite curing agent (hydrochloric acid/phosphoric acid/p-toluene sulfonic acid/water 58/4/6/3) was used for curing phenolic foams, and petroleum ether was a blowing agent.[33] A phenolic foam was made by foaming and curing a foamable phenolic resin composition that comprised a phenolic resin, a blowing agent, an acid catalyst, and an inorganic filler. The blowing agent comprised an aliphatic hydrocarbon containing from 1 to 8 carbon atoms.[34] The acid curing catalyst initiated polymerization of the phenolic resin.[34]

Important properties to be considered before selection of binder material for syntactic foams are low viscosity, readily controlled gelation time, small exothermal effect during curing, low curing shrinkage, good adhesion, wettability, and compatibility with modifiers and fillers.[35] Appropriate matrix selection contributes to strength, stiffness, and bonds untreated hollow spheres towards matrix, improves the matrix rheology and improves curing temperature, time and degree of curing, hydrolytic stability in deep water application, and cost.[35]

Expandable microspheres/epoxy foams with different densities and microstructures were prepared by changing the foaming temperature and the precuring extent.[36] The microstructure of foams had a homogeneous distribution of cells at high precuring extent and high foaming temperature, while small cell size at high precuring extent and low foaming temperature.[36]

Syntactic foams are lightweight composite materials for spacecraft protection capable of attenuating shock waves.[37] The syntactic foam had a density of 0.64 ± 0.02 g/cm^3 and 55 vol% of microspheres.[37] The matrix comprised of epoxy resin and curing agent and had density in the cured state of 1.15 ± 0.01 g/cm^3.[37] The glass microspheres had an average diameter of 83 μm and the wall thickness of 1 μm.[37]

The addition of filler to the epoxy matrix may affect curing process kinetics and resulting composite properties.[38] The carbon foam particles, regardless of their size distribution, made the curing process more complex and affected curing reaction parameters (accelerated cure).[38]

In epoxy composites reinforced with fluorinated single-walled carbon nanotubes, the curing agent should be carefully adjusted; otherwise, the fluorine on the carbon nanotube surfaces can transform into free radicals and break the chains of the diglycidyl ether of bisphenol F, thereby decreasing crosslink density, glass transition temperature, and decomposition temperature of the epoxy matrix.[39]

11.5 DILUENTS

Diluents, as examples below show, may play different roles in the foaming processes, leading to essential modifications of the foam structure. Carbon dioxide dissolves in polyols and isocyanates and has a dilution effect on the reaction system, which could slow down the reaction rate of polymerization by hindering the molecular collisions according to the theory of diffusion-controlled reactions in the presence of a diluent.[40]

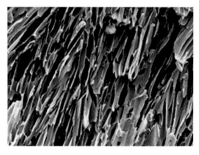

Figure 11.1. SEM images of channel-like structures in polyacrylonitrile foam. [Adapted, by permission, from Wu, Q-Y; Wan, L-S; Xu, Z-K, *Polymer*, **54**, 284-91, 2013.]

The polyacrylonitrile foams were prepared using thermally induced phase separation in the presence of dimethyl sulfone as a crystallizable diluent.[27] The channel-like structures stem from the spherulitic orientation of dimethyl sulfone crystals in the polymer matrix (Figure 11.1).[41] The dimethyl sulfone crystals are capable of acting as *in situ* formed templates, which subsequently enable to shape the final pore structure of polyacrylonitrile foams.[41] PAN foams are useful as precursors for pyrolyzing into carbon monoliths with different macroporous structures.[41] The macroporous carbon monoliths can be used in various fields such as redoxase immobilization, filtration, adsorption, and catalyst support.[41]

Tris(1-chloro-2-propyl) phosphate was used as a diluent/plasticizer in polyurethane foams.[42] This is a frequent application of plasticizers in polyurethanes. Many formulations have too high a viscosity for easy processing, and therefore, they require inert diluent that does not interfere with cure characteristics. At the same time, the selection of plasticizer should include potential benefits of more elastic networks, but it should limit the potential for plasticizer to exude from three-dimensional networks.

The glass transition temperature of polymer materials was studied in a batch microcellular foaming process by carbon dioxide or nitrogen.[43] The glass transition temperature of polymers drops to the room temperature in this process, and it is governed by the following relationship:[43]

$$T_g = T_{go} \exp[-M_p^{-1/3} \rho^{-1/4} \alpha w] \qquad [11.1]$$

where:
- T_g glass transition temperature of polymer material containing diluents or gas
- T_{go} glass transition temperature of original polymer material
- M_p molecular weight of polymer material
- ρ specific gravity of polymer material
- α material constant between gas and polymer
- w weight gain ratio for gas absorption in %.

Chow's equation for polymer-diluent mixtures has been used to correlate the decrease of glass transition temperature.n The sorption of CO_2 into the polymer matrix (poly(lactic-co-glycolic) acid) promoted plasticization resulting in a lower glass transition temperature.[44]

Figure 11.2. Intercalation/exfoliation process of montmorillonite in polystyrene and polypropylene nanocomposites. [Adapted, by permission, from Istrate, O M; Chen, B, *Polym. Int.*, **63**, 2008-16, 2014.]

11.6 EXFOLIATED ADDITIVES

Figure 11.3. Mechanism for the exfoliation of montmorillonite in the presence of blowing agent during an extrusion process. [Adapted, by permission, from Istrate, O M; Chen, B, *Polym. Int.*, **63**, 2008-16, 2014.]

Laminar fillers such as graphene and nanoclay dispersed in a polymer matrix could be found in three forms: aggregated, intercalated, and exfoliated.[45]

A blowing agent was used to enhance exfoliation in polymer nanocomposite.[46] Polypropylene or polystyrene form polymer matrix for this composite.[46] Nanomer® I.44P (a dimethyl dihydrogenated tallow ammonium chloride modified montmorillonite) is nanofiller.[46] Azodicarbonamide was used as a blowing agent with ZnO as the activator.[46] Figure 11.2 shows the steps of intercalation and exfoliation, and Figure 11.3 summarizes the mechanism of the process.[46] The degree of exfoliation may be controlled by the melt-

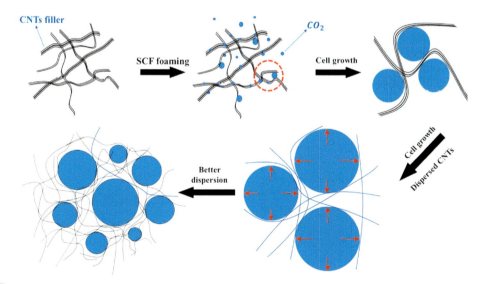

Figure 11.4. Mechanism of CNTs dispersion during SCF foaming process. Due to gas expansion, agglomerates break up. Equibiaxial flow occurs on the surface of bubbles and in struts as bubbles expand, which is extensional in nature and entails a greater viscosity than shear flow, leading to imparting a greater force on agglomerates. [Adapted, by permission, from Huang, A; Lin, J; Chen, S; Fang, H; Wang, H; Peng, X, *Compos. Sci. Technol.*, **201**, 108519, 2021.]

compounding parameters.[46] The process just described may be a convenient and inexpensive method of production of nanocomposites.[46]

The foaming process can effectively separate the aggregated conductive fillers by causing the polymer matrix to swell, resulting in a more uniform and expanded conductive path and higher electrical conductivity.[47] Figure 11.4 shows the mechanism of increased filler (such as CNTs) exfoliation and dispersion during the foaming process.[47]

11.7 FIBERS

The cell density of cellulose fiber-reinforced polypropylene foam increased with the increase of the cellulose content.[48] The void fraction at the 40 wt% cellulose was higher than that at the 20 wt% cellulose.[48] The cellulose fiber had a nucleating effect in the poly(lactic acid) foams-blown by supercritical carbon dioxide and nitrogen.[49] At low fiber loadings, the addition of treated bagasse and palm oil fibers to the natural rubber foam caused a decrease in the average cell size and an increase in the number of foam cells, but when its load increased above 5 wt%, the cell size and the cell number had the opposite trends.[50] The addition of fibers enhanced the sound absorption efficiency of natural rubber foams.[50] Cellulosic fibers incorporated into poly(lactic acid) in co-rotating twin-screw extruder and subsequently microcellular injection molded using dissolved nitrogen acted as crystal nucleating agents.[51] They increased crystallization temperature and crystallinity and decreased crystallization half time.[51] A finer and more uniform cell structure was achieved in the cellulosic fiber composite foams, which was attributed to the nucleating effect of the fibers and the increased melt strength.[51]

Microcellular injection molding of polypropylene and glass fiber composites was performed using supercritical nitrogen as the physical blowing agent.[52] The cell morphology and the glass fiber orientation of parts were influenced by cooling and shear effects.[52] The mechanical properties of glass fiber composites were enhanced by improved cell morphology, dispersion state, and orientation of the glass fiber.[52] The optimum amount of glass fibers to achieve these improvements was 11.8 wt%.[52]

Polymethylmethacrylate/multiwalled carbon nanotube composite foams were fabricated using a fully automatic and controllable supercritical carbon dioxide foaming system.[53] Foaming reduced the electromagnetic reflection by 60% and enhanced the specific electromagnetic absorption by 96%.[53] The electromagnetic percolation threshold was cell shape-dependent.[53] Electromagnetic absorption was improved by increasing cell density and decreasing cell size.[53]

Foaming increased the percolation threshold, reduced DC and AC conductivities, widened the insulator-conductor transition window, and reduced the dissipation factor of the polystyrene/multiwalled carbon nanotubes composites.[55] These results were attributed to deterioration of conductive network and inferior dispersion and distribution of carbon nanotubes caused by the presence of foam cells in nanocomposites.[55]

The cream time, rise time, tensile, compressive strength at room temperature, glassy and rubbery state moduli, glass transition temperature, and shape memory increased with an increased amount of multiwalled carbon nanotubes in polyurethane foam.[56]

When shear and extensional forces were applied to poly(ethylene-co-octene)/multiwall carbon nanotube nanocomposites, which were prepared by a melt blending process, they facilitated the orientation of nanotubes in the composite.[57] During foaming with supercritical carbon dioxide carbon, nanotubes oriented on cell walls.[57] The strain hardening, due to the presence of carbon nanotubes, acted as a self-reinforcing element, protecting cells from destruction during their growth.[57] A dramatic decrease in the open cell content and a high cell density were observed in composition, unlike in the copolymer foamed without the presence of carbon nanotubes.[57]

The dispersion of nanoparticles in PVC medium was increased by functionalization of multiwalled carbon nanotubes in sodium hypochlorite solution, and the morphological properties of foams containing functionalized nanoparticles were improved.[58] The density of nanocomposite foams decreased with functionalized carbon nanotubes.[41]

In high-pressure foam injection molding, polystyrene/carbon-fiber exhibited both translational and rotational displacements in the close proximity to the growing cells, caused by the melt's biaxial stretching.[54] These displacements were a strong function of the cell size, the initial cell-fiber distance, and the initial fiber angle.[58] The fiber length had little or no effect on the fibers' orientation and translation.[58]

Cellulose fibers in poly(lactic acid)/poly(butylene succinate) increased the number of closed cells and made cell distribution more uniform.[59] Composites had shear thinning behavior, caused by the orientation of fibers in the flow direction.[59] A lower content of cellulose fibers in the blend contributes to a lower melt viscosity in the whole range of shear rates.[59] The crystallinity of composite foams had an increased degree of crystallinity when cellulose fibers were present.[59]

Ultra-porous and lightweight epoxy foam composites loaded with carbon fibers were obtained using 15 wt% acetone (the density of epoxy foam was 0.05 g cm^{-3}).[60]

11.8 FILLERS

Numerous fillers, such as calcium carbonate, organoclay nanofillers, nanosilica, halloysite, nanocrystalline cellulose, graphene oxide, hydroxyapatite, wood flour, bark powder, starch, and talc were compounded into compositions subjected to foaming, and their effects are discussed below.

Styrene-butadiene-styrene and calcium carbonate nanoparticles were melt-blended with polystyrene and extrusion-foamed with supercritical carbon dioxide to toughen and reinforce polystyrene foams.[61] With the addition of 5% $CaCO_3$ nanoparticles, the cell diameter was reduced by 72.2%, and the compressive modulus was improved by 379.2%.[61] Figure 11.5 shows that a gradual increase in the calcium carbonate addition causes a proportional decrease in the cell size but only up to 5 wt%.[61] There are two reasons for the cell diameter decrease and cell density increase: nucleation by nanoparticles, which reduces the critical energy barrier for bubble nucleation and significantly enhanced system viscosity with the addition of filler, which increases surface tension, which the bubble needs to overcome to expand, resulting in smaller cells.[61] In iPP/nano calcium carbonate, the bubble size of the microcellular samples could be effectively decreased while the cell density increased, especially at high carbon dioxide concentration and backpressure, low mold temperature and injection speed, and high filler content.[62] Low-density polyethylene containing nano-calcium carbonate (0.5-7 phr), foamed with azodicarbonamide had the lowest mean cell diameter of 27 μm and the largest cell density of 8x10^8 cells/cm^3 at a concentration of nanofiller at 5 phr.[63]

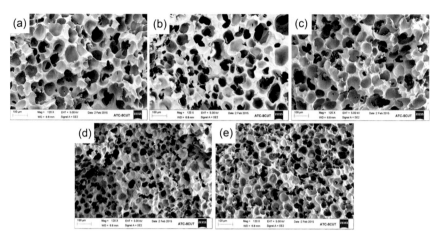

Figure 11.5. SEM images of the extrusion-foamed samples: (a) PS/SBS, (b) 1% $CaCO_3$, (c) 3% $CaCO_3$, (d) 5% $CaCO_3$, and (e) 10% $CaCO_3$. [Adapted, by permission, from Jing, X; Peng, X-F; Mi, H-Y; Wang, Y-S; Zhang, S; Chen, B-Y; Zhou, H-M; Mou, W-J, J. Appl. Polym. Sci., 135, 42508, 2016.]

A blowing agent was developed to contain nanoparticles of an inorganic carbonate and an acid or its salt.[64] The acid (salt) can be liquefied by heating to react with the carbonate to release carbon dioxide.[64] The blowing agent, optionally pelletized in a suitable polymer carrier, is useful as a heat-activated blowing agent additive in polymer compositions.[64] The polymer matrix is made out of polyolefin or polystyrene, carbonate is calcium carbonate, and the acid salt is sodium citrate.[64]

Nanocomposite foams were fabricated from 60/40 wt% ethylene vinyl acetate/natural rubber blends using azodicarbonamide and two nanofillers (sodium montmorillonite and organoclay).[65] Nanofiller acted as a blend compatibilizer (sodium montmorillonite more effective).[65] Addition of nanofiller (sodium montmorillonite intercalated with bis-(hydroxyethyl terephthalate)) led to a drastic decrease of melt viscosity, which suppressed foaming.[66] Nanoclays had better interaction with polypropylene in PP/EPDM blends and were located primarily in the polypropylene phase.[67] Addition of small amount of nanofiller improved foam structure.[67] Ethylene-vinyl acetate/montmorillonite nanocomposites were injection molded by a conventional and microcellular method using nitrogen as a blowing agent.[68] Cell size decreased as the clay loading increased.[68] The surface free energy of nanoclays is a key parameter in controlling the localization of the particles in the polyurethane matrix.[69] The surface free energy determines the dispersion of the nanoclays before foaming and the available surface area for heterogeneous nucleation during foaming.[69] The nanoclay particles, which get trapped at the polyurethane-air interface during foaming, yield improved mechanical properties.[69] The interfacial adsorption stabilizes the cell wall during cell growth, leading to smaller cells.[69] The optimal foam density (0.1-0.2 g cm^{-3}) and cell size (<100 nm) are required to obtain a polymer foam with an effective thermal conductivity lower than that of air.[70] To maintain cell size below 100 nm, nucleation must increase its density to 10^{15} cm^{-3}.[70] It is possible to achieve this by combining high homogeneous and heterogeneous nucleation mechanisms in nanostructured polymeric materials.[70]

11.8 Fillers

Poly(ether imide) foams were extruded using a single screw extruder with pressurized carbon dioxide used as a blowing agent.[71] Nanosilica particles (0.08-0.6 wt%) increased only the density of foam but did not increase the nucleation rate and the pore size distribution, most likely because of their poor dispersibility and agglomerated state in the single screw extruder.[71] Nanosilica modified by vinyltriethoxysilane was used in polystyrene composite foamed by supercritical carbon dioxide.[72] The size of the pores decreased, and the cell density increased with the decrease in nanosilica size and the increase in silica loading.[72] The surface treatment of the nanosilica particles had a substantial effect on the decrease of the cell size and the increase of the cell density.[72] The isocyanate-functionalized silica nanoparticles were incorporated into polyurethane during the synthesis of foam from polypropylene glycol and toluene diisocyanate using water as the blowing agent.[73] The shape recovery and the strain energy storage significantly increased with reduced hysteresis loss due to the incorporation of modified silica.[73] Modification improved substantially silica dispersion.[73]

By producing blowing agent concentration or temperature gradient inside the sample, polymeric foams with a gradient of cell size can be produced.[74] Anodized aluminum oxide, AAO, film modified with a fluorinated silane was used as the substrate.[74] Polystyrene, polymethylmethacrylate, and poly(lactic acid) were compressed onto the modified aluminum oxide film and then foamed by a batch foaming process using supercritical carbon dioxide as a blowing agent.[74] Figure 11.6 shows results for polystyrene, but similar results were obtained for other polymers with a difference for the ranges of cell sizes.[74] By changing the properties of anodized aluminum, cell sizes can be modified.[74]

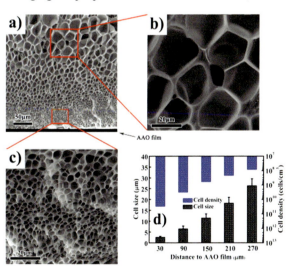

Figure 11.6. a) SEM images of PS foams on AAO film with pores of ~150 nm (saturating pressure of 13.8 MPa and foaming temperature of 100°C) and b) top and c) bottom partially enlarged SEM images, and d) average cell size and cell density at different positions. [Adapted, by permission, from Yu, J; Song, L; Chen, F; Fan, P; Sun, L; Zhong, M; Yang, J, *Mater. Today Commun.*, **9**, 1-6, 2016.]

Biodegradable poly(butylene succinate)/halloysite nanotubes (1-7 wt%) composites were foamed using supercritical carbon dioxide as a physical blowing agent.[75] The cell size decreased, and both cell density and volume expansion ratio increased with the addition of halloysite.[75] The smallest cell size and the highest cell density, and the volume expansion ratio were obtained with 5 wt% halloysite.[75] Bionanocomposites halloysite/plasticized starch were prepared by melt-extrusion using water as a natural blowing agent.[76] The addition of halloysite has two benefits: nucleation and a role of barrier agent increasing the proportion of small cells.[76] The plasticized starch with a blend of glycerol and sorbitol containing 6 wt% of halloysite, extruded at 117°C, had a cellular structure and the mechanical properties required for scaffold applications.[76] A pressure-sensitive adhesive foam comprising a non-syntactic foam blend of styrenic block copolymer and acrylic copolymer, which has an elongation greater than 600%.[77]

Polypropylene/nano-crystalline cellulose composite foams were produced by extrusion compounding combined with injection molding, using azodicarbonamide as foaming additive.[78] The nano-crystalline cellulose reduced cell size by 42-71% and increased cell density by 5-37 times.[78] Low-density green polyurethane foams were reinforced with nano-crystalline cellulose and blown using water.[79] Addition of nanofiller in small amounts helps to increase modulus at low temperature.[79]

The foaming time and foam setting time decreased with an increase in graphene oxide concentration because the acidic functional groups present on graphene oxide catalyzed the –OH with –OH condensation, which was responsible for the foaming.[80]

Ethylene-vinyl acetate copolymer/wood-flour foams were obtained with azodicarbonamide.[81] The shape of foams containing 20% wood-flour were more uniform with increasing content of stearic acid (5 wt% is an optimum).[81] With increasing content of stearic acid, more gas remained in the matrix, and the average cell size and density increased.[81]

A scaffold was developed by combining microwave irradiation and gas foaming (sodium bicarbonate).[82] Chitosan matrix was reinforced with hydroxyapatite.[82] Rapid heating with microwave irradiation enhanced gas blowing and increased pH, which provided better crosslinking.[82] So obtained scaffold had superior properties for bone tissue engineering.[82]

Extrusion foaming using water as a blowing agent offers a green process of preparing polymer foams.[83] Water has limited solubility in synthetic polymers, but thermoplastic starch is an effective water carrier.[83] The polypropylene/thermoplastic starch blend foams with low density and high open-cell content are promising candidate absorbent materials for oil spill cleanup application.[83]

Thermally insulating polymeric foams improve energy-management efficiency.[84] To limit transmission of infrared radiation through such foams, strong IR-shielding carbon particles are often incorporated to block IR radiation.[84] But, carbon particles can also dramatically increase solid-phase conduction, which counteracts efforts of decreasing thermal conductivity.[84] Microcellular foam of superior thermal insulation (~28.5 mW·m^{-1}K^{-1}) was produced using ethylene-norbornene-based cyclic olefin copolymers.[84] Unlike the traditional carbon-filled approach, the incorporation of more norbornene segments (33-58 mol%) to cyclic olefin copolymer structure greatly improved its ability to block thermal radiation without increasing its solid thermal conductivity.[84] Using the supercritical CO_2 and n-butane as physical blowing agents, cyclic olefin copolymer foams were fabricated with tunable morphology.[84] The void fraction of these foams ranged from 50 to 92%, with a high degree of closed-cell content (>98%).[84] In cyclic olefin copolymer foams with given cellular structures (e.g., the void fraction of 90%, the cell size of 100-200 μm and cell density of ~10^7 cells/cm^3), their total thermal conductivity decreased from 49.6 to 37.9 mW·m^{-1}K^{-1} with increasing norbornene content from 33 to 58%, which was attributed to high-norbornene cyclic olefin copolymer's strong ability to attenuate thermal radiation.[84]

The 3D structure of foam facilitates the dispersion of fillers in the matrix and reduces the amount of filler.[85] Monomer casting polyamide-6/melamine foam/graphene oxide/paraffin wax composites were synthesized using melamine foam as a three-dimensional template and adsorbing graphene oxide and paraffin wax on the melamine foam skeleton.[85] The addition of both graphene oxide and paraffin wax significantly improved tribological properties of composites.[85] The improvement of tribological performance was attributed to the synergistic lubrication and anti-wear effects of paraffin wax and graphene oxide.[85]

11.9 FIRE RETARDANTS

The halogen-free flame retardant poly(vinyl alcohol) foam was prepared by continuous extrusion using water as a blowing agent.[86] Melapur MP 116 (melamine phosphate) was used as a flame retardant.[65] Melamine phosphate acted as a heterogeneous nucleating agent.[86] Tris(1-chloro-2-propyl) phosphate was used as a fire retardant in polyurethane foam.[87] The flame-retardant rigid polyurethane foams were produced using a combination of dimethyl methylphosphonate/[bis(2-hydroxyethyl)amino]-methyl-phosphonic acid dimethyl ester/expandable graphite.[88] The trinary synergistic effect was achieved in inhibiting flame and forming a phosphorus-rich char layer, thus forming stable flame-retardant action.[88]

Extruded polystyrene foams contained brominated styrene-butadiene polymer as a flame retardant.[89] The blowing agent was a mixture of carbon dioxide, ethanol, water, and a C4-C5 hydrocarbon.[89] The blowing agent mixture helps to overcome the tendency of the brominated styrene-butadiene to form very small cells.[89] The formulation allows to obtain good quality, low density extruded foam.[89]

The oxidative thermal decomposition of a non-fire retardant and a fire retardant polyurethane foams was investigated using thermogravimetric analysis and differential scanning calorimetry in both air and nitrogen.[90] Oxidative thermal decomposition of polyurethane foam was heating rate dependent.[90] Fire retardant additives form char residue, lowering decomposition rate, increasing decomposition temperature range, and improving thermal stability.[90]

Ultra-lightweight and porous cellulose foam, useful in a broad range of applications (e.g., structural materials, sound insulation, or thermal insulation), was produced with borate crosslinking to improve mechanical and fire retarding properties.[91] The pulp foam had a low density of 13.3-16.4 mg/cm^3, high porosity (>98%), and high compressional strength (up to 74.1 kPa, which was 28 times higher than that of pristine foam without borate).[91] The pulp foam containing boron had low thermal conductivity (about 0.045 W/m·K), improved flame retarding properties, and excellent self-extinguishing behavior (with boron content of 3.45 wt% completely incombustible).[91]

Phosphorus and nitrogen-containing flame retardant (1,4-bis(diethyl methylenephosphonate) piperazine) were used in flame retarded rigid polyurethane foams.[92] With the addition of 10 wt% expanded graphite and 15 wt% flame retardant, PU foam passed UL-94 V-0 with the LOI value of 25.7%.[92] The density and compression strength of pure RPUF were 45.42 kg m^{-3} and 213 kPa.[92]

11.10 FOAM STABILIZERS

The cell stabilizer (SureCell T-50) for cellular PVC is all acrylic foam cell promoter.[93] It provides an even expansion and superior cell uniformity across wide sheet dies with less shrinkage.[94] Cellular voids are reduced, and cellular PVC has a lower gloss that is easy to coat because of good ink and paint adhesion.[94] The stabilizer permits processing at a lower temperature (reduced energy consumption and lower risk of decomposition and yellowing).[94] It easily substitutes into existing formulations.[94]

To produce flexible polyolefin foam for thermal insulation, a cell stabilizer is required.[95] Cell stabilizer prevents gas of blowing agent from escaping from the polymer melt immediately after injection.[95] Examples of suitable cell stabilizers include stearic acid amide, glycol monostearate, and fatty acids of glycine.[95] The total quantity of cell stabilizer in the foam composition mixture was 2-5 wt%, based on the total quantity of polymers and additives.[95]

11.11 NUCLEATING AGENTS

Uniform formation of small cells is frequently one of the major goals of formulation and product development. For these reasons, nucleating agents are important additives used in foaming technologies. Major findings are discussed below.

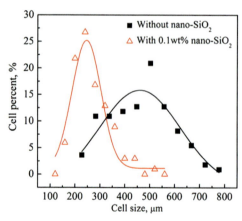

Figure 11.7. The influence of nano-SiO$_2$ on cell size distribution of PET foams. [Adapted, by permission, from Fan, C; Wan, C; Gao, F; Huang, C; Xi, Z; Xu, Z, Zhao, L; Liu, T, *J. Cellular Plast.*, **52**, 3, 277-98, 2016.]

Nucleation of poly(ethylene terephthalate) based on homogeneous nucleation suffers from a high nucleation energy barrier, which results in insufficient nucleation.[96] The addition of nucleating agent decreases the nucleation energy barrier and increases the density of nucleation.[96] Excessive nucleation may have an adverse effect on strain hardening.[96] Figure 11.7 shows that in the presence of a nucleating agent, cells are smaller and have narrower distribution.[96] In polystyrene foaming using supercritical carbon dioxide, the size of the pores decreases, and the cell density increases with the decrease in the nanosilica size and the increase of silica loading.[97] The surface treatment of the nanosilica particles by vinyltriethoxysilane have even more pronounced effect on the decrease of the cell size and the increase of the cell density than unmodified nanosilica.[97] Spherical ordered mesoporous silica particles were modified by selective grafting of polystyrene brushes on the outside surface, by which the mesoporous structure inside particles was maintained.[98] Such prepared particles exhibited an excellent heterogeneous effect on polystyrene foaming.[98]

PTFE acted as a crystal nucleating agent, and the PTFE fibrils induced PET crystallization along the PTFE fibrils surface.[99] This phenomenon was most evident when the con-

11.11 Nucleating agents

tents of PTFE was 1 wt%.[99] PTFE fibrils entanglement network promoted crystallization and enhanced the melt strength effectively, which improved foaming behavior.[99]

In microcellular injection molded polycarbonate parts, sodium chloride, and active carbon were used as nucleating agents and water as a foaming agent.[100] Water vapor-foamed polycarbonate, containing cubic sodium chloride, had a smooth surface comparable to that of solid parts, whereas polycarbonate foamed in the presence of active carbon had desirable mechanical properties.[100]

The extrusion foaming of poly(3-hydroxybutyrate-co-3-hydroxyvalerate) with a chemical blowing agent based on sodium bicarbonate and citric acid applied calcium carbonate as nucleation agent.[101] Higher screw speed generated higher pressure drop at the die exit, which enhanced cell nucleation and expansion.[101] The heterogeneous nucleation by the $CaCO_3$ decreased cell size and increased cell density.[101] Nano-calcium carbonate was also used in microcellular foaming of low-density polyethylene with the use of azodicarbonamide.[102] The formulation containing 5 phr nano-$CaCO_3$ had microcellular foam with the lowest mean cell diameter of 27 μm and the largest cell density of 8×10^8 cells/cm^3.[102]

In foam extrusion of thermoplastic cellulose acetate, HFO 1234ze was used as a low global warming blowing agent and talc as a nucleating agent.[103] A homogeneous, fine foam morphology with closed cells was developed with talc.[103] Depending on the blowing agent and talc contents, the average cell sizes ranged from 1 to 0.12 mm, and foam density ranged between 100 and 400 kg/m^3.[103]

Foaming isotactic polypropylene by foam injection molding requires nucleating agents to prevent the formation of large cell structures, which diminish the homogeneity of foams. Talc (2 wt%) and benzene trisamide-based nucleating agents can be used to control cell nucleation during the foaming.[104] Trisamide-based nucleating agents formed supramolecular nanostructures in the polymer melt, acting first as nucleating sites for foam formation and then as nuclei for the polymer crystallization.[104] Only 0.02 wt% of an additive is required to obtain a remarkable reduction in the cell sizes.[104]

Cell nucleation of thermoplastic polyolefin foam blown with nitrogen was studied by batch foaming simulation system with a visualization component.[105] At a low nitrogen concentration, talc had a significant effect on the cell density, but the cell density became insensitive to the talc concentration at a high nitrogen concentration.[105]

A cellulose fiber was used as a nucleating agent for poly(lactic acid) blown with supercritical gases, but a decrease in the strain at break followed.[106] A high talc content was necessary to achieve good quality foam sheets, but it led to a higher brittleness of the material.[106]

Liquid nucleating agent (perfluoroalkane) was used in rigid polyurethane foams blown with HFC-365mfc.[107] The addition of perfluoroalkane induced smaller cell sizes, most likely because of lower surface tension of polyol and perfluoroalkane mixture, resulting in a high nucleation rate.[107]

The main limitation of supercritical CO_2 used in the microcellular foaming industry is its solubility.[108] Four kinds of amines (formamide, ethylenediamine, polyamide dendrimer, and polyhedral oligomeric silsesquioxane) were introduced into polyvinylalcohol to improve foaming behavior.[108] Formamide and polyhedral oligomeric silsesquioxane plasticized PVAl, and formamide and ethylenediamine had a nucleating effect and improved the foaming effect.[108] Figure 11.8 illustrates the effects of amines with data.[108]

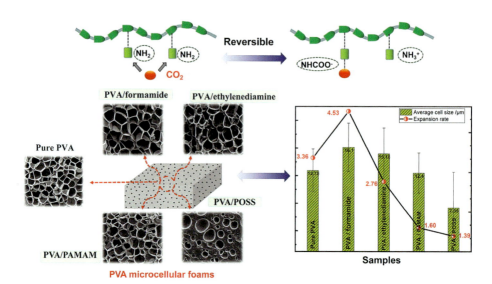

Figure 11.8. Effect of amines on PVAl foaming results. PAMAM – polyamide dendrimer; POSS – polyhedral oligomeric silsesquioxane. [Adapted, by permission, from Xiang, A; Yin, D; He, Y; Li, Y; Tian, H, *J. Supercritical Fluids*, in press, 105156, 2021.]

Nanocellular polymers were produced by extrusion using two nucleating agents (needle-like sepiolite and polymethylmethacrylate-polybutylacrylate-polymethylmethacrylate block copolymer).[109] Highly anisotropic nanocellular structures were obtained.[109] Cell sizes of 100-300 nm and anisotropy ratios as high as 2.77 were achieved.[109] Copolymer nanostructuration controlled nucleation and sepiolite increased anisotropy.[109] The alignment of the copolymer micelles and sepiolite in the extrusion direction promoted coalescence in the extrusion direction, leading to highly anisotropic nanocellular structures.[109]

Spring-like sandwich flexible foam composite with a 3D resilient concave-convex structured fabric core was obtained using two gradient-structured styrene-ethylene/butene–styrene copolymer-g-maleic anhydride-filled polyurethane foam by two-step foaming.[110] When the amount of SEBS-g-MAH was increased from 0 wt% to 4 wt%, the average pore sizes decreased from 0.13 mm to 0.09 mm, because SEBS-g-MAH increased viscosity of polyether polyol and acted as nucleating agent.[110] During polymer heterogeneous nucleation, the Gibbs free energy required for free foam nucleation decreased and prompted nucleation at the interface.[110]

From the above discussion, it is pertinent that nucleating agents are very important additives of foaming systems. It was not sufficiently stressed here that nucleating additives not only affect cell sizes but can also speed up production through nucleation of polymer crystallization. To learn more about additives and their applications, two specialized monographic sources can be consulted.[111,112]

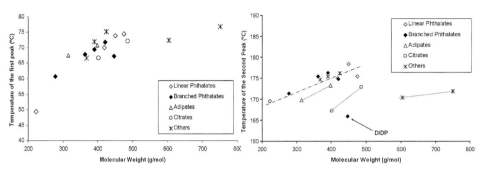

Figure 11.9. Temperature of the first peak (swelling) versus the molecular weights of plasticizers. [Adapted, by permission, from Zoller, A; Marcilla, A, *J. Appl. Polym. Sci.*, **121**, 3314-21, 2011.]

Figure 11.10. Temperature of the second peak (maximum decomposition rate of azodicarbonamide) versus the molecular weights of plasticizers. [Adapted, by permission, from Zoller, A; Marcilla, A, *J. Appl. Polym. Sci.*, **121**, 3314-21, 2011.]

11.12 PLASTICIZERS

Based on the studies of 20 plasticizers in azodicarbonamide foaming of PVC, two processes were found to impact foaming: the interactions of plasticizer with resin (swelling) and plasticizer effect on the decomposition of azodicarbonamide.[113] The heat evolved in the swelling process (swelling is an exothermic process) is almost linearly decreasing function of molecular weight of plasticizer (the relationship between molecular weight and heat evolved is best projected when compared within the same chemical group of plasticizers; for example, the heat in the case of adipates is slightly lower than that of phthalates) (Figure 11.9).[113] The maximum temperature of the azodicarbonamide decomposition (also exothermic process) is an increasing function of the molecular weight of the plasticizer (Figure 11.10).[113]

Studies of phthalate plasticizers, which are most frequently used in foams, give more background information on the subject of incorporation and selection of plasticizers for foaming formulations. Foams of good quality are prepared with plasticizers having smaller molecular weight because they are more compatible with the PVC-VA resins.[114] These plastisols undergo earlier gelation with elevated viscosities and fully developed properties at the temperatures of the azodicarbonamide decomposition.[114] Formulations based on less compatible plasticizers have late gelation with lower viscosity and slower bubble formation, thus resulting in poorer quality foams.[114]

Extensional viscosity is a fundamental characteristic of the material, responsible for the behavior of a system when undergoing the extensional stress produced by the released gases.[115] Depending on the structure and consequently, on the compatibility of a plasticizer, each plastisol develops its properties and structure accordingly.[115] Due to its behavior during the curing and foaming processes, formulations with some plasticizers may not be able to withstand the pressure formed by the released gases during the foaming process yielding foams of poorer quality.[115] The extensional viscosity is a very important property of PVC plastisols.[115] High extensional viscosity favors the formation of foams with a homogeneous bubble size distribution, providing that the plastisol has developed all the required properties at the moment of the gas generation.[115]

In plasticization of polycaprolactone with ethyl lactate for foaming with supercritical carbon dioxide, the premixing was found to be essential for obtaining a good quality foam.[116]

Production of expanded polystyrene using D-limonene as a plasticizer allows reduction of the amount of pentane in the process, yet permits generating beads with similar cell size as if greater quantities of pentane had been used.[117]

The presence of plasticizer (epoxidized soybean oil, Phosflex 71B HP, or bisphenol A bis-(diphenyl phosphate)) in combination with thermoplastic copolyester elastomer (55 wt% poly(tetramethylene oxide) soft segment and poly(butylene terephthalate) hard segment gave low-density foams, which exhibited less cracks.[118]

11.13 POLYMERIC MODIFIERS

Styrene-butadiene-styrene and calcium carbonate nanoparticles were melt-blended with polystyrene for extrusion-foaming using supercritical carbon dioxide to simultaneously toughen and reinforce polystyrene foams.[119] The cell structure was related to the system viscosity, by which toughening agent (SBS) also exerted a positive influence on foam quality in addition to nano-calcium carbonate that acted as nucleating agent.[119] Styrene-butadiene block copolymer was used as toughening additive and talc as a nucleating agent in the production of poly(lactic acid) foam with supercritical nitrogen.[106] The toughening additive acted as a nucleating agent, due to which the amount of talc could be reduced.[106] Polypropylene foam sheets containing impact modifier (Hytrel IM-1) were manufactured with excellent impact strength and low density, using supercritical carbon dioxide, for application in electronic packaging.[120]

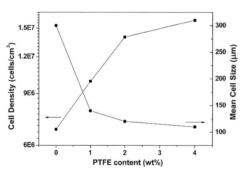

Figure 11.11. The cell size and cell density of liner polypropylene and linear polypropylene/PTFE blend foams obtained at 160°C. [Adapted, by permission, from Wang, K; Wu, F; Zhai, W; Zheng, W, *J. Appl. Polym. Sci.*, **129**, 2253-60, 2013.]

Poly(lactic acid) has poor melt strength, and for this reason, it is difficult to foam because of severe cell coalescence during foaming.[121] The addition polytetrafluoroethylene particles is known to stabilize its foams against bubble coalescence and collapse.[121] The foams are stable even when held under molten conditions for an extended period of time.[121] The foam stability is attributed to an interfacial mechanism: because of their low surface energy, the PTFE particles adsorb on the inner surface of the foam bubbles at a high surface coverage and endow the bubbles with an interfacial "shell" that prevents coalescence (see Figures 6.6 and 6.7 for the illustration of mechanism and demonstration of results).[121] Figure 11.11 shows the effect of PTFE particles on the development of cell size and cell density in the polypropylene foams.[122]

11.14 RETARDERS

Retarders are added to the formulation to slowdown crosslinking. Considering that the quality of foam depends on the balance and timing of development of rheological properties (e.g., melt strength) and decomposition rate of blowing agent (or gas formation rate), the retarders can provide good means to regulate this process. In sulfur-cured EPDM, rubber phthalimide plays the role of retarder, which can affect the timing of rheology development and blowing agent decomposition.[123] Unlike in mineral building materials (e.g., cementitious foams), where retarders are very common, in plastic foams, more options are still to be developed in the future since this method of regulating process kinetics seems to be very convenient.

11.15 SURFACTANTS

Phenolic foams were produced using petroleum ether as a blowing agent. Polysorbate 80 (non-ionic) was a surfactant (12%) added to maintain uniform cell structure.[124] NIAX L-6900 (silicone) was used as a surfactant in products of rigid urethane foams using blowing agents, such as methyl formate and C5-C6 hydrocarbons.[87,107] The hydrophile-lipophile balance value of this surfactant is 8.47.[107] NIAX silicone L-3627 is a low emission silicone surfactant for high resilience molded polyurethane foam.[125] It promotes optimum cell opening and provides stabilization.[125] NIAX silicone L-3629 surfactant was developed for use in molded polyurethane foam, where lower emissions were needed.[125] It provided excellent cell structure, decreased surface voids, and reduced vent collapse.[125] Evonik manufactures a series of silicone-based surfactants (Tegostab) used as stabilizers for rigid foam polyurethanes.

Surfactant (sodium dodecyl sulfate) facilitated the formation of cells by lowering the surface tension of the mixture of carbon fiber/solvent/surfactant/epoxy.[126] Inhomogeneous distribution of surfactant particles increased locally cell size during the foaming process and affected coalescence.[126]

REFERENCES

1　Moscoso-Sanchez, F J; Mendizabal, E; Jasso-Gastinel, C F; Ortega-Gudino, P; Robledo-Ortiz, J R; Gonzalez-Nunez; R; Rodrigue, D, *J. Cellular Plast.*, **51**, 5-6, 489-503, 2015.
2　Radovanovic, R; Jaso, V; Pilic, B; Stoiljkovic, D, *Hem. Ind.*, **68**, 6, 701-7, 2014.
3　Stevens Russel, **US20150267024**, *Lanxess Limited*, Sep. 24, 2015.
4　Limerkens, D; Lindsay, C I; Nijs, C; Woutters, S A, **EP20032621**, *Huntsman International LLC*, Mar. 11, 2009.
5　Delevati, G; Oviedo, M A S; Anderle, F M; Renck, O W; Viveiro, J A E, **US20190315948A1**, *Braskem SA*, Oct. 17, 2019.
6　Hamad, W N F W; Teh, P L; Yeoh, C K, *Polym.-Plast. Technol. Eng.*, **52**, 754-60, 2013.
7　Zoller, A; Marcilla, A, *J. Appl. Polym. Sci.*, **128**, 354-62, 2013.
8　Ghanbari, A; Tabarsa, T; Mashkour, M, *Carbohydrate Polym.*, **197**, 305-11, 2018.
9　Wypych, G, **PVC Degradation & Stabilization**, 4th Ed., *ChemTec Publishing*, Toronto 2020.
10　Malave, J A K; Cavalier, K; Pascal, J-P, **US20200165402A1**, *Solvay SA*, May 28, 2020.
11　Ariff, Z M; Afolabi, L O; Rodriguez-Perez, M A, *J. Mater. Res. Technol.*, **9**, 5, 9929-40, 2020.
12　Lee, Y; Jang, M G; Choi, K H; Han, C; Kim, W N, *J. Appl. Polym. Sci.*, **133**, 43557, 2016.
13　Long, Y; Sun, F; Liu, C; Xie, X, *RSC Adv.*, **6**, 23726, 2016.
14　Yu, B; Williams, D J, **WO2014133986**, *Honeywell International Inc.*, Sep. 4, 2014.
15　Burdeniuc, J J; Vincent, J L; Miller, T J, **WO2014066596**, *Air Products and Chemicals, Inc.*, May 1, 2014.
16　Albers, R; Klasen, P; Schneider, H-H; Vogel, V, **US20140162006**, *Bayer Intellectual Property GmbH*, Jun. 12, 2014.

17 Song, S A; Oh, H J; Kim, B G; Kim, S S, *Compos. Sci. Technol.*, **76**, 45-51, 2013.
18 Strachota, A; Cerný, M; Chlup, Z; Depa, K; Šlouf, M; Sucharda, Z, *Ceramics Int.*, **41**, 10A, 13561-71, 2015.
19 Lubczak, R; Szczęch, D; Lubczak, J, *Polym. Testing*, **93**, 106884, 2021.
20 Sun, Y; Chu, Y; Wu, W; Xiao, H, *Carbohydrate Polym.*, **255**, 117489, 2021.
21 Sui-chieh, J; Yu, X S, **WO2014039218**, *Nike International Ltd.*, Mar. 13, 2014.
22 Lee, S Y, **WO2014003376**, *Fine Chemical Co. Ltd.*, Jan. 3, 2014.
23 Bollmann, H; Kamm, A; Huprikar, A G; Holwitt, U, **US20130197118**, Aug. 1, 2013.
24 Olang, F N, **US10120183694**, *Owens Corning Intellectual Capital LLC*, Jul. 19, 2012.
25 Yang, C; Zhe, X; Zhang, M; Wang, M; Wu, G, *Radiat. Phys. Chem.*, **131**, 35-40, 2017.
26 Li, G; Qi, R; Lu, J; Hu, X; Luo, Y; Jiang, P, *J. Appl. Polym. Sci.*, **127**, 3586-94, 2013.
27 Li, H; Chuai, C; Igbal, M; Wang, H; Kalsoom, B B; Khattak, M; Khattak, M Q, *J. Appl. Polym. Sci.*, **122**, 973-80, 2011.
28 Yanez-Flores, I G; Ibarra-Gomez, R; Rodriguez-Fernandez, O S; Gilbert, M, *Eur. Polym. J.*, **36**, 2235-41, 2000.
29 Tiwari, A P; Chhetri, K; Kim, H Y, *J. Energy Storage*, **33**, 102080, 2021.
30 Ben-Daat, D; Baldwin, J; Bock, K M; Sieradzki, P, **US20180281260A1**, *Toray Plastics America Inc*, Oct. 4; 2018.
31 Zhou, J; Kohl, C; Früh, T, **US20190382547A1**, *Arlanxeo Deutschland GmbH*, Dec. 19, 2019.
32 Restrepo-Zapata, N C; Osswald, T A; Hernandez-Ortiz, J P, *Polym. Eng. Sci.*, **55**, 2073-88, 2015.
33 Ma, Y; Wang, J; Xu, Y; Wang, C; Chu, F, *J. Appl. Polym. Sci.*, **132**, 42730, 2015.
34 Coppock, V; Ruud Zeggelaar, R; Hiroo Takahashi, H; Toshiyuku Kato, T, **US8765829**, *Kingspan Holdings Limited*, Jul. 1, 2014.
35 Afolabi, L O; Mutalib, N A A; Ariff, Z M, *J. Mater. Res. Technol.*, **8**, 5, 3843-51, 2019.
36 Wang, L; Yang, X; Zhang, J; Zhang, C; He, L, *Composites Part B: Eng.*, **56**, 724-32, 2014.
37 Rostilov, T A; Ziborov, V S, *Acta Astronautica*, **178**, 900-7, 2021.
38 Szeluga, U; Olszowska, K; Tsyntsarski, B, *Wear*, **466-467**, 203558, 2021.
39 Bao, F; Wang, J; Wang, J; Zeng, S; Guo, X, *J. Mater. Res. Technol.*, **9**, 6, 12391-403, 2020.
40 Yang, Z; Liu, T; Hu, D; Xu, Z; Zhao, L, *J. Supercritical Fluids*, **147**, 254-62, 2019.
41 Wu, Q-Y; Wan, L-S; Xu, Z-K, *Polymer*, **54**, 284-91, 2013.
42 Long, Y; Sun, F; Liu, C; Xie, X, *RSC Adv.*, **6**, 23726-36, 2016.
43 Hwang, Y D; Cha, S W, *Polym. Testing*, **21**, 3, 269-75, 2002.
44 Álvarez, I; Gutiérrez, C; García, M T, *J. Supercritical Fluids*, **164**, 104886, 2020.
45 Lan, B; Li, P; Luo, Y; Gong, P, *Polymer*, **212**, 123159, 2021.
46 Istrate, O M; Chen, B, *Polym. Int.*, **63**, 2008-16, 2014.
47 Huang A; Lin, J; Chen, S; Fang, H; Wang, H; Peng, X, *Compos. Sci. Technol.*, **201**, 108519, 2021.
48 Kuboki, T, *J. Cellular Plast.*, **50**, 2, 113-28, 2014.
49 Geissler, B; Feuchter, M; Laske, S; Fasching, M; Holzer, C; Langecker, G R, *J. Cellular Plast.*, **52**, 1, 15-35, 2016.
50 Tomyangkul, S; Pongmuksuwan, P; Harnnarongchai, W; Chaochanchaikul, K, *J. Reinforced Plast. Compos.*, **35**, 8, 672-81, 2016.
51 Ding, W D; Jahani, D; Chang, E; Alemdar, A; Park, C B; Sain, M, *Compos. Part A: Appl. Sci. Manuf.*, **83**, 130-9, 2016.
52 Xi, Z; Sha, X; Liu, T; Zhao, L, *J. Cellular Plast.*, **50**, 5, 489-505, 2014.
53 Alkuh, M S; Famili, M H N; Shrivan, M M M; Moeini, M H, *Mater Design*, **100**, 73-83, 2016.
54 Shaayegan, V; Ameli, A; Wang, S; Park, C B, *Compos. Part A: Appl. Sci. Manuf.*, **88**, 67-74, 2016.
55 Arjmand, M; Mahmoodi, M; Park, S; Sundararaj, U, *J. Cellular Plast.*, **50**, 6, 551-62, 2014.
56 Kang, S M; Kwon, S H; Park, J H; Kim, B K, *Polym. Bull.*, **70**, 885-93, 2013.
57 Zhai, W; Wang, J; Chen, N; Naguib, H E; Park, C B, *Polym. Eng. Sci.*, **52**, 2078-89, 2012.
58 Ghasemi. I; Farsheh, A T; Masoomi, Z, *J. Vinyl Addit. Technol.*, **18**, 161-7, 2012.
59 Vorawongsagul, S; Pratumpong, P; Pechyen, C, *Food Packaging Shelf Life*, **27**, 100608, 2021.
60 Breiss, H; El Assal, A; Harmouch, A, *Mater. Res. Bull.*, **137**, 111188, 2021.
61 Jing, X; Peng, X-F; Mi, H-Y; Wang, Y-S; Zhang, S; Chen, B-Y; Zhou, H-M; Mou, W-J, *J. Appl. Polym. Sci.*, **135**, 42508, 2016.
62 Xi, Z; Chen, J; Liu, T; Zhao, L; Turng, L-S, *Chinese J. Chem. Eng.*, **24**, 180-9, 2016.
63 Abbasi, M; Khorasani, S N; Bagheri, R; Esfahani, J M, *Polym. Compos.*, **32**, 1718-25, 2011.
64 Lapirre, R M, **US8563621**, *Polyfil Corporation*, Oct. 22, 2013.
65 Lopattananon, N; Julyanon, J; Masa, A; Kaesaman, A; Thongpin, C; Sakai, T, *J. Vinyl Addit. Technol.*, **21**, 134-46, 2015.
66 Scamardella, A M; Vietri, U; Sorrentino, L; Lavorgna, M; Amendola, E, *J. Cellular Plast.*, **48**, 6, 557-76,

2012.
67 Keramati, M; Ghasemi, I; Karrabi, M; Azizi, H, *Polym. J.*, **44**, 433-8, 2012.
68 Hwang, S-s; Liu, S-p; Hsu, P P; Yeh, J-m; Yang, J-p; Chen, C-l, *Int. Commun. Heat Mass Transfer*, **29**, 383-9, 2012.
69 Van Hooghten, R; Gyssels, S; Estravis, S; Rodriguez-Perez, M A; Moldenaers, P, *Eur. Polym. J.*, **60**, 135-44, 2014.
70 Forest, C; Chaumont, P; Cassagnau, P; Swoboda, B; Sonntag, P, *Prog. Polym. Sci.*, **41**, 122-45, 2015.
71 Aktas, S; Gevgilili, H; Kucuk, I; Sunol, A; Kalyon, D M, *Polym. Eng. Sci.*, **54**, 2064-74, 2014.
72 Zakiyan, S; Hossein, M; Famili, N; Ako, M, *J. Mater. Sci.*, **49**, 6225-39, 2014.
73 Kang, S M; Kim, M J; Kwon, S H; Park, H; Jeong, H M; Kim B K, *J. Mater. Res.*, **27**, 22, 2837-43, 2012.
74 Yu, J; Song, L; Chen, F; Fan, P; Sun, L; Zhong, M; Yang, J, *Mater. Today Commun.*, **9**, 1-6, 2016.
75 Wu, W; Cao, X; Lin, H; He, G; Wang, M, *J. Polym. Res.*, **22**, 177, 2015.
76 Schmitt, H; Creton, N; Prashantha, K; Soulestin, J; Lacrampe, M-F; Krawczak, P, *J. Appl. Polym. Sci.*, **132**, 41341, 2015.
77 Satrijo, A; Erickson, K P; Ma, J; Marshall, D A; Martin, M C; Peloquin, R L; Wang, S, **WO2013155362**, *3M Innovative Properties Company*, Oct. 17, 2013.
78 Yousefian, H; Rodrigue, D, *J. Appl. Polym. Sci.*, **132**, 42845, 2015.
79 Mosiewicki, M A; Rojek, P; Michalowski, S; Aranguren, M I; Prociak, A, *J. Appl. Polym. Sci.*, **132**, 41602, 2015.
80 Narasimman, R; Vijayan, S; Prabhakaran, K, *J. Mater. Sci.*, **50**, 8018-28, 2015.
81 Kim, J-H; Kim, G-H, *J. Appl. Polym. Sci.*, **131**, 40894, 2014.
82 Beskardes, I G; Demirtas, T T; Durukan, M D; Gümüsderelioglu, M, *J. Tissue Eng Regen. Med.*, **9**, 1233-46, 2015.
83 Xu, M Z; Bian, J J; Han, C Y; Dong, L S, *Macromol. Mater. Eng.*, **301**, 149-59, 2016.
84 Zhang, R; Kim, E S; Lee, P C; *Chem. Eng. J.*, **409**, 128251, 2021.
85 Huang, S; Pan, B; Xie, M; Wang, H, *Tribol. Int.*, **154**, 106726, 2021.
86 Guo, D; Bai, S; Wang, Q, *J. Cellular Plast.*, **51**, 2, 145-63, 2015.
87 Al-Moameri, H; Zhao, Y; Ghoreishi, R; Suppes, G J, *J. Appl. Polym. Sci.*, **132**, 42454, 2015.
88 Xi, W; Qian, L; Huang, Z; Cao, Y; Li, L, *Polym. Deg. Stab.*, **130**, 97-102, 2016.
89 Kram, S L; Lee, S; Stobby, W G; Morgan, T A, **WO2012082332**, *Dow Global Technologies LLC*, Jun. 21, 2012.
90 Pau, D S W; Fleischmann, C M; Delichatsios, *Fire Safety J.*, **111**, 102925, 2020.
91 He, S; Liu, C; Chi, X; Peng, H, *Chem. Eng. J.*, **371**, 34-42, 2019.
92 Zhang, Z; Li, D; Xu, M; Li, B, *Polym. Deg. Stab.*, **173**, 109077, 2020.
93 Surecell T-50 All Acrylic Foam Cell Promoter for Cellular Vinyl Products. DOW 875-00001-0111.
94 Anon. *Plast. Addit. Comp.*, **10**, 6, 11, 2008.
95 Renders, M J P J; Duzak, T; Aksoy, M, **US20200283593A1**, *Thermaflex International Holding BV*, Sep. 10, 2020.
96 Fan, C; Wan, C; Gao, F; Huang, C; Xi, Z; Xu, Z, Zhao, L; Liu, T, *J. Cellular Plast.*, **52**, 3, 277-98, 2016.
97 Zakiyan, S E; Famili, M H N; Ako, M, *J. Mater. Sci.*, **49**, 6225-39, 2014.
98 Yang, J; Huang, L; Zhang, Y; Chen, F; Zhong, M, *J. Appl. Polym. Sci.*, **130**, 4308-17, 2013.
99 Jiang, C; Han, S; Chen, S; Zhou, H; Wang, X, *Polymer*, **212**, 123171, 2021.
100 Peng, J; Sun, X; Mi, H; Jing, X; Peng, X-F; Turng, L-S, *Polym. Eng. Sci.*, **55**, 1634-42, 2015.
101 Szegda, D; Sitthi Duangphet, S; Jim Song, J; and Karnik Tarverdi, K, *J. Cellular Plast.*, **50**, 2, 145-62, 2014.
102 Abbasi, M; Khorasani, S N; Bagheri, R; Esfahani, J M, *Polym. Compos.*, **32**, 1718-25, 2011.
103 Zepnik, S; Hendriks, S; Kabasci, S; Radusch, H-J, *J. Mater. Res.*, **28**, 17, 2394-2400, 2013.
104 Stumpf, M; Spoerrer, A; Schmidt, H-W; Altstaedt, V, *J. Cellular Plast.*, **47**, 6, 519-34, 2011.
105 Kim, S G; Lee, J W S; Park, C B; Sain, M, *J. Appl. Polym. Sci.*, **118**, 1691-703, 2010.
106 Geissler, B; Feuchter, M; Laske, S; Fasching, M; Holzer, C; Langecker, G R, *J. Cellular Plast.*, **52**, 1, 15-35, 2016.
107 Lee, Y; Jang, M G; Choi, K H; Han, C; Kim, W N, *J. Appl. Polym. Sci.*, **133**, 43557, 2016.
108 Xiang, A; Yin, D; He, Y; Li, Y; Tian, H, *J. Supercritical Fluids*, in press, 105156, 2021.
109 Bernardo, V; Martin-de Leon, J; Rodriguez-Perez, M A, *Mater. Lett.*, **255**, 126587, 2019.
110 Li, T-T; Dai, W; Huang, S-Y; Lou, C-W, *Compos. Part B: Eng.*, **197**, 108171, 2020.
111 Wypych, A; Wypych, G, **Databook of Nucleating Agents**, 2nd Ed., *ChemTec Publishing*, Toronto, 2021.
112 Wypych, G, **Handbook of Nucleating Agents**, 2nd Ed., *ChemTec Publishing*, Toronto. 2021.
113 Zoller, A; Marcilla, A, *J. Appl. Polym. Sci.*, **121**, 3314-21, 2011.
114 Zoller, A; Marcilla, A, *J. Appl. Polym. Sci.*, **121**, 1495-505, 2011.
115 Zoller, A; Marcilla, A, *J. Appl. Polym. Sci.*, **128**, 354-62, 2013.

116 Salerno, A; Fanovich, M A; Pascual, C D, *J. Supercritical Fluids*, **95**, 394-406, 2014.
117 Gibeault, J-P, **US8772362**, *Nexkemia Petrochimie Inc.*, Jul. 8, 2014.
118 Roozemond, P; Zheng, L; van Hemelrijck, E, **US20190375889A1**, *DSM IP Assets BV*, Dec. 12, 2019.
119 Jing, X; Peng, X-F; Mi, H-Y; Wang, Y-S; Zhang, S; Chen, B-Y; Zhou, H-M; Mou, W-J, *J. Appl. Polym. Sci.*, **133**, 43508, 2016.
120 Kuo, C-C; Liu, L-C; Liang, W-C; Liu, H-C; Chen, C-M, *Composites Part B: Eng.*, **79**, 1-5, 2015.
121 Lobos, J; Iasella, S; Rodriguez-Perze, M A; Velankar, S S, *Polym. Eng. Sci.*, **56**, 9-17, 2016.
122 Wang, K; Wu, F; Zhai, W; Zheng, W, *J. Appl. Polym. Sci.*, **129**, 2253-60, 2013.
123 Restrepo-Zapata, N C; Osswald, T A; Hernandez-Ortiz, J P, *Polym. Eng. Sci.*, **55**, 2073-88, 2014.
124 Ma, Y; Wang, J; Xu, Y; Wang, C; Chu, F, *J. Appl. Polym. Sci.*, **132**, 42730, 2015.
125 Anon. *Focus on Surfactants*, **2010**, 8, 3, 2010.
126 Breiss, H; El Assal, A; Harmouch, A, *Mater. Res. Bull.*, **137**, 111188, 2021.

12

EFFECT OF FOAMING ON PHYSICAL-MECHANICAL PROPERTIES OF FOAMS

12.1 COMPRESSION SET, STRENGTH, AND MODULUS

The compression set was measured by the reduction in foam thickness after material was aged under compressed conditions. The higher the compression set, the lower the resilience of the foam. Figure 12.1 shows that the compression set was decreased for EVA/wood flour composite foams when the concentration of wood particles was increased, which means that wood particles had a reinforcing effect on foam cell walls, which reduced cell collapse under load.[1] The disadvantage of EVA in its use in midsole is its poor compression set.[5] Polyurethane displays a distinctly better compression set even at low densities.[5] Some footwear manufacturers blend EVA with more durable polymers (e.g., polyolefin elastomer) or replace EVA altogether to reduce compression set and enhance resilience and durability.[6]

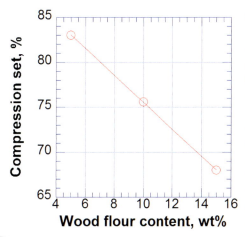

Figure 12.1. Effect of wood-flour content on compression set of EVA/wood-flour foams containing 5%, 10%, and 15% wood-flour. [Adapted, by permission, from Kim, J-H; Kim, G-H, *J. Appl. Polym. Sci.*, **131**, 40894, 2014.]

Two silicone foams were thermally aged under compressive strain in open-air containers or in high humidity environments.[2] More compression set was induced when foams were aged in the presence of moisture.[2] The residues from tin octoate, the catalyst used to accelerate the cure of the foams, were implicated in promoting aging.[2] Hydrolysis and rearrangement were either catalyzed by tin(II) species or by a combination of trace amounts of water and octanoic acid, a product of the hydrolysis of tin octoate.[2] PDMS foams prepared with high catalyst content displayed the enhanced compression set.[4]

Polyolefin elastomers, when crosslinked, exhibit significantly enhanced melt strength and may be used to produce high-quality foam products with exceptional heat aging, compression set, and weather resistance properties.[3] Increasing crosslinker loadings in PDMS foams improved compression set resistance.[4]

Based on the above examples, it is easy to come to the conclusion that reinforcement of cell walls improves foam resilience and reduction of catalytic residues, which may

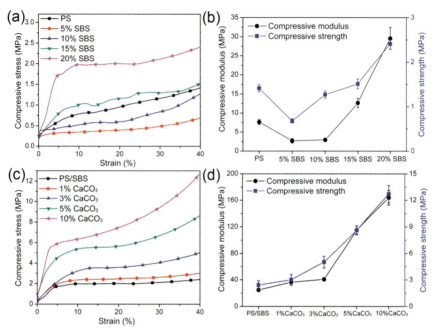

Figure 12.2. Compressive properties of the (a,b) extrusion-foamed PS/SBS blends and (c,d) extrusion-foamed PS/SBS/CaCO$_3$ nanocomposites: (a,c) compressive stress–strain curves and (b,d) statistical results for the compressive modulus and strength. [Adapted, by permission, from Jing, X; Peng, X-F; Mi, H-Y; Wang, Y-S; Zhang, S; Chen, B-Y; Zhou, H-M; Mou, W-J, *J. Appl. Polym. Sci.*, **133**, 43508, 2016.]

increase hydrolysis rate, improves compression set over time, and minimizes the effect of the action of environmental elements.

Compressive strength or compression strength is the capacity of a material or structure to withstand loads tending to reduce the size of the sample. Having this in mind, it is easy to conclude that foams generally have lower compression strength than solid materials.

The addition of the rubbery-phase (styrene-butadiene-styrene) to polystyrene foams increased their compressive modulus by 289.5%.[7] With the further addition of 5% rigid CaCO$_3$ nanoparticles, the compressive modulus was improved by 379.2%.[7] Figure 12.2 shows data characteristic of these improvements.[7] The elastic recovery and compressive strength of the nanocomposite foams made out of natural rubber/ethylene vinyl acetate blend decreased with increasing sodium montmorillonite and organoclay content, whereas the opposite trend was observed for conventional composite foams with china clay.[10] Nanofiller played the role of compatibilizer, whereas china clay also worked as the nucleating agent.[10] The compressive strength of chitosan hydrogels was improved around five- to six-fold when hydroxyapatite was incorporated into hydrogel composites designed as bone scaffolds.[11] The biopolyurethane nanocomposite foams reinforced with carrot nanofibers had a narrow cell size distribution, and the compressive strength and modulus were significantly elevated.[14] The highest compressive strength and modulus were reached with 0.5 phr fibers.[14]

12.1 Compression set, strength, and modulus

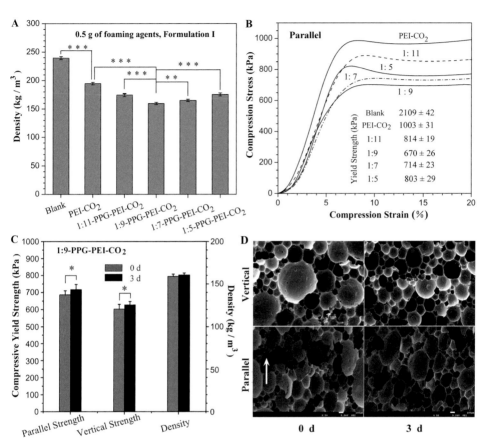

Figure 12.3. Properties and morphology of polyurethane foams. (A) The foams blown by 1:9-PPG-PEI–CO$_2$ possessed the lowest density among all the foams. (B) The compressive yield strength in the foam rise direction (parallel) varied mainly depending on the sample density. (C and D) The anisotropic mechanical strength and anisotropic morphology observed in 1:9-PPG-PEI–CO$_2$ blown foams, with the white component aged for 0 or 3 d. The foam rise direction (parallel) is indicated by a white arrowed line. The white component aging enhanced the mechanical strength (C), lowered and homogenized the pore size of the foams (D), but kept the densities statistically the same (C). Scale bars in (D): 100 μm. [Adapted, by permission, from Jing, X; Peng, X-F; Mi, H-Y; Wang, Y-S; Zhang, S; Chen, B-Y; Zhou, H-M; Mou, W-J, J. Appl. Polym. Sci., 133, 43508, 2016.]

The compressive strength is generally related to the foam density (Figure 12.3B).[7] However, the white component (PPG-PEI quickly absorbs CO$_2$ and forms a white solid (called here the white component) in about 1 min) aging for 3 days promoted the dispersion of 1:9 PPG-PEI.CO$_2$, thereby lowering and homogenizing the pore size of the resultant foams (Figure 12.3D).[7] This morphological change did not affect the foam density but improved the compressive strength (indicated by * in Figure 12.3C).[7] The anisotropic morphology with the major axes of the elliptical pores, pointing to the foam rise direction (Figure 12.3D) is caused by the horizontal confinement of the foam expansion along the container wall.[7] Therefore, the compressive strength at the foam rise direction was larger than across it, both before and after the white component aging (Figure 12.3C).[7] These

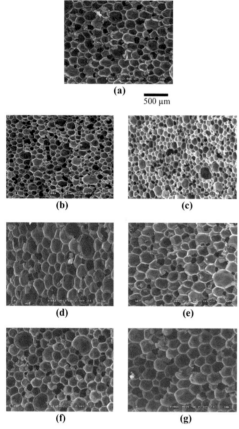

Figure 12.4. Scanning electron micrographs of PUR foams with varying additive species and concentrations (phr): (a) without additive, (b) perfluoroalkane 1, (c) perfluoroalkane 2, (d) propylenecarbonate 1, (e) propylenecarbonate 2, (f) acetone 1.0, (g) acetone 2. [Adapted, by permission, from Lee, Y; Jang, M G; Choi, K H; Han, C; Kim, W N, *J. Appl. Polym. Sci.*, **133**, 43557, 2016.]

observations clearly confirm that not only the foam density but also the foam morphology affects its mechanical strength.[7]

In the studies of starch/halloysite foams for biomedical applications, the largest compressive stress (0.85 MPa) was observed at a porosity of 4%.[9] The lowest compressive stress (0.45 MPa) was recorded at a porosity of 28%.[9] Low porosity foams tend to have thicker cell walls and higher solid fraction, and hence they are able to better resist deformation.[9] A maximum compressive strength and specific compressive strength of 5.2 MPa and 21.3 MPa g^{-1} cm^{-3}, respectively, were achieved at a very low graphene oxide concentration of 0.25 wt% in the carbon composite foams, which corresponded to an increase of 189 and 133%, respectively.[12] In general, the compressive strength of foam materials depends on their bulk density, cell size, and type of reinforcement used.[12] The increase in compressive strength with the incorporation of graphene oxide is due to the reinforcement effect of the graphene produced by the *in situ* reduction of graphene oxide.[12]

Studies of the influence of different types of liquid nucleating agents used in the production of polyurethane foams with an HFC-365mfc show that the compressive strength of the PUR foams prepared with perfluoroalkane was higher than the PUR foams prepared with the propylene carbonate and acetone.[13] The PUR foams synthesized with perfluoroalkane showed a smaller average cell diameter, and for this reason, they had a better compressive strength (see Figure 12.4).[13]

A decreased polystyrene melt flow index lowered the foam compressive strength, and an increase in the CO_2 concentration increased the compressive strength.[15] They both had an influence on cell size.[15]

12.2 CRYSTALLINITY

The foaming behavior of PPS/PES blends was closely related to the blend morphology, crystallinity, and the mass-transfer rate of the CO_2 in each polymer phase.[16] The degree of crystallinity of polymer blend decreased with the increase of PES content (PPS has a poor impact toughness due to the rigid backbone chain and a high crystallinity).[16] The incorporation of PES destroys the continuity and decreases the crystalline ability of the PPS phase, leading to a reduced crystallinity of the blend.[16] The diffusion of CO_2 in PPS/PES blends is hindered by crystal barriers in the semi-crystalline PPS component.[16] The equilibrium concentration and desorption diffusivity of CO_2 increased as the PES content increased.[16] Higher saturation pressure permitted the formation of foams with smaller cell size and larger cell density in the microcellular PPS/PES blends due to the higher gas concentration.[16] This, in turn, resulted in the improved mechanical performance of the blend, which could be tailored to the requirements of high-performance applications.[16]

The crystallinity of 20% was reached in poly(lactic acid) samples after CO_2 saturation, and high crystallinity of about 38.2% was achieved for the foamed poly(lactic acid).[17] Non-foamed poly(lactic acid) had a low elongation at break between 3.6 and 15.1%.[17] Poly(lactic acid) foam presented a significant increase in elongation at break by up to 15.1 times compared with that of the non-foamed sample.[17] Microcellular foaming improved tensile strength of poly(lactic acid) foams by 53.1%.[17]

The degree of crystallinity of poly(lactic acid)/maleic-anhydride-grafted-polypropylene blends had a profound influence on controlling the cellular morphology of their foams.[18] A faster crystallization rate and a higher degree of crystallinity helped in solidifying the cell walls during the foaming processes.[18] A pressure-induced flow method was applied to neat PLLA.[21] As a result, the PLLA's crystallinity and its spherulite size increased from the original 15.3% and 5.0 nm to 42.1% and 28.5 nm, respectively.[21] High-strength and low-density PLLA foams with uniform and controlled cellular morphology were produced.[21] Crystallinity of polylactide composite produced by injection molding was remarkably enhanced by introducing foaming and additives.[22] The presence of nanoclay and talc, presence of pressurized N_2, foaming action, and mold opening all enhanced the PLA's crystallization kinetics during injection molding, leading to a threefold increase in the final crystallinity.[22] Long-chain branched polylactide had substantially higher crystallinity than the linear polylactide.[23] The addition of nanoclay markedly reduced the crystallinity of both types of polylactide.[23] The presence of dispersed nanoparticles strongly hinders chain mobility, resulting in lower crystallinity.[23] Too high a crystallinity will depress the expansionary ability of PLA foam because of the increased stiffness.[25] The degree of crystallinity induced during blowing agent saturation needs to be controlled by selecting the proper saturation temperature and time.[25] Both linear and branched PLA with different crystallizability can produce double crystal melting peak with reasonably high final crystallinity.[25]

Poly(lactic acid) was foamed with supercritical CO_2.[26] Crystallinity decreased with increasing annealing temperature or saturation pressure.[26] Ring-banded spherulites were formed by crystallization under compressed CO_2.[26] Uniform closed-cell structure was obtained by controlling foaming temperature in the range of 80-100°C at which only ring-banded spherulites were formed.[26] Cellular structure was affected by both crystallinity and crystal morphology.[26] Pressure also influenced crystallization behavior; complex spheru-

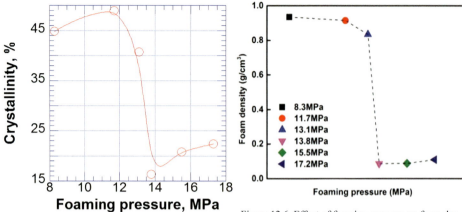

Figure 12.5. Crystallinity of PP foams under different foaming pressures. [Data from Fu, D; Chen, F; Kuang, T; Li, D; Peng, X; Chiu, D Y; Lin, C S; Lee, L J, *Mater. Design*, **93**, 509-13, 2016.]

Figure 12.6. Effect of foaming pressure on foam density (the foaming temperature and saturation time at 155°C and 2 h). [Adapted, by permission, from Fu, D; Chen, F; Kuang, T; Li, D; Peng, X; Chiu, D Y; Lin, C S; Lee, L J, *Mater. Design*, **93**, 509-13, 2016.]

lites with ring-banded texture in the periphery were obtained under saturation pressure of 8 MPa, resulting in the formation of a number of stamen-like cell structures.[26] Higher pressure led to imperfect crystals with more close-packed lamellae.[26]

The foaming process of poly(vinylidene fluoride-co-hexafluoropropylene) using supercritical carbon dioxide leads to a lower crystallinity and higher melting temperature.[19] Usually, uniform cells with fine cell size occur in low-crystallinity polymers, while non-uniform structures are observed in high-crystallinity polymers.[19] One means to overcome the melt strength issue and reduce cost is to form a PVDF alloy with other polymers, such as acrylics, to reduce the crystallinity and improve the foaming process.[27]

Linear polypropylene is very difficult to foam due to its low melt strength and high crystallinity.[20] The crystallinity of foams changes with foaming pressure.[20] Below 13.8 MPa, the crystallinity of PP foams remained similar, but when the pressure reached 13.8 MPa, an apparent drop in crystallinity was observed (see Figure 12.5).[20] At around the same pressure, the foam density drastically decreased (Figure 12.6).[20]

The crystallinity degree of polypropylene containing carbon nanotubes was reduced after foaming, yielding lower mechanical properties of the foam cell walls.[24] The cell wall properties can be estimated based on the degree of crystallinity obtained from DSC measurements of the foamed samples.[24]

Rapid cooling affected iPP polycrystallinity, generating both crystals and nodular mesophase structures.[28] By increasing the cooling rate, crystalline phases tend to become more disordered; therefore, mesophases may appear due to the rapid reduction of chain mobility.[28] Cellular growth during foaming can induce biaxial stretching forces.[28] This stress could induce localized differences in the resulting material's crystallinity due to possible orientation and generation of more ordered crystals.[28] The biaxial stretching caused by the rapid CO_2 expansion results in the generation of more ordered crystalline phases.[28]

12.3 DEFORMATION RECOVERY

The deformation recovery is caused by cell wall rebounding and the air pressure inside the foam cells.[29] The deformation recovery is impacted by the repeated loading-unloading (low-speed impact), causing a significant stress softening phenomenon of the bumper foams.[29] The softening greatly depends on the deformation history of the material.[29]

The fracture behavior and the recovery ratio of insulating foams were significantly affected by low temperatures.[30] The recovery ratio for all tested materials significantly decreased at -163°C.[30] A polyurethane foam specimen after an experiment at the temperature of -163°C was easily broken into pieces even under the application of small loads.[30]

The recovery of poly(ε-caprolactone) foams correlated with their compression modulus.[31]

Polyether foam having open cells had a 95% height recovery at the time of shorter than 3 seconds.[32] The recovery rate in less than 5 seconds indicates rapid recovery, such as typical of resilient foam types.[32] The recovery rate longer than 5 seconds is indicative of a slow recovery foam often referred to as "viscoelastic" or "memory" foam.[32] High airflow is usually a desirable performance trait for flexible foam because it relates to improved recovery characteristics and durability.[32]

12.4 DENSITY

Table 12.1 gives a range of densities for different foams.

Table 12.1. Foam densities

Foam type	Density, g/cm^3	Ref.
Polyurethane foam blown with CO$_2$ adducts of polypropylene glycol-grafted polyethyleneimines	0.06-0.24	8
Ethylene vinyl acetate/natural rubber containing sodium montmorillonite, organoclay, China clay	0.25-0.5, 0.3-0.5, 0.5-0.55	10
Chitosan-based super-porous hydrogels	0.059-0.091	11
Graphene-reinforced carbon composite foam	0.125-0.3	12
Rigid PU foams blown with HFC-365mfc in the presence of liquid nucleating agents such as acetone, propylene carbonate, and perfluoroalkane	0.0943 (acetone) 0.1152 (no additive)	13
Water blown rigid PU; thermal conductivity (mW/mK) 33/53/85	0.2/0.4/0.6	33
Glass fiber-reinforced phenolic foam (microwave foamed)	0.02-0.04 (good thermal insulator)	34
Carbon foams derived from phenol-formaldehyde resin	0.34-0.51	35
Ultralow density carbon foam with embedded carbon nanotube whiskers for high-performance microwave absorption materials	0.0094	36
Anisotropic, crosslinked high-density polyethylene foamed with azodicarbonamide	0.2	37
Atactic polypropylene foamed with supercritical carbon dioxide, radiation crosslinked	0.22-0.59	38

Table 12.1. Foam densities

Foam type	Density, g/cm³	Ref.
Ethylene co-monomer in random copolymers of polypropylene; fast/slow CO_2 pressure release at 140°C	0.06/0.81	28
Extruded polystyrene foam blown with trans HFO-1234ze and trans HCFO-1233zd	<0.05	39
Polyether-based TPU for midsole footwear	0.28-0.44	40
Ultra low-density biodegradable shape memory polymer foams with tunable physical properties	0.005	41
Natural rubber foams; amount of blowing agent (sodium bicarbonate): 4/8/12/16 phr	0.34/0.28/0.24/0.18	42

The above examples show that density decrease due to foaming is usually larger than the density gain due to the addition of various fillers. The nucleating action of fillers is responsible for this phenomenon.

12.5 ELASTIC MODULUS

The addition of azodicarbonamide leads to a loss of elasticity of the polymeric matrix (polypropylene) due to a significant decrease in collapse stress.[43] Foamed samples exhibit smaller values of both elastic modulus and collapse stress than their corresponding precursors due to density reduction.[43] Elastic modulus decreases with the increase in open-cell content.[43]

Polyurethane foam and cell unit exhibit three behaviors due to external compression; linear elasticity, plateau (plastic collapse), and densification.[44] Linear elasticity is controlled by cell wall bending and cell face stretching, which contribute to elastic moduli.[44] Figure 12.7 illustrates deformation mechanisms and their potential effect on the material's elasticity.[44] The thin membrane over the face of the cell contributes to elastic moduli as the cell fluid is compressed.[44] Permanent deformation occurred with an impact energy of 1,288J/m², which caused a plastic hinge on the microstructure cell.[44] Therefore, it means that the occurrence of plastic hinges affects cell wall bending strength, which determines elastic modulus of the specimen subjected to impact and compression tests.[44] The critical impact energy that degraded the impact and compressive performance of specimen was between 1,092 and 1,228 J/m².[44] When the impact stress exceeded the yield strength, the elastic modulus decreased.f Decrease of the elastic modulus of the insulation material can cause changes in the structural response of the system, which can compromise the structural integrity of the system owing to sloshing impact.[44] After 80 repetitions of a 1,500 J/m² impact, the elastic modulus decreased by 44.5% compared to that of the intact specimen.[44]

For open-cell polyurethane foam used as cancerous bone surrogates made of urethanes, epoxies, and structural fillers, a mean elastic modulus of 3 GPa was estimated for high-density foams, 2.7 GPa for medium-density foams, and 1 GPa for low-density foams.[45]

Elastic modulus of polyurethane foams at room temperature was in the range of 21-65 MPa varying with foam density in the range of 0.18-0.09 g cm⁻³.[45]

Elastic modulus of LLDPE foam was in the range of 37-112 MPa increasing with the volume fraction of polymer increase (elastic modulus of pure LLDPE was 450 MPa).[47]

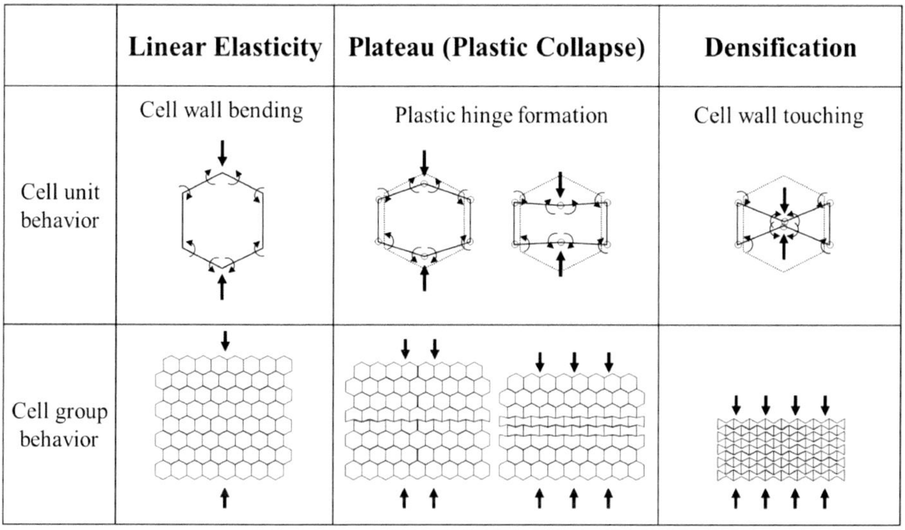

	Linear Elasticity	Plateau (Plastic Collapse)	Densification
Cell unit behavior	Cell wall bending	Plastic hinge formation	Cell wall touching
Cell group behavior			

Figure 12.7. Schematic diagram of the deformation mechanism in terms of cell unit and group under the compression. The hexagon represents one cell composed of the strut (line) and strut joint (angular point). Linear elasticity is controlled dominantly by a cell wall (strut) bending. As the stress increased, a plastic hinge occurs at the corner of the cell. A black circle indicates plastic hinge formation. At this time, the cell is plastically collapsed. Plastic deformation of the cells propagates to adjacent cell layers. When the cell walls touch each other, and the cell group can no longer be deformed, the stress increases rapidly, which is densification. [Adapted, by permission, from Kim, S; Kim, J-D; Lee, J-M, *Int. J. Mech. Sci.*, **194**, 106188, 2021.]

Polystyrene foam had elastic modulus in compression state of 1.24 MPa and elastic modulus in tension state of 16.1 MPa.[48]

The elastic modulus of PVC foam (30-150 MPa) correlated with foam density in the range of 0.19-0.04 g cm^{-3} (elastic modulus of non-foamed PVC was 2,700 MPa).[49]

12.6 ELONGATION

Table 12.2 shows the effect of foaming on elongation.

Table 12.2. Elongation

Foam type	Elongation,% solid	Elongation,% foamed	Ref.
Polypropylene (azodicarbonamide 0.4/0.8/2 wt%)	150	120/80/55	50
Polyethylene (azodicarbonamide 0.4/0.8/2 wt%)	105	95/80/50	50
Poly(vinyl chloride) (azodicarbonamide 0.4/0.8/2 wt%)	220	180/170/145	50
Poly(lactic acid) blown with compressed CO_2	3.6-15.1	44-88	51
Poly(vinylidene fluoride)	100-400	50-80	52

Table 12.2. Elongation

Foam type	Elongation,% solid	Elongation,% foamed	Ref.
Pressure-sensitive adhesive foams from styrenic and acrylic copolymers		>600	53
EPDM foam produced using azodicarbonamide and urea for sealing materials		470-835	54
Oil-resistant polyurethane shoe soles made using all-propylene oxide polyols		>400	55
Polyurethane foam containing 0/5/20; solid foamed	682	682/690/493 475/529/212	56
PU foam vs density of 0.15/0.2/0.25 TPU foam for athletic shoes for the same densities		277/309/317 280/300/302	57
Ethylene vinyl acetate foamed with azodicarbonamide (different densities of foamed material): 0.24/0.205/0.141		324/285/261	58
Polyetherimide (water vapor-induced phase separation)	60	20.9	59

12.7 EMI SHIELDING

Graphene-reinforced carbon composite foams were prepared by thermo-foaming of graphene oxide dispersions in molten sucrose to form solid organic foams, followed by carbonization at 900°C.[12] The hydrogen bonding interaction between sucrose hydroxyls and the functional groups on graphene oxide caused a decrease in the melting point of sucrose from 185 to 120°C at the graphene oxide concentration of 1.25 wt%.[12] The EMI shielding effectiveness of the graphene-reinforced carbon composite foams increased with an increase in the GO concentration up to 0.15 wt% because of a decrease in foam density.[12] The maximum EMI shielding effectiveness of 38.6 dB was achieved at graphene oxide concentrations of 0.15 wt%.[12]

The foamed composites of poly(butylene terephthalate) and carbon fiber had better electrical conductivity and EMI shielding properties than those of solid materials.[60] A partial foaming with a density gradient from the surface to the center of polycarbonate/carbon nanotube has been performed.[62] The gradient foamed materials were very advantageous for EMI shielding application since the foamed structure at the surface had a low dielectric constant and limited the reflection of the EMI signal while the presence of highly conductive solid in the middle ensured a high absorption of the electromagnetic radiation.[62] Cell formation in polypropylene/carbon fiber foam increased the fibers inter-connectivity by biaxial stretching of the matrix.[63] It also changed the fiber orientation.[63] The introduction of foaming reduced density of injection-molded samples by 25%, lowered the volume fraction of the percolation threshold from 8.5 to 7 vol% of carbon fibers, enhanced the through-plane conductivity by up to a maximum of six orders of magnitude, increased the dielectric permittivity, and resulted in the increase of the specific EMI shielding effectiveness by up to 65%.[63]

The phenolic foams with the open-cell structure, containing 2 wt% multiwalled carbon nanotubes and 7 wt% Fe_3O_4 and having a thickness of 2 mm, had EMI shielding

12.7 EMI shielding

effectiveness of ~34 dB at 8-12 GHz.[61] The obtained open-cell foams exhibited the highest EMI shielding effectiveness of ~62 dB at 8-12 GHz.[61] EMI shielding properties of foams increased with increasing foam thickness.[61]

The process used a water-miscible compound to modify graphene oxide, such as isophorone diamine, that also functioned as both a carrier and a curing agent in the preparation of epoxy composite foams produced without the use of a solvent in EMI shielding applications.[64]

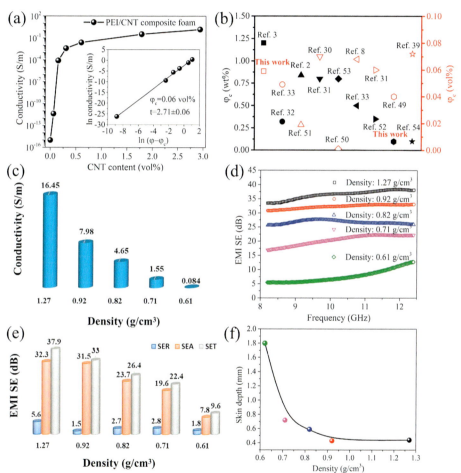

Figure 12.8. (a) The electrical conductivity of the composite foam vs. CNT content (The inset shows a ln-ln plot of the conductivity vs. φ-φ_c of the foam. Lines correspond to the least square fitting of the experimental data); (b) Percolation threshold from this work and the previously reported polymer/CNT composites; (c) Conductivity of PEI/CNT2.94 foam of different density; (d) EMI SE for the PEI/CNT2.94 foam vs. density; (e) Comparison of SEA and SER at the frequency of 11.5 GHz for the PEI/CNT2.94 foam of different density; (f) Skin depth of PEI/CNT2.94 vs. density at the frequency of 11.5 GHz. [Adapted, by permission, from Feng, D; Liu, P; Wang, Q, Composites Part A: Appl. Sci. Manuf., **124**, 105463, 2019.]

Lightweight TPU/graphite membranes with excellent mechanical and EMI shielding properties were fabricated with CO_2.[65] Nanographite effectively improved cellular structure, strength, rigidity, ductility and enhanced the EMI shielding performance for both solid (PU – 0.5 dB; composite with 20% graphite – 7.5 dB) and foamed (PU foam – 0.3 dB, composite foam with 20% nanographite – 6.2 dB) TPU composite.[65]

Because of the synergistic effect of high carbon black, CB, the content of 25-30 wt% in polypropylene, microcellular foam structure, and thickness, the PPCB foams with the conductivity of 0.6-1.3 S/m exhibited EMI shielding of 10-31 dB in the broadband frequency range of 6-18 GHz.[66]

Microporous polyetherimide/carbon nanotube segregated composite foams exhibited an excellent conductivity and EMI shielding effectiveness of 7.98 S/m and 32.3 dB, respectively.[67] Figure 12.8 illustrates the effect of foam properties on EMI shielding performance of polyetherimide foams.[67]

12.8 EXPANSION RATIO

Table 12.3 gives some data on the expansion ratio of cellular foams. Expansion ratio or porosity was calculated by dividing the density of unexpanded formulation by the density of a foam.[72]

Table 12.3. Expansion ratio

Foam type	Expansion ratio	Ref.
PC/ABS and PP produced with nitrogen in MuCell process (expansion ratio of up to 2 gives good mechanical performance)	1.29-2.7 & 1.29-3	68
PP foams produced by addition of thermoplastic starch containing 20% water	1.9-3.6	69
Open-cell polycarbonate foamed with nitrogen in MuCell process	1-8	70
PVC plastisol foamed by azodicarbonamide for flooring applications	4.25+5.25	71
Organic cellulose ester blown with HFO 1234ze (depending on the concentration of blowing agent from 2-3 wt% and talc from 0-0.8 wt%)	1.82-3.98	72
Poly(ethylene terephthalate) by extrusion foaming using CO_2	9.5	73
Commercial polyolefin-based foams produced by Sekisui	13.5	74
Insulation foams	40	76
EPDM foam for sealing material	>2	77
Recycled scrap ethylene-vinyl acetate foam	1.5-2.2	78
Ethylene/alpha-olefin/non-conjugated polyene interpolymer	1.33-1.37	79
The expansion ratio of UHMWPE of foam produced by supercritical carbon dioxide depended on temperature 100/120/140°C	1.2/8.8/18.2	80

The extrusion foaming studies of long-chain branched polylactide using CO_2 as a physical blowing agent show that branching promotes the foam's expansion ratio.[75]

The cell wall thickness was a function of the expansion ratio and cell density.[8] When the cell density reached 10^{22} cells/cm^3, even if the expansion ratio is only 2, the cell wall thickness was about 0.1 nm, which was below the thickness of a polymer chain.[8] Cell wall rupture and cell coalescence may have been inevitable during cell growth.[81]

12.9 FLEXURAL MODULUS

Flexural modulus data are included in table 12.4.

Table 12.4. Flexural modulus

Foam type	Flexural modulus, MPa	Ref.
Epoxy foamed using sodium bicarbonate (0 to 20 phr)	2500-2000	82
Medium density polypropylene foams blown with azodicarbonamide (5-15 wt%)	1060 (control) 1300-1400	43
Polypropylene with 20% glass: non-foamed/MuCell/IQ Foam	7700/6500/7000	83
PP/nanocrystalline cellulose foam produced by extrusion combined with injection molding using azodicarbonamide (0.5-5 wt%)	1180 (control) 1600-1900	84
Vinyl ester syntactic foams containing 60% cenospheres	3000 (control) 4500	85
Impregnated (polyester or epoxy) vs. non-impregnated polyurethane foam	184.55; 235.86; 146	86
Polystyrene foam for containers	1655	87
Poly(lactic acid)/poly(butylene succinate)/cellulose fiber (5/10/15) composite for hot cups packaging application	1522/1594/1650	88
Poly(lactic acid) glass fiber reinforced: 0/5/10/15/20 wt%	3/3.8/4.3/5.1/6.1	89

A higher flexural modulus of elasticity than for the compact material can be achieved up to an expansion ratio of approximately 2, which correlates with a density reduction of approximately 40%.[68]

12.10 GLASS TRANSITION TEMPERATURE

Blowing agent (p,p'-oxybis benzene sulfonyl hydrazide) induces several differences to the vulcanization reaction of EPDM.[90] It decreases reaction temperatures while increasing reaction heats.[90] It eliminates the exothermic peak before vulcanization and decreases the fully cured resin's glass transition temperature.[90] The presence of the blowing agent reduces the operational window.[90]

The glass transition temperature of the polymer/gas mixture (polycarbonate/carbon dioxide) has a good linear relationship with the absorbed CO_2 concentration.[91] Foaming of samples takes place above the T_g of the polymer/gas mixture, which is well below the glass transition temperature of the polymer.[91] Experimental data for carbon dioxide in poly(styrene-co-acrylonitrile) show that solubility of carbon dioxide follows Henry's law above the glass transition temperature, and the diffusivity is a relatively strong function of temperature.[92]

The glass transition temperature of polymers drops to room temperature in the batch microcellular foaming process.[93] The glass transition temperature of ABS (T_g = 120°C) and PS (T_g = 100°C) decreased to 30°C, when the values of the weight gain ratio (due to absorption of gas) were 8.76% and 7.62%, respectively.[93] There was a little difference between the effects of argon, helium, and nitrogen on the glass transition temperature of polymer containing the same amount of any of these gases.[93] The glass transition tempera-

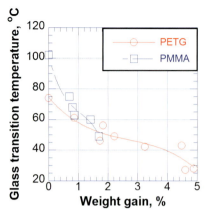

Figure 12.9. Glass transition temperature of glycol modified poly(ethylene terephthalate), PETG, and polymethylmethacrylate, PMMA, containing different concentrations of carbon dioxide. [Adapted, by permission, from Yoon, J D; Cha, S W, *Polym. Testing*, **20**, 287-93, 2001.]

ture of polymers containing Ar, He, and N_2 gases dropped to 70°C with very low gas absorption compared to carbon dioxide.[93] When a gas dissolves into the polymer, it weakens attraction forces between polymer molecules and changes the entanglement density, making the molecular movements more likely.[93]

Figure 12.9 shows the effect of the amount of carbon dioxide dissolved in glycol-modified poly(ethylene terephthalate), PETG, and polymethylmethacrylate, PMMA, on their glass transition temperatures.[94] Cha-Yoon model (presented in the paper) can be used to predict processing temperature because glass transition temperature and melting temperature are closely related in all thermoplastic matrices.[94]

The glass transition temperature of polymethylmethacrylate decreased in carbon dioxide from 380.5 to 311.4K within the pressure range of 0.1-12 MPa but increased gradually to 323.1K from 12 to 22 MPa.[95] Bimodal cell structure was developed when the temperature of depressurization was in the vicinity of the glass transition temperature.[97]

A disalt of malic acid (e.g., disodium malate) was used in the production of a polyurethane foam to lower the glass transition temperature of the polyurethane foam.[96]

The glass transition temperature of PET was 84.9°C, and because crystal perfection proceeded *via* migration and rejection of the structural defects at the crystallites induced by slow crystallization, the crystallinity increased rapidly with the rise of isothermal temperature, especially above the glass transition temperature.[98]

With high-pressure CO_2, the gas molecules are absorbed by the polymer matrix until saturation, causing a decrease in glass transition temperature of the polymer.[99] Polymers that contain carbonyl groups interact strongly with CO_2, increasing its solubility, including polymethylmethacrylate, which absorbs CO_2 more readily than in polymers without C=O groups such as polystyrene or polyvinylchloride.[99] On depressurization, as the CO_2 concentration decreases, the glass transition temperature begins to rise again, and the pores formed by the nucleation of the CO_2 bubbles become permanent, giving rise to the formation of a microcellular foam.[99] Rapid depressurization results in rapid cooling leading to the collapse of the pore walls.[99] Faster cooling rate involves a rapid increase in polymer matrix viscosity and glass transition temperature, which prevents the break of pores walls resulting in smaller cell size.[99]

12.11 IMPACT STRENGTH

Impact strength data are included in table 12.5.

Table 12.5. Impact strength

Foam type	Impact strength, kJ/m² polymer	Impact strength, kJ/m² foam	Ref.
Microcellular polycarbonate using MuCell technology with different grades	6-52	18-42	100
Polypropylene with azodicarbonamide (1/5/10 wt%)	7	4.1/3.5/2.8	43
Polypropylene foamed with supercritical carbon dioxide (presence of cells smaller than 10 μm in the PP enhances the impact strength)	12.3	12.5-17.3	102
Polyoxymethylene foam containing 3 wt% azodicarbonamide	4.61-5.38	9.74	103
Shish-kebab structured iPP foamed with CO_2 (Izod, J/m) nanoscale pores	3.2	23.9	104
Microcellular polylactate CO_2 blown (Izod, J/m) (foaming time 3/5/15 s)	21	94/85/76	108

Microcellular foams provide better impact strength than the standard foam products.[101] Figure 12.10 shows that cell size in PMMA microcellular foams affects impact resistance.[105]

The impact strength of highly oriented PS foam, having a relative density of 0.3, was ~1.5 times higher than for non-foamed PS.[106] The molecular chains can be oriented by shearing of bubbles by oriented foaming, which leads to improvement of tensile strength along the oriented direction.[106] The cells oriented parallel to the impact direction result in poor impact properties.[106]

Acrylonitrile-butadiene-styrene foamed by thermally activated microspheres had higher impact strength than its solid precursor.[107]

Propylene/1-butene copolymer was foamed by supercritical CO_2.[109] Enhanced impact strength (from 3834.3 to 6699.65 MJ/m²) resulted from high 1-butene content.[109] This improvement was attributed to a decrease in the regularity of the segment and pore size distribution.[109] The crystallinity of the material decreased with the decrease in the proportion of the crystalline regions in the material, and the overall rigidity of the material also decreased.[109]

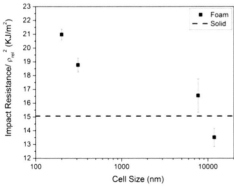

Figure 12.10. Mechanical behavior at high strain rates divided by the square of the relative density for both PMMA solid and for PMMA microcellular and nanocellular foams. [Adapted, by permission, from Notario, B; Pinto, J; Rodriguez-Perez, M A, *Polymer*, **63**, 116-126, 2015.]

12.12 RELATIVE PERMITTIVITY

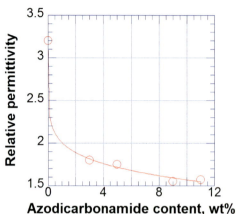

Figure 12.11. Relative permittivity of bismaleimide foam vs. amount of azodicarbonamide used for its foaming. [Data from Xie, X; Gu, A; Liu, P; Liang, G; Yuan, L, *Polym. Adv. Technol.*, **22**, 1731-7, 2011.]

High-performance foams with excellent dielectric properties were produced from toughened bismaleimide resin blown with azodicarbonamide.[110] The dielectric properties were not only dependent on the properties of the basic component (e.g., polymer) but also on the material microstructure.[110] Foams were composed of a polymer and the gas present in the cells, therefore the dielectric properties of foams depended on both parts.[110] Because air has very low relative permittivity, it is expected that the relative permittivity of foam should be much lower than that of a pure resin.[110] The higher the porosity of foams, the lower the relative permittivity of the foam.[110] Figure 12.11 shows that with the increased amount of azodicarbonamide the relative permittivity of foam decreases, meaning that the foam has lower dielectric properties than the solid polymer.[110]

Bisphenol A epoxy resin was foamed with different concentrations of polymethylhydroxysiloxane.[111] The higher the amount of the blowing agent, the lower the relative permittivity.[111]

Epoxy resin foamed with sodium bicarbonate and acetic acid was used as a catalyst.[82] Addition of catalyst increased foaming density and reduced dielectric constant.[82] Relative permittivity of polycarbonate improved upon foaming and the addition of graphene.[112]

Low dielectric constant dielectrics possess a broad prospect for application in microelectronic devices.[113] The most efficient method for fabricating low dielectric constant dielectrics is the incorporation of pores into the materials.[113] Polyarylene ether nitrile foams possess ultralow dielectric constant and excellent mechanical properties.[113] The porous structure of PEN foams is composed of finger-like pores and cellular pores (Figure 12.12).[113] Due to the existence of these pores, the density of the PEN foams is as low as 0.158 g·cm^{-3}, and the homologous void fraction is up to 87.5%.[113] As a result of the high void fraction, the PEN foam demonstrates an ultralow dielectric constant of 1.25 at 1 kHz.[113]

A high dielectric constant value is required for energy storage applications, such as dielectric capacitors.[114] Polyvinyl-formaldehyde foams were doped with 1-4 wt% multiwalled carbon nanotubes.[114] The 4 wt% PVF/MWCNTs shows a high dielectric constant (ε') value of 165 and a low loss tangent (tan δ) value of 0.2 at 50°C and 50 Hz.[114] This was attributed to Maxwell-Wagner-Sillars interfacial polarization existing between the polymer and MWCNTs and/or polarization of the MWCNTs.[114]

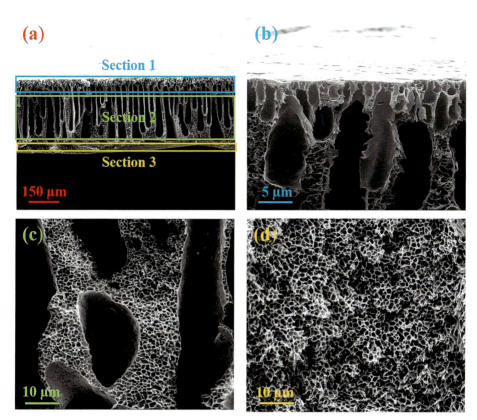

Figure 12.12. Microporous morphology of PEN foams: (a) cross-section, (b) enlarged Section 1, (c) enlarged Section 2, (d) enlarged Section 3. [Adapted, by permission, from Wang, L; Liu, X; Liu, C; Liu, X, *Chem. Eng. J.*, **384**, 123231, 2020.]

12.13 RESISTIVITY (ELECTRICAL)

The volume resistivity of epoxy foams was not significantly affected by cell size or morphology. It was only slightly higher for higher density foams having more closed cells.[111]

The compression-molded solid samples of polystyrene containing multiwalled carbon nanotubes had a percolation threshold equal to 0.7 wt%.[115] The injection samples foamed using azodicarbonamide had insulation behavior even up to 3.5 wt%.[115] Foam cells deteriorate conductive networks and prevent a sharp insulator-conductor transition.[115]

Figure 12.13 illustrates differences in the behavior of micrometer and nanometer foams.[116] The resistivity of foam (polymethylmethacrylate foamed with carbon dioxide) increased by two orders of magnitude when the pore size decreased from 1460 nm to 710 nm, regardless of a slight decrease in relative density (that should reduce the resistivity).[116] An increase in the electrical resistivity when cell size shifted from the micro to the nanoscale was caused by an increase in tortuosity of the solid phase.[116] Materials with cell

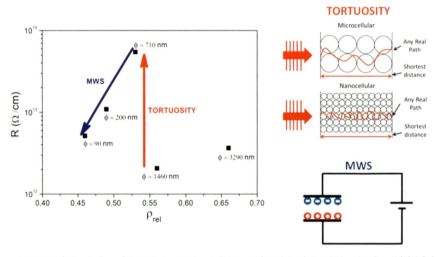

Figure 12.13. (Left) Evolution of the DC electrical resistivity as a function of the relative density. (Right) Scheme of the tortuosity (top) and interfacial polarization phenomenon (bottom). The numbers in the figure of DC resistivity vs. relative density correspond to the cell size of each foam. [Adapted, by permission, from Notario, B; Pinto, J; Verdejo, R; Rodriguez-Perez, M A, *Polymer*, **107**, 302-5, 2016.]

sizes below 700 nm showed a reduction of resistivity when the cell size was reduced.[116] The reason for this behavior was a conductive mechanism at low frequencies related to the polarization phenomena (or MWS) caused by the accumulation of charges on the opposing cell walls as depicted in the diagram.[116] MWS occurs in heterogeneous materials at the interfaces, leading to the separation of charges.[116]

Foams of blends containing HDPE, LDPE, and EVA, filled with carbon black and foamed with azodicarbonamide, had "switching behavior," meaning that at increased temperatures, they showed a rapid increase in resistivity.[117] They displayed a positive temperature effect.[117]

The volume resistivity decreased by ~7 orders of magnitude at 10 wt% carbon nanotube loading compared to pure polyetherimide, and tensile strength increased from 115 to 123 MPa.[118] PEI nanocomposite with 5 wt% of expanded graphite had electrical volume resistivity decreased by ~3 orders of magnitude and lower percolation threshold, which was related to large particle size and partial aggregation of untreated graphite particles within the PEI matrix.[119]

12.14 RHEOLOGY

Analysis of different grades of poly(ethylene terephthalate) showed that the long-chain branched polymers had higher melt viscosity and greater elasticity than the linear ones.[73] Only one long-chain branched poly(ethylene terephthalate) had the extensional viscosity characteristics that led to strain hardening.[73] This type was suitable for foaming.[73] Branching improves strain hardening of polymers.[73]

The cell structure evolution was found to be highly correlated to the system viscosity.[7] System viscosity can be successfully modified by additives such as toughening agents or nanofillers.[7]

In formulations having higher melt viscosity, the initial bubble expansion speed was lower, resulting in the smaller bubble size. On more advanced stages of bubble expansion, the diffusion rate becomes an important factor and the influence of melt viscosity is diminished.[120] The melt viscosity is controlled by temperature, shear rate, type, concentration, and distribution of foaming agents, and properties of polymer matrix and additives used in the formulation.[120] Melt viscosity is reduced by a dissolved gas.[120]

During the curing stage of polyurethane foams, the viscoelasticity of the liquid became dominant, and the rheological behavior was Maxwell-type viscoelasticity with relaxation time exponentially increasing with time.[121]

The presence of LDPE in the LLDPE matrix only slightly enhanced the linear viscoelasticity under small deformations, but 20 wt% LDPE was sufficient to significantly improve the non-linear behavior under large deformations to the same level as for pure LDPE.[122] Steep stress growth caused by strain hardening may dramatically increase the pressure of large bubbles and then effectively curb the coalescence of neighboring bubbles, thus facilitating good foamability.[122] Addition of a small amount of LDPE significantly improved the foamability of LLDPE/LDPE blends.[122]

12.15 SHAPE MEMORY

Shape fixity, shape recovery, and strain energy storage significantly increased with reduced hysteresis loss when silica nanoparticles were incorporated into polyurethane foam.[123] Silica particles reinforce polyurethane, if dispersed well, and act as multifunctional crosslinks, elastic energy storage, and relaxation retarder.[123] The best shape memory performance was obtained with 2% silica.[123] The shape fixity and shape recovery of polyurethane foam increased with the increasing amount of multiwalled carbon nanotubes.[124] Nanotubes reinforced the foam, increased the rubber elasticity, and retarded stress relaxation during stretching in the rubbery state.[124]

Shape memory foams can respond to specific external stimulus changing their configuration and then remember the original shape.[125] Figure 12.14 gives a schematic representation of the thermomechanical cycle for the shape memory foam.[125]

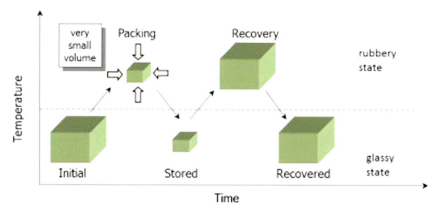

Figure 12.14. Termomechanical cycle of shape memory foam. [Adapted, by permission, from Santo, L, Progress in Aerospace Sciences. Shape memory polymer foams, *Elsevier*, 2016, pp. 60-65.]

Partial crystallization and vitrification are the basis of the shape memory effect in shape memory epoxy resins.[126] It is extremely difficult to detect and repair micro-cracks in polymers due to thermal or UV radiation.[126] Micro-cracks can be rebound to heal the damaged area by the self-healing mechanism.[126]

Micro-gravity did not affect the ability of the samples to recover their shape when tested in spacecraft, but there were limits in the case of the heating system design because of differences in heat transfer on the earth and on the orbit.[127]

Heat, electricity, light, magnetism, moisture, and even a change in pH value were used as external stimuli to cause macroscopic deformation of shape memory foam.[128] Smart materials developed based on the shape memory foams included deployable structures, morphing structures, biomaterials, smart textiles and fabrics, automobile actuators, and self-healing composite systems.[128]

Conductive biomass-derived d-glucaric acid-chitosan/single-walled carbon nanotube polymer composite layer and a crystallized paraffin layer were used to fabricate ultra-lightweight polyurethane foams with shape memory and adjustable EMI shielding functions, of which the shape change *via* external stimulus enables the foams to adjust EMI shielding efficiency autonomously.[129] The foam coated with synthesized conductive composite layers showed an exceptional EMI shielding effectiveness of 56 dB at an ultra-low density of 0.03 g/cm^3 with only 0.171 vol% SWNT.[129] The foam also exhibited ultra-high durability over 2000 compression-recovery cycles.[129] Introduction of paraffin layer as a reversible network resulted in shape memory foams with a high fixity ratio >95% and recovery ratio >90%.[129]

Two-way reversible shape memory polymer foams with porous 3D structures of interconnected pores were made using salt-leaching technology based on benzoyl peroxide thermo-crosslinked poly(ethylene-co-vinyl acetate).[130] Pore size was regulated using different NaCl particle sizes.[130] Large pore size exhibited ideal two-way shape memory behavior under prestretching strain at the crosslinking temperature of 200°C.[130]

12.16 SHEAR MODULUS

Shear modulus data are included in table 12.6.

Table 12.6. Shear modulus

Foam type	Shear modulus, MPa control	Shear modulus, MPa foam	Ref.
Poly(ε-caprolactone)/starch biodegradable foams	87.07 Pa	92.84-150.16 Pa	31
PET foam at different temperatures: 0, 20,40, 80°C	19.09/18.93/14.63/4.46		131
PUR foam at different temperatures: 0, 20,40, 80, 120°C	7.10/6.23/5.28/4.10/1.46		131
PET insulating foam	32	12-24	131
Effect of irradiation on shear modulus of irradiated silicone rubber; irradiation dose:150/300500	0.3	0.4/0.5/0.6	134

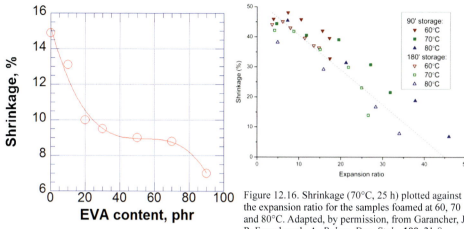

Figure 12.15. Shrinkage of CPE and its blends with EVA. [Adapted, by permission, from Zhang, B S; Zhang, Z X; Lv, X F; Lu, B X; Xin, Z X, *Polym. Eng. Sci.*, **52**, 218-24, 2012.]

Figure 12.16. Shrinkage (70°C, 25 h) plotted against the expansion ratio for the samples foamed at 60, 70 and 80°C. Adapted, by permission, from Garancher, J-P; Fernyhough. A, *Polym. Deg. Stab.*, **100**, 21-8, 2014.]

The shear modulus decreased with increasing cell size and cell wall thickness variation.[132]

12.17 SHRINKAGE

Foams made out of chlorinated polyethylene rubber blown by compression molding using azodicarbonamide suffer from extensive shrinkage.[135] Blending with ethyl vinyl acetate improves the performance of foam (Figure 12.15).[135]

The dimensional stability of polylactic foams, obtained using carbon dioxide, was related to the expansion ratio and the strain-induced crystallinity developed during foaming.[136] The shrinkage leads to a decrease in the size of the cells and the overall expansion ratio.[136] Shrinkage at 70°C (above the glass transition temperature of polylactide) virtually disappeared with crystallinity above 40%.[136] The "frozen-in" chains move back into a lower energy state resulting in shrinkage.[136] The cell walls in an expanded thermoplastic foam can be considered as analogous to the biaxially stretched films.[136] Upon re-heating above the glass transition temperature, the shrinkage of the cell walls forces the whole foamed structure to shrink.[136] With a substantial crystallinity, shrinkage can be suppressed.[136] Shrinkage is decreased by the expansion ratio increase (Figure 12.16).[136] During expansion, the local stretching of the amorphous phase increased in the cell walls.[136] With increased expansion ratio, the crystallinity of the samples increased too, generating a reinforcing crystalline network able to resist shrinkage upon re-heating above the glass transition temperature.[136]

1-Chloro-3,3,3-trifluoropropene used as a blowing agent leads to a reduced shrinkage, low lambda value, long cream time, and fast penetration time of polyurethane foams.[137]

A large number of –OH groups existing on organic montmorillonite surface hydrogen bonded with TPU soft/hard segments after foaming, which stabilized TPU molecular chains and reduced TPU foam's shrinkage.[138] Strong interaction between TPU's soft segments with organic montmorillonite inhibited TPU molecular chain's relaxation and reduced shrinkage of TPU supercritical CO_2 foams.kk TPU foams' shrinkage increased with increasing cell size at the same cell density.[138] With organic montmorillonite loading increased, a sufficient amount of hydrogen bonding could stabilize TPU oriented molecular chains and finally reduce TPU foams' shrinkage.[138] TPU foams with a large expansion ratio (more than 7 folds) and less shrinkage ratio (less than 30%) could be manufactured with 1 wt% organic montmorillonite loading.[138]

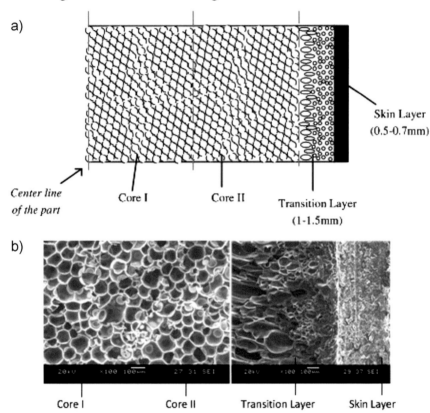

Figure 12.17. a) A schematic showing the overall structure of the foam injection-molded sample using mold opening, and b) representative SEM micrographs, obtained at temperature of 300°C, showing the skin layer and the transition and core regions of the injection-molded samples using mold opening. [Adapted, by permission, from Jahani, D; Ameli, A; Saniei, M; Ding, W; Park, C B; Naguib, H E, *Macromol. Mater. Eng.*, **300**, 48-56, 2015.]

12.18 SOUND ABSORPTION

Relative to their solid polycarbonate, the sound wave transmission loss was increased by up to 2.5-fold, and their thermal conductivity was decreased by up to fivefold by varying expansion ratio.[70] The open cell structure in the core region improved the sound wave absorption (Figure 12.17).[70] In the core region, the cells were interconnected with an open-cell structure in most cases.[70] Pinholes in the cell walls may be created by biaxia

12.18 Sound absorption

stretching caused by expansion and polymer shrinkage of the thin walls during the solidification.[70] The mold opening caused the melt to flow one-dimensionally, and cells to grow further in the thickness direction and to experience additional shear and extensional stresses.[70] The low melt strength of linear PC can cause cell wall rupturing.[70] The degree of mold opening, the melt temperature, and the injection flow rate were the most influential parameters in creating a high open-cell content.[70]

Sound absorption efficiency of natural rubber foams was improved by using treated bagasse and oil palm fibers.[139] The addition of natural fiber increased sound absorption values, especially at medium and high frequencies.[139] Improved interfacial bonding between the natural rubber and natural fibers enhanced sound absorption efficiency.[139]

Polyurethane foam is commonly used in the automobile industry as a sound absorption material because of its high sound absorption efficiency.[140] Magnesium hydroxide filler helps to improve the acoustic property as illustrated in Figure 12.18.[140] The filler improves the sound absorption efficiency by a combination of enhanced damping motion and increased number of partially opened pores.[140] The optimum of open porosity is achieved at the filler content of 1.0 wt% for the highest noise reduction coefficient.[140]

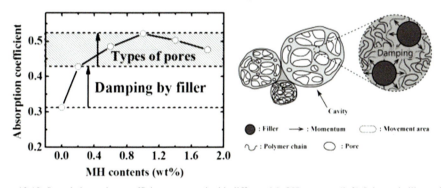

Figure 12.18. Sound absorption coefficient measured with different MgOH contents (left). Schematic illustration of mechanical damping effects of polyurethane foams loaded with inorganic fillers under sound waves (right). [Adapted, by permission, from Sung, G; Kim, J W; Kim, J H, *J. Ind. Eng. Chem.*, **44**, 99-104, 2016.]

A blend of linear low-density polyethylene (LLDPE) with poly(ethylene-octene) elastomer (POE) and foamed POE were used as the film and foam layers, respectively (Figure 12.19).[141] The sound absorption coefficient of the microlayer foam/film structure improved by 2-3 times compared with conventional single-layer material.[141] When the thickness ratio of the foam layer to film layer was low, the characteristic sound absorption peaks generally appeared at the medium-to-low frequency range.[141] When the foam/film thickness ratio increased, the characteristic sound absorption peaks shifted to medium-to-high frequencies.[141] This shows that the thickness

Figure 12.16. Schematic diagram of microlayers structure subjected to a sound wave. [Adapted, by permission, from Sun, X; Liang, W, *Compos. Part B: Eng.*, **87**, 21-26, 2016.]

and/or thickness ratio of the microlayer foam/film sheet can be designed to modulate the sound absorption profile and to maximize the absorption efficiency in the target frequency range for the particular application.[141]

PU composite foams, including silicone-acrylic filler particles, were fabricated to examine the sound absorption properties.[142] The best sound absorption performance was achieved with 2 wt% silicone-acrylic filler particles, giving the smallest cavity size and the highest partial-open pore ratio.[142]

12.19 SURFACE ROUGHNESS

The surface roughness of electromagnetic materials has an important impact on the service life and reliability of the electromagnetic components, especially the fitting properties, wear resistance, fatigue strength, and vibration.[143] Information on wall thicknesses and surface roughness are of particular interest to tissue engineering.[144]

The formation of surface irregularities was affected by the distribution of the CO_2 phase in the extrudate, especially at the exit section of the die.[145] The gas-phase entrained into the polymer melt migrated to the surface of the die and gave rise to a lubrication zone, acting as an apparent slip layer.[145] When the continuity of the flow boundary condition was interrupted (e.g., change of the stick condition to a slip condition or *vice versa*), the time-dependent fluctuations in pressure, flow rate and wall slip velocity developed, causing bulk irregularities.[145]

The surface roughness of the polypropylene composites was significantly reduced due to the microfoaming and presence of an endothermic foaming agent (surface roughness was reduced by 70% compared to non-foamed composites).[146] This may be due to the slow nucleation process by endothermic foaming agent (Hydrocerol BIH20).[146]

Surface roughness is controlled by mold filling pattern. Three methods exist to smooth the surface of microcellular parts:[147]
- use of co-injection or gas counter pressure
- use of hot mold surfaces or coated mold surfaces
- surface improvements resulting from processing, mold, material, or part design.

12.20 SURFACE TENSION

The foam surface area is inversely proportional to the surface tension.[148] The lower the surface tension, the better the foamability (from the perspective of surface energy).[148] The critical Weber number, which is a dimensionless ratio of inertial force (causing the deformation) and surface tension (restoring the bubble sphericity), has been used as a criterion of the bubble breakup.[148] The Weber number is given by the following equation:

$$We = \frac{\rho v^2 l}{\sigma} \qquad [12.1]$$

where:
- ρ fluid density
- v velocity of a fluid
- l characteristic dimension (droplet diameter)
- σ surface tension

Low surface tension of polyol and perfluoroalkane mixture prevented the coalescence of bubbles and promoted the formation of dense and small cells.[13] The perfluoroalkane was well-mixed in polyol and consequently produced a higher nucleation rate.[13]

Surfactant (sodium dodecyl sulfate) addition (0.15 wt%) to carbon fiber/epoxy foam facilitated formation of cells by lowering the surface tension of the mixture.[149] When the content of sodium dodecyl sulfate increased, large inhomogeneous cells were formed at the sides and inside of the foam samples, which was explained by inhomogeneous distribution of sodium dodecyl sulfate particles in resin, increasing the cell sizes locally during the foaming process, and so the coalescence phenomena.[149]

12.21 TEAR STRENGTH

Tear strength data are included in table 12.7.

Table 12.7. Tear strength

Foam type	Tear strength, N/mm	Ref.
CPE/EVA blend foamed with azodicarbonamide (EVA=0, 20, 90 wt%)	2.7, 3.8, 4.7	135
PVC flooring foamed using azodicarbonamide=0.8, 1, 1.2 wt%	14, 7.8, 6	71
NBR/carbon nanotube=0, 5, 15 phr	10, 13, 15 kJ/m^2	150
SBS/PS/SBR foamed with azodicarbonamide, crosslinked with dicumyl peroxide=0.1; 0.15; and 0.2 phr	2.6; 3.2; 3.7 kg/cm	151

PU foams of similar densities exhibited higher tear strength (18.3-35.9%; on average 27% greater) than TPU foams.[57]

12.22 TENSILE MODULUS

Tensile modulus data are included in table 12.8.

Table 12.8. Tensile modulus

Foam type	Tensile modulus, MPa solid	Tensile modulus, MPa foam	Ref.
Linear medium density polyethylene foamed with azodicarbonamide=0.15, 0.5, 1 phr	350	250, 100, 50	152
Medium density polypropylene foamed with azodicarbonamide=5 wt%	1150	400	43
iPP foamed with CO_2 (modulus decreases proportionally to relative density of foam)			103
Polyoxymethylene foam obtained with 1 phr azodicarbonamide	1088	831	113
Polymethylmethacrylate foamed with carbon dioxide	3000	850	153
PET vs. crystallinity: 22/25/30	158/165/185		98

Crystalline phase contents in poly(ethylene terephthalate) foam had a positive correlation with the tensile modulus, but the shape ratio of crystals had no significant influence on tensile modulus.[98]

12.23 TENSILE STRENGTH

Tensile strength data are included in table 12.9.

Table 12.9. Tensile strength

Foam type	Tensile strength, MPa solid	Tensile strength, MPa foam	Ref.
Polypropylene blown with 0.8% azodicarbonamide	15	12	50
Polyethylene blown with 0.8% azodicarbonamide	8.5	7.2	50
Polyvinylchloride blown with 0.8% azodicarbonamide	8	6.3	50
Polycarbonate unimodal/bimodal foams having relative density of 0.45		27/33	91
Polylactide blown with CO_2	52.4	11.9-17	17
Polycaprolactone blown with CO_2	11.02	8.02	50
Polycarbonate blown with N_2	55.1	47.1	154
Linear medium density polyethylene blown with 1 phr azodicarbonamide	22	2	152
Polypropylene copolymer	20	7	155

Thermal and vibration treatment on rigid polyurethane foam had its tensile strength reduced, and fracture elongation decreased due to the chemical degradation of polyurethane matrix and physical breaking of the foam structure.[156] The tensile strength and fracture elongation of rigid polyurethane foam both decreased with the increase of treatment time.[156]

12.24 THERMAL CONDUCTIVITY

Thermal conductivity data are included in table 12.10.

Table 12.10. Thermal conductivity

Foam type	Thermal conductivity, W/mK solid	Thermal conductivity, W/mK foam	Ref.
Building walls (requirement)		<0.027	8
Refrigerators (requirement)		<0.022	8
Underground steel pipes (requirement)		<0.03	8
Ethylene vinyl acetate/natural rubber blown with azodicarbonamide		0.147	10

Table 12.10. Thermal conductivity

Foam type	Thermal conductivity, W/mK solid	Thermal conductivity, W/mK foam	Ref.
Cell size of PS foam vs. thermal conductivity: 40, 80, 100, 140, 200, 250 μm	0.009, 0.0052, 0.0048, 0.0042, 0.0054, 0.007		157
Open-cell polycarbonate obtained by MuCell (relative density of foam 0.2 obtained with mold opening of 12.8 mm)	0.2	0.04	70
Foamed concrete		0.25-0.7	158
Recycled PET-based foam has reduced thermal conductivity by cell size and cell orientation		0.025	159
Extruded PS blown with *trans*HFO-1234ze		0.0192	160
Phenolic closed-cell foam blown with cycloisopentane		0.0201	161
Polyurethane foam obtained with 1-chloro-3,3,3-trifluoropropene		0.0218-0.024	137
Ethylene-norbornene cyclic olefin copolymers vs norbornene content of 33/36/51/58%	49.6/45.9/40.4/37.9		162

REFERENCES

1 Kim, J-H; Kim, G-H, *J. Appl. Polym. Sci.*, **131**, 40894, 2014.
2 Labouriau, A; Robison, T; Meincke, L; Wrobleski, D; Taylor, D; Gill, J, *Polym. Deg. Stab.*, **121**, 60-8, 2015.
3 Sun, X; Liang, W, *Composites Part B: Eng.*, **87**, 21-6, 2016.
4 Robinson, M W C; Swain, A C; Khan, N A, *Polym. Deg. Stab.*, **116**, 88-93, 2015.
5 Bollmann, H; Kamm, A; Huprikar, A G; Holwitt, U, **US20130197118**, Aug. 1, 2013.
6 Swigart, J F; Jiang, W Y, **US20130340280**, Columbia Sportswear North America, Inc., Dec. 26, 2013.
7 Jing, Y; Peng, X-F; Mi, H-Y; Wang, Y-S; Zhang, S; Chen, B-Y; Zhou, H-M; Mou, W-J, *J. Appl. Polym. Sci.*, **133**, 43508, 2016.
8 Long, Y; Sun, F; Liu, C; Xie, X, *RSC Adv.*, **6**, 23726-23736, 2016.
9 Schmitt, H; Creton, N; Prashantha, K; Soulestin, J; Lacrampe, M-F; Krawczak, P, *J. Appl. Polym. Sci.*, **132**, 41341, 2015.
10 Lopattananon, N; Julyanon, J; Masa, A; Kaesaman, A; Thongpin, C; Sakai, T, *J. Vinyl Addit. Technol.*, **21**, 134-46, 2015.
11 Beskardes, I G; Demirtas, T T; Durukan, M D; Gümüþderelioglu, M, *J. Tissue Eng. Regen. Med.*, **9**, 1233-46, 2015.
12 Narasimman, R; Vijayan, S; Prabhakaran, K, *J. Mater. Sci.*, **50**, 8018-28, 2015.
13 Lee, Y; Jang, M G; Choi, K H; Han, C; Kim, W N, *J. Appl. Polym. Sci.*, **133**, 43557, 2016.
14 Zhou, X; Sethi, J; Geng, S; Berglund, L; Frisk, N; Aitomäki, Y; Sain, M M; Oksman, K, *Mater. Design*, **110**, 526-31, 2016.
15 Han, B; Breindel, R M; Brammer, S T; Delaviz, Y; Boudreaux, C J; Fabian, B; Annan, N, **WO2014152410**, Owens Corning Intellectual Capital, Llc, Sep. 25, 2014.
16 Ma, Z; Zhang, G; Shi, X; Yang, Q; Li, J; Liu, Y; Fan, X, *J. Appl. Polym. Sci.*, **132**, 42634, 2015.
17 Ji, G; Wang, J; Zhai, W; Lin, D; Zheng, W, *J. Cellular Plast.*, **49**, 2, 101-17, 2013.
18 Wang, X; Liu, W; Li, H; Du, Z; Zhang, C, *J. Cellular Plast.*, **52**, 1, 37-56, 2016.
19 Tang, M; Wang, T-C, *J. Taiwan Inst. Chem. Eng.*, in press, 2016.
20 Fu, D; Chen, F; Kuang, T; Li, D; Peng, X; Chiu, D Y; Lin, C S; Lee, L J, *Mater. Design*, **93**, 509-13, 2016.
21 Kuang, T; Chen, F; Chang, L; Zhao, Y; Fu, D; Gong, X; Peng, X, *Chem. Eng. J.*, **307**, 1017-25, 2017.
22 Ameli, A; Nofar, M; Jahani, D; Rizvi, G; Park, C B, *Chem. Eng. J.*, **262**, 78-87, 2015.
23 Najafi, N; Heuzey, M-C; Carreau, P J; Therriault, D; Park, C P, *Eur. Polym. J.*, **73**, 455-65, 2015.
24 Wan, F; Tran, M-P; Leblanc, C; Béchet, E; Plougonven, E; Léonard, A; Detrembleur, C; Noels, L; Thomassin, J-M; Nguyen, V-D, *Mech. Mater.*, **91**, 1, 95-118, 2015.

25. Park, C B; Nofar, M, **WO2014158014**, *Synbra Technology B.V.*, Oct. 2, 2014.
26. Yang, Y; Li, X; Zhang, Q; Yu, P, *J. Supercritical Fluids*, **145**, 122-32, 2019.
27. Zerafati, S; Stabler, S M, **EP2449012**, *Arkema, Inc.*, May 9, 2012.
28. Pin, J-M; Tuccitto, A V; Lee, P C, *Polymer*, **212**, 123123, 2021.
29. Yang, X; Xia, Y; Zhou, Q, *Mater. Design*, **32**, 3, 1167-76, 2011.
30. Park, S-B; Lee, C-S; Choi, S-W; Kim, J-H; Bang, C-S; Lee, J-M, *Composite Structures*, **136**, 258-69, 2016.
31. Ogunsona, E; D'Souza, N A, *J. Cellular Plast.*, **51**, 3, 245-68, 2015.
32. Hager, S L; Britt, B; Mcvey, S; Moore, M; Uthe, P; Reese, J R, **WO2013043645**, *Bayer Materialscience Llc*, Mar. 28, 2013.
33. Andersons, J; Kirpluks, M; Cabulis, U, *Construction Build Mater.*, **260**, 120471, 2020.
34. Choe, J; Kim, M; Kim, J; Lee, D G, *Composite Structures*, **152**, 239-46, 2016.
35. Gong, Q; Zhan, L; Zhang, Y-z; Wang, Y-i, *Carbon*. **110**, 523, 2016.
36. Shi, H-G; Wang, T; Cheng, J-B; Wang, Y-Z, *J. Alloys Compounds*, in press, 158090, 2021.
37. Bernardo, V; Laguna-Gutierrez, E; Lopez-Gil, A; Rodriguez-Perez, M A; *Mater. Design*, 114, 83-91, 2017.
38. Yang, Zhe, X; Zhang, M; Wang, M; Wu, G, *Radiat. Phys. Chem.*, **131**, 35-40, 2017.
39. Bowman, J M; Williams, D J, **WO2014015315**, *Honeywell International Inc.*, Jan. 23, 2014.
40. Baghdadi, H A, **US20150038605**, *Nike, Inc.*, Feb. 5, 2015.
41. Singhal, P; Wilson, T S; Cosgriff-Hernandez, E; Maitland, D J, **US20140142207**, *Lawrence Livermore National Security, Llc, The Texas A&M University System,* May 22, 2014.
42. Samsudin, M S F; Ariff; Z M; Ariffin, A, *MaterialsToday: Proc.*, **17**, Part 3, 1133-42, 2019.
43. Saiz-Arroyo, C; Rodriguez-Perez, M A; Tirado, J; Lopez-Gil, A; de Saja, J A, *Polym. Int.*, **62**, 1324-33, 2013.
44. Kim, Jeong-Dae Kim, Jae-Myung Lee, J-M, *Int. J. Mech. Sci.*, **194**, 106188, 2021.
45. Belda, R; Palomar, M; Giner, E, *Mater. Sci. Eng.: C*, **120**, 111754, 2021.
46. Hwang, B-K; Kim, S-K; Kim, J-H; Lee, J-M, *Int. J. Mech. Sci.*, **180**, 105657, 2020.
47. Cao, S; Liu, T; Jones, A; Tizani, W, *Mechanics Mater.*, **136**, 103081, 2019.
48. Tang, N; Lei, D; Huang, D; Xiao, R, *Polym. Testing*, **73**, 359-65, 2019.
49. Liu, Y; Rahimidehgolan, F; Altenhof, W, *Polym. Testing*, **91**, 106836, 2020.
50. Garbacz, T; Palutkiewicz, P, *Cellular Polym.*, **34**, 4, 189-214, 2015.
51. Ji, G; Wamg, J; Zhai, W; Lin, D; Zheng, W, *J. Cellular Plast.*, **49**, 2, 101-17, 2013.
52. Ebnesaijad, S, **Fluoroplastics**, *William Andrew*, 2015, pp. 412-31.
53. Satrijo, A; Erickson, K P; Ma, J; Marshall, D A; Martin, M C; Peloquin, R L; Wang, S, **WO2013155362**, *3M Innovative Properties Company*, Oct. 17, 2013.
54. Iwase, T; Kawata, J; Kousaka, T; Takahashi, N, **US20130267619**, *Nitto Denko Corporation*, Oct. 10, 2013.
55. Corinti, E; Zanacchi, L; Benvenuti, A, **EP2582739**, *Dow Global Technologies LLC*, Apr. 24, 2013.
56. Li, X; Wang, G; Yang, C; Zhao, J; Zhang, A, *Polym. Testing*, **93**, 106891, 2021.
57. Ramirez, B J; Gupta, V, *Int. J. Mech. Sci.*, **150**, 29-34, 2019.
58. Nishi, T; Yamagichi, T; Shibata, K; Hokkirigawa, *Biotribology*, **22**, 100128, 2020.
59. Sun, X; Zhao, X; Ye, L, *Compos. Part A: Appl. Sci. Manuf.*, **126**, 105579, 2019.
60. Hwang, S-S, *Composites Part B: Eng.*, **98**, 1-8, 2016.
61. Li, C; Chen, L; Ding, J; Zhang, J; Li, X; Zheng, K; Zhang, X; Tian, X, *Carbon*, **104**, 90-105, 2016.
62. Monnereau, L; Urbanczyk, L; Thomassin, J-M; Pardoen, T; Bailly, C; Huynen, I; Jérôme, C; Detrembleur, C, *Polymer*, **59**, 117-23, 2015.
63. Ameli, A; Jung, P U; Park, C B, *Carbon*, **60**, 379-91, 2013.
64. Jorba, D A; Thomas, E L, **US20120322917**, *Massachussetts Institute of Technology*, Dec. 20, 2012.
65. Li, X; Wang, G; Yang, C; Zhao, J; Zhang, A, *Polym. Testing*, **93**, 106891, 2021.
66. Li, Y; Lan, X; Wu, F; Liu, J; Zheng, W, *Compos. Commun.*, **22**, 100508, 2020.
67. Feng, D; Liu, P; Wang, G, *Composites Part A: Appl. Sci. Manuf.*, **124**, 105463, 2019.
68. Heim, H-P; Tromm, M, *J. Cellular Plast.*, **52**, 3, 299-319, 2016.
69. Xu, M Z; Bian, J J; Han, C Y; Dong, L S, *Macromol. Mater. Eng.*, **302**, 149-59, 2016.
70. Jahani, D; Ameli, A; Saniei, M; Ding, W; Park, C B; Naguib, H E, *Macromol. Mater. Eng.*, **300**, 48-56, 2015.
71. Radovanovic, R; Jaso, V; Pilic, B; Stoiljkovic, D, *Hem. Ind.*, **68**, 6, 701-7, 2014.
72. Zepnik, S; Hendriks, S; Kabasci, S; Radusch, H-J, *J. Mater. Res.*, **28**, 17, 2394-2400, 2013.
73. Fan, C; Wan, C; Gao, F; Huang, C; Xi, Z; Xu, Z; Zhao, L; Liu, T, *J. Cellular Plast.*, **52**, 3, 277-98, 2016.
74. Arencon, D; Antunes, M; Realinho, V; Velasco, J I, *Polym. Testing*, **43**, 163-72, 2015.
75. Nofar, M, *Mater. Design*, **101**, 24-34, 2016.
76. Okolieocha, C; Raps, D; Subramaniam, K; Altstädt, V, *Eur. Polym. J.*, **73**, 500-19, 2015.
77. Iwase, T; Kawata, J; Kousaka, T; Takahashi, N, **US20130267619**, *Nitto Denko Corporation*, Oct. 10, 2013.
78. Yu, S-c J; Shen, X, **WO2014039218**, *Nike International Ltd, Nike, Inc.*, Mar. 13, 2014.

References

79. Lipishan, C; Clayfield, T E; Delatte, S M, **EP2925797**, *Dow Global Technologies LLC*, Oct. 7, 2015.
80. Liu, J; Qin, S; Wang, G; Gao, Y, *Polym. Testing*, **93**, 106974, 2021.
81. Yeh, S-K; Demewoz, N M; Kurniawan, V, Polym. Testing, 93, 107004, 2021.
82. Hamad, W N F W; Teh, P L; Yeoh, C K, *Polym.-Plast. Technol. Eng.*, **52**, 754-60, 2013.
83. Gómez-Monterde, J; Hain, J; Maspoch, M L, *J. Mater. Proc. Technol.*, **268**, 162-70, 2019.
84. Yousefian, H; Rodrigue, D, *J. Appl. Polym. Sci.*, **132**, 42845, 2015.
85. Labella, M; Zeltmann, S E; Shunmugasamy, V C; Gupta, N; Rohatgi, P K, *Fuel*, **121**, 240-9, 2014.
86. Marsavina, L; Sadowski, T; Knec, M; Negru, R, *Int. J. Non-linear Mech.*, **45**, 10, 969-75, 2010.
87. Tippet, J; Daniels, L; Gonzalez, J, **US20120187019**, *Fina Technology, Inc.*, Jul. 26, 2012.
88. Vorawongsagul, S; Pratumpong, P; Pechyen, C, *Food Packaging Shelf Life*, **27**, 100608, 2021.
89. Wang, G; Zhang, D; Wan, G; Li, B; Zhao, G, *Polymer*, **181**, 121803, 2019.
90. Restrepo-Zapata, N C; Osswald, T A; Hernandez-Ortiz, J P, *Polym. Eng. Sci.*, **55**, 2073-88, 2015.
91. Ma, Z; Zhang, G; Yang, Q; Shi, X; Shi, A, *J. Cellular Plast.*, **50**, 1, 55-79, 2014.
92. Balashova, I M; Danner, R P, *Fluid Phase Equilibria*, **428**, 92-4, 2016.
93. Hwang, Y D; Cha, S W, *Polym. Testing*, **21**, 269-75, 2002.
94. Yoon, J D; Cha, S W, *Polym. Testing*, **20**, 287-93, 2001.
95. Li, R; Zhang, Z; Fang, T, *J. Supercritical Fluids*, **110**, 110-6, 2016.
96. Landers, R; Hubel, R, **US20150031781**, *Evonik Industries AG*, Jan. 29, 2015.
97. Xu, L-Q; Huang, H-X, *J. Supercritical Fluids*, **109**, 117-85, 2016.
98. Yao, S; Hu, D; Xi, Z; Zhao, L, *Polym. Testing*, **90**, 106649, 2020.
99. Álvarez, I; Gutiérrez, C; García, M T, *J. Supercritical Fluids*, **164**, 104886, 2020.
100. Bledzki, A K; Rohleder, M; Kirschling, H; Chate, A, *J. Cellular Plast.*, **46**, 415-40, 2010.
101. Giles, H F; Wagner, J R; Mount, E M, **Extrusion, Foam Extrusion**. *William Andrew*, 2005. pp. 513-7.
102. Bao, J-B; Junior, A N; Weng, G-S; Wang, J; Fang, Y-W; Hu, G-H; *J. Supercritical Fluids*, **111**, 63-73, 2016.
103. Mantaranon, N; Chirachanchai, S, *Polymer*, **96**, 54-62, 2016.
104. Mi, H-Y; Chen, J-W; Geng, L-H; Chen, B-Y; Jing, X; Peng, X-F, *Mater. Lett.*, **167**, 274-7, 2016.
105. Notario, B; Pinto, J; Rodriguez-Perez, M A, *Polymer*, **63**, 116-26, 2015.
106. Bao, J-B; Liu, T; Zhao, L; Hu, G-H; Niao, X; Li, X, *Polymer*, **53**, 25, 5982-93, 2012.
107. Al Jahwari, F; Huang, Y; Naguib, H E; Lo, J, *Polymer*, **98**, 270-81, 2016.
108. Matuana, L M, *Bioresource Technol.*, **99**, 9, 3643-50, 2008.
109. Wen, S; Yu, L; Phule, A D; Zhang, Z X, *J. Supercritical Fluids*, **165**, 104987, 2020.
110. Xie, X; Gu, A; Liu, P; Liang, G; Yuan, L, *Polym. Adv. Technol.*, **22**, 1731-7, 2011.
111. Abraham, A A; Chauhan, R; Srivastava, A K; Katiyar, M; Tripathi, D N, *J. Polym. Mater.*, **28**, 2, 267-74, 2011.
112. Gedler, G; Antunes, M; Velasco, J I; Ozisik, R, *Mater. Design*, **90**, 906-14, 2016.
113. Wang, L; Liu, X; Liu, C; Liu, X, *Chem. Eng. J.*, **384**, 123231, 2020.
114. Abdel-Baset, T A; Hekal, E A; Azab, A A; Anis, B, *Physica B: Condensed Matter*, in press, 412666, 2021.
115. Arjmand, M; Mahmoodi, M; Park, S; Sundararaj, U, *J. Cellular Plast.*, **50**, 6, 551-62, 2014.
116. Notario, B; Pinto, J; Verdejo, R; Rodriguez-Perez, M A, *Polymer*, **107**, 302-5, 2016.
117. Li, J-x; Zhang, G; Li, Z-s; Wang, X-l; Liu, X-q, *Chem. Res. Chinese Uni.*, **24**, 2, 215-9, 2008.
118. Sun, X; Ye, L; Zhao, X, *Eur. Polym. J.*, **116**, 488-96, 2019.
119. Abbasi, H; Antunes, M; Velasco, J J, *Prog. Mater. Sci.*, **103**, 319-73, 2019.
120. Liu, P S; Chen, G F, **Porous Materials. Producing Polymer Foams**. *Butterworth Heinemann*, 2014, pp. 345-82.
121. Takeda, S; Ohashi, M; Ichihara, M, *J. Volcanology Geothermal Res.*, **393**, 106771, 2020.
122. Zhang, H; Liu, T; Li, B; Xin, Z, *Polymer*, in press, 123209, 2021.
123. Kang, S M; Kim, M J; Kwon, S H; Park, H; Jeong, H M; Kim, B K, *J. Mater. Res.*, **27**, 22, 2837-43, 2012.
124. Kang, S M; Kim, M J; Kwon, S H; Park, H; Jeong, H M; Kim, B K, *Polym. Bull.*, **70**, 885-93, 2013.
125. Santo, L, **Progress in Aerospace Sciences. Shape memory polymer foams**, *Elsevier*, 2016, pp. 60-65.
126. Kumar, K S S; Biju, R; Nair, C P R, *Reactive Funct. Polym.*, **73**, 2, 421-30, 2013.
127. Santo, L; Quadrini, F; Villadei, W; Mascetti, G; Zolesi, V, *Procedia Eng.*, **104**, 50-56, 2015.
128. Leng, J; Lan, X; Liu, Y; Du, S, *Prog. Mater. Sci.*, **56**, 7, 1077-11-35, 2011.
129. Zhu, S; Zhou, Q; Wang, M; Ye, C, *Compos. Part B: Eng.*, **204**, 108497, 2021.
130. Hui, J; Fu, Y; Qiu, Y; Ni, Q-Q, *Matter. Lett.*, **264**, 127343, 2020.
131. Garrido, M; Correia, J R; Keller, T, *Constr. Build. Mater.*, **76**, 150-7, 2015.
132. Chen, Y; Das, R; Battley, M, *Int. J. Solids Struct.*, **52**, 150-64, 2015.
133. Meller, M; Li, J; Dolega, J, **EP2671911**, *Armacell Enterprise GmbH*, Dec. 11, 2013.
134. Jia, D; Yan, S; Peng, Y; Wan, Q, *Polym. Deg. Stab.*, in press, 109410, 2021.
135. Zhang, B S; Zhang, Z X; Lv, X F; Lu, B X; Xin, Z X, *Polym. Eng. Sci.*, **52**, 218-24, 2012.

136 Garancher, J-P; Fernyhough. A, *Polym. Deg. Stab.*, **100**, 21-8, 2014.
137 Balbo, B M; Wang, Y; Hwang, C H; Xi, B, **WO2014037476**, *BASF SE*, Mar. 13, 2014.
138 Lan, B; Li, P; Luo, X; Gong, P, *Polymer*, **212**, 123159, 2021.
139 Tomyangkul, S; Pongmuksuwan, P; Harnnarongchai, W; Chaochanchaikul, K, *J. Reinfor. Plast. Compos.*, **35**, 8, 672-81, 2016.
140 Sung, G; Kim, J W; Kim, J H, *J. Ind. Eng. Chem.*, **44**, 99-104, 2016.
141 Sun, X; Liang, W, *Compos. Part B: Eng.*, **87**, 21-26, 2016.
142 Baek, S H; Kim, J H, *Compos. Sci. Technol.*, **198**, 108325, 2020.
143 Li, J; Wang, A; Zhang, H; Ma, Z, Zhang, G, *Compos. Part A: Appl. Sci. Manuf.*, **140**, 106144, 2021.
144 Rodenburg, C;Viswanathan, P; Battaglia, G, Ultramicroscopy, 139, 13-19, 2014.
145 Aktas, S; Gevgilili, H; Kucuk, I; Sunol, A; Kalyon, D M, *Polym. Eng. Sci.*, **54**, 2064-74, 2014.
146 Bledzki, A K; Faruk, O, *Compos. Part A: Appl. Sci. Manuf.*, **37**, 9, 1358-67, 2006.
147 Dougherty, E P; Edgett, K; Turng, L-S; Lacey, C; Lee, J; Gorton, P J; Sun, X, **WO2012034036**, *Playtex Products, Inc.*, Mar. 12, 2012.
148 Wang, J; Nguyen, A V; Farrokhpay, S, *Adv. Colloid Interface Sci.*, **228**, 55-70, 2016.
149 Breiss, H; El Assal, A; Harmouch, Z, *Mater. Res. Bull.*, **137**, 111188, 2021.
150 Chen, Y; Zhang, Y; Xu, C; Cao, X, *Carbohydrate Polym.*, **130**, 149-54, 2015.
151 Shih, R-S; Kuo, S-W; Chang, F-C, *Polymer*, **52**, 3, 752-9, 2011.
152 Moscoso-Sanchez, FJ; Mendizabal, E; Jasso-Gastinel, C F; Ortega-Gudino, P; Robledo-Ortiz, J R; Gonzalez-Nunez, M; Rodrigue, D, *J. Cellular Plast.*, **51**, 5-6, 489-503, 2015.
153 Zeng, C; Hossieny, N; Zhang, C; Wang, B; Walsh, S M, *Compos. Sci. Technol.*, **82**, 29-37, 2013.
154 Peng, J; Sun, X; Mi, H; Jing, X; Peng, X-F; Turng, L-S, *Polym. Eng. Sci.*, **55**, 1634-42, 2014.
155 Vo, C V; Bunge, F; Duffy, J; Hood, L, *Cellular Polym.*, **30**, 3, 137-55, 2011.
156 Zuo, K; Xu, J; Xie, S; Chen, J, *J. Supercritical Fluids*, in press, 105161, 2021.
157 He, Y; Wu, J; Qiu, D; Yu, Z, *Int. J. Mech. Sci.*, **193**, 106164, 2021.
158 Leung, C K Y; Cheung, K F; Zhu, H; Lin, S W, **US20130216802**, *Nano and Advanced Materials Institute Limited*, Aug. 22, 2013.
159 Meller, M; Li, J; Dolega, J, **EP2671911**, *Armacell Enterprise GmbH*, Dec. 11, 2013.
160 Bowman, J M; Williams, D J, **WO2014015315**, *Honeywell International Inc.*, Jan 23, 2014.
161 Rochefort, M; Ripley, L; Holland, P; Coppock, V, **WO2014044715**, *Kingspan Holdings (Irl) Limited*, Mar. 27, 2014.
162 Zhang, R; Kim, E S; Romero-Diez, S; Lee, P C, *Chem. Eng. J.*, **409**, 128251, 2021.

13

ANALYTICAL TECHNIQUES USEFUL IN FOAMING

Many analytical methods are used to analyze and improve foaming processes. Many such applications do not differ much from applications in other areas of applied polymer chemistry. Here, we give some specific examples which are particularly useful in the foaming processes.

13.1 CELL DENSITY

The most popular method of determination of cell density (more than 300 hits on Science Direct for foams) is based on ImageJ.[1] The method of sample preparation involved the following operations:[1] Foam samples were cut and photographed with a camera. Also, SEM images were recorded.[1] Both optical and SEM images were processed and analyzed using the software ImageJ to obtain the average value of the cell diameter and the cell size distribution.[1] The images selected for analysis had a resolution chosen in such a manner that 60-200 cells were visible, and the smallest cells were at least 10 pixels across.[1] Corrections specified in the ASTM standard D3576[2] were applied to obtain the cell size for a 3D distribution of cells from a plane image.[1] Cell density was calculated using the following equation:[1]

$$N = \frac{6}{\pi \langle D^3 \rangle}\left(1 - \frac{\rho_f}{\rho_s}\right) \qquad [13.1]$$

where:
- D — cell diameter
- $\pi \langle D^3 \rangle / 6$ — the mean volume of cell
- ρ_f — foam density
- ρ_s — density of solid pellets

ImageJ is an open-source Java image processing software, which can be downloaded from many sources, including US government site.[3] Very active user community offers many variations (flavors) of the software and many plug-ins.

Images for analysis are converted to gray-scale images.[4] A threshold value is set in ImageJ to distinguish a particle in the image.[4] A minimum particle size and circularity are selected from the analytical module.[4] The obtained data are imported into Microsoft Excel for further analysis.[4]

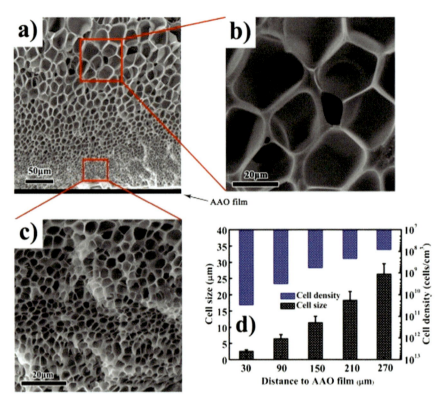

Figure 13.1. a) SEM images of PS foams on anodized aluminum oxide film with pores of ~150 nm (saturating pressure of 13.8 MPa and foaming temperature of 100°C) and b) top and c) bottom partially enlarged SEM images, and d) average cell size and cell density at different positions. [Adapted, by permission, from Yu, J; Chen, F; Fan, P; Sun, L; Zhong M; Yang, J; *Mater. Today Commun.*, **9**, 1-6, 2016.]

13.2 CELL SIZE

Similar to cell density, the cell size is determined by the image analysis. The image can be acquired by many available digital cameras and image capture cards, as well as scientific image viewing equipment. It is possible to follow the cell growth rate by the time-lapse acquisition. These images are then processed by image analysis software, such as ImageJ or commercial image processing software, such as, for example, Image-Pro Plus from Media Cybernetics.[5] Their website shows essential steps of image analysis.[6] Gatan Microscopy Suite® is software for (scanning) transmission electron microscope experimental control and analysis.[7] It is used with various research instruments, such as scanning electron microscopes, fast STEM spectrum imaging, EELS analysis, and many others.[7,8] Figure 13.1 shows an example of results that can be obtained using SEM images.

A numerical model was developed to simulate bubble nucleation and growth during depressurization of thermoplastic polymers saturated with supercritical blowing agents.[9] The model has the ability to predict cell size distribution based on the specific process conditions and polymer properties.[9] Experimental data obtained by foaming acrylate copolymers with carbon dioxide compare well with model predictions of average cell size, porosity, and cell size distribution.[9]

13.3 DENSITY

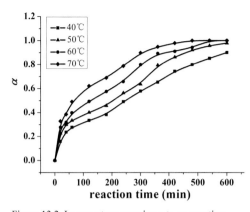

Figure 13.2. Isocyanate conversion rate vs. reaction time at different temperatures. [Adapted, by permission, from Huo, S-P; Nie, M-C; Kong, Z-W; Wu, G-M; Chen, J, *J. Appl. Polym. Sci.*, **125**, 152-7, 2012.]

Gas adsorption methods give information regarding micropores (<2 nm) and mesopores (2-50 nm), while mercury porosimetry is the standard technique used for the analysis of meso- and larger macropores (0.004-950 μm).

The evaluation of the closed-cell or open-cell content of foams is based on the volume determination by means of gas pycnometer. The analysis is performed according to ISO 4590[10] using Penta-Foam.[11] Foam sample is cut to a form of cube or cylinder, its geometric volume measured with a micrometer screw.[11] The sample is then analyzed in a pycnometer at a pressure of about 0.25 bar.[11] The instrument can be used in the corrected analysis procedure in which cells opened by cutting a specimen are eliminated from the final result.[11] All required information for this procedure and sample preparation is included in ISO standard.[10]

13.4 DIFFERENTIAL SCANNING CALORIMETRY

DSC measurements of poly(lactic acid)/polycarbonate blends foamed using supercritical carbon dioxide were used for determination of enthalpy of crystallization on heating, ΔH_c, and melting enthalpy, ΔH_m, heats of fusion of pure crystalline phases of both polymers, and crystallinity of their blend.[12] The cold crystallization and melting behavior of various poly(lactic acid)/maleic-anhydride-grafted-polypropylene blends were studied using DSC to determine glass transition temperature, cold crystallization temperature, and melting temperature of blends and pure polymer.[13]

Determination of crystallinity and glass transition temperature are the two most important applications of DSC in the analysis of foams.

13.5 FOURIER TRANSFORM INFRARED

Fourier transform infrared-attenuated total reflectance spectroscopic analysis was used to study the kinetics of the formation of the polyurethane foam.[14] The crosslinking reaction was followed by monitoring the change of absorbance of isocyanate group at 2270 cm^{-1}.[14] The isocyanate conversion rate was calculated from the following equation:[14]

$$\alpha = \frac{\dfrac{A_0/\varepsilon l_0}{A_0^r/\varepsilon^r l_0} - \dfrac{A_t/\varepsilon l_t}{A_t^r/\varepsilon^r l_t}}{\dfrac{A_0/\varepsilon l_0}{A_0^r/\varepsilon^r l_0}} = \frac{\dfrac{A_0}{A_0^r} - \dfrac{A_t}{A_t^r}}{\dfrac{A_0}{A_0^r}} = 1 - \frac{A_t A_0^r}{A_t^r A_0} \qquad [13.2]$$

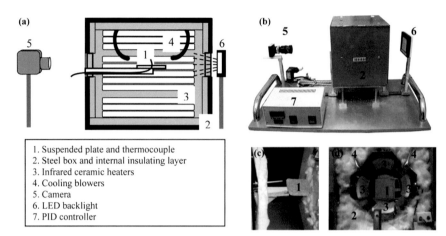

Figure 13.3. Optical expandometer. [Adapted, by permission, from Solorzano, E; Antunes, M; Saiz-Arroyo, C; Rodriguez-Perez, M A; Velasco, J I; de Saja, J A, *J. Appl. Polym. Sci.*, **125**, 1059-67, 2012.]

where:

A_0 absorbance of NCO group at time 0
A_t absorbance of NCO group at time t
A_0^r absorbance of benzene ring at time 0
A_t^r absorbance of benzene ring at time t
l path length of the beam
ε molar attenuation

Figure 13.2 shows an example of data of the isocyanate conversion rate vs. reaction time.[14] These FTIR data can be further recalculated to determine reaction rate, kinetics of the reaction, and the reaction rate constants.[14]

13.6 OPTICAL EXPANDOMETRY

Figure 13.3 shows the diagram of the optical expandometer, which is an instrument designed to determine *in situ* the free expansion kinetics of chemically foamed thermoplastic foams.[15] The camera monitors a free-foaming of material placed inside a furnace. Images are acquired under special illumination to facilitate image processing.[15] The method permits measurement of the foam evolution without affecting the material's expansion and collapse.[15] Novadep has developed an optical expandometer to visualize and quantify processes "in-situ," such as foaming at high temperatures (1000°C).[16]

13.7 POLARIZING OPTICAL MICROSCOPE

Spherulitic morphology of foam samples can be observed using a polarizing optical microscope.[13] The method was used to study the effect of maleic-anhydride-grafted-polypropylene on the formation of poly(lactic acid) spherulites in blends of both polymers.[13] The presence of maleic-anhydride-grafted-polypropylene increased the formation of spherulites because branched chains provided more heterogeneous nucleating points.[1] POM micrographs were also used to inspect the morphology of layer composites containing alternate foamed and unfoamed layers.

13.8 Scanning electron microscopy

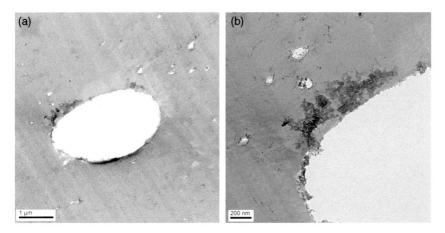

Figure 13.4. TEM images of the foamed MWCNT/PS nanocomposites: (a) low-magnification and (b) high magnification. [Adapted, by permission, from Arjmand, M; Mahmoodi, M; Park, S; Sundararaj, U, *J. Cellular Plast.*, **50**, 6, 551-62, 2014.]

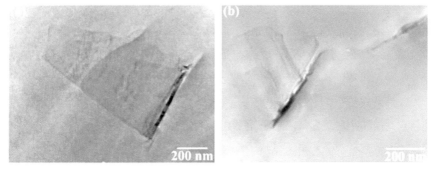

Figure 13.5. TEM micrographs of graphene nanoparticles present in PC-GnP nanocomposite foams prepared *via* 1 step foaming: (a) X1PC05 foams exhibited thinner and more close GnP nanoparticles when compared to (b) X1PC2 foams. [Adapted, by permission, from Gedler, G; Antunes, M; Borca-Tasciuc, T; Velasco, J I; Ozisik, R, *Eur. Polym. J.*, **75**, 190-9, 2016.]

13.8 SCANNING ELECTRON MICROSCOPY

SEM is a very important technique used to observe the morphology of foams, especially important in the case of microfoams, which require high magnifications. Preparation of sample should restrict artifacts, which can be easily induced in relatively elastic foam materials. Foams are usually placed in the liquid nitrogen for 3 hours before microtoming to avoid damage to morphological structures.[18] The surface of foam samples for SEM viewing is coated by sputtering with palladium, gold, or their mixture.[18] As discussed in Sections 13.1 and 13.2, micrographs can be digitized and used with dedicated software to analyze selected features of morphology.[18] Study of polypropylene having fiber-reinforced skin was conducted using image analysis to determine the influence of skin thickness and reinforcement on mechanical properties of the foam.[18]

The cell density and the number of cells per unit volume of poly(lactic acid)/polycarbonate foam were estimated by analyzing the SEM micrograph using the Image-Pro Plus software.[12] Similar studies were done on poly(phenylene sulfide)/poly(ether sulfone) blend foams and many other foams.[19]

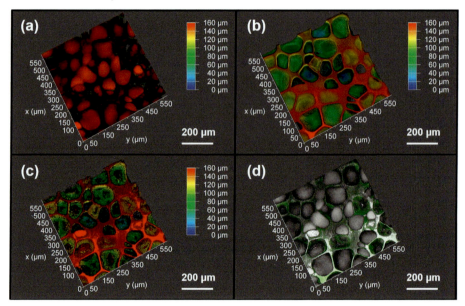

Figure 13.6. CLSM micrographs of Nile Blue A stained TPU foam: (a) transmission mode, (b) fluorescence mode, (c) reflection mode and (d) multi-channel overlay mode. [Adapted, by permission, from Ma, H; Gong, P; Zhai, S; Li, G, *Chem. Eng. Sci.*, **207**, 892-902, 2019.]

13.9 TRANSMISSION ELECTRON MICROSCOPY

High magnification capabilities of TEM give results that cannot be obtained by other microscopy methods. Figure 13.4 illustrates this for polystyrene/multiwalled carbon nanotube foam.[20] The dark areas on the perimeters of foam cells are made out of carbon nanotubes.[20] Their agglomerates were formed during the cell growth step.[20] Figure 13.5 shows that the thickness of graphene dispersed in polycarbonate foam can be assessed using TEM.[21]

13.10 CONFOCAL LASER SCANNING MICROSCOPY

Confocal Laser Scanning Microscopy, CLSM, is a 3D characterization method.[22] Compared with SEM, POM, and TEM, the main advantage of CLSM is to perform an optical section and to provide micrographs of a bulk of a sample without sample destruction.[22] The micrographs obtained from CLSM provide reliable information on the internal foam morphology.[22] CLSM works by focusing on a single 2D plane at a selected depth.[22] Micrographs at different depths can then be obtained by changing the position of the focal plane, which is called an optical section.[22] Transmission, reflection, and fluorescence modes can be simultaneously used to get a comprehensive 3D structure (Figure 13.6).[22]

13.11 X-RAY ANALYSIS

Crystalline structures of components of foamed polymers in the blend (poly(lactic acid) and maleic-anhydride-grafted-polypropylene) can be clearly distinguished using x-ray diffraction.[13] This, in turn, helps in identification that spherulites were made from poly(lactic acid).[13] The poly(phenylene sulfide)/poly(ether sulfone) blend has its crystalline structure dependent on poly(phenylene sulfide) because poly(ether sulfone) is amorphous.[19] The analysis of blends by wide-angle x-ray diffraction shows that poly(ether sulfone) interferes with the crystallization of poly(phenylene sulfide) by partial destruction of the continuity of poly(phenylene sulfide) phase because the diffraction peak of poly(phenylene sulfide) becomes much weaker at higher concentrations of poly(ether sulfone).[19]

At a low pressure of nitrogen in poly(lactic acid) (30 bars), the dissolved gas promoted the crystal nucleation. The crystals were freely growing to the larger sizes without any interference by other crystals until a very late stage. As a result, a closely-packed crystalline structure was obtained. When nitrogen pressure was higher (60 bars), more gas was dissolved in PLA, which led to a larger number of nucleated crystals. The multitude of growing crystals increased the polymer chain entanglement, which hindered the crystal growth and, that is why a loosely packed crystalline structure and a relatively low degree of crystallinity were observed (Figure 13.7).[23]

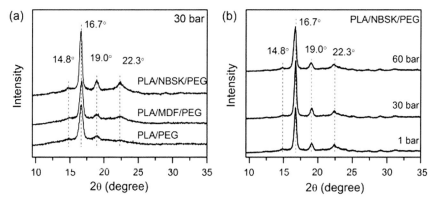

Figure 13.7. WAXD diffraction patterns of (a) PLA materials at 30 bar and (b) PLA/NBSK (northern bleached softwood kraft)/PEG (polyethyleneglycol) composites under various N$_2$ pressures cooled at a cooling rate of 2°C/min. [Adapted, by permission, from Ding, W D; Jahani, D; Chang, E; Alemdar, A; Park, C B; Sain, M, *Compos. Part A: Appl. Sci. Manuf.*, **83**, 130-9, 2016.]

REFERENCES

1 Lobos, J; Iasella, S; Rodriguez-Perez, M A; Velankar, S S, *Polym. Eng. Sci.*, **56**, 9-17, 2016.
2 ASTM D3576-20 Standard Test Method for Cell Size of Rigid Cellular Plastics.
3 Rasband, W S; ImageJ, U. S. National Institutes of Health, Bethesda, Maryland, USA, http://imagej.nih.gov/ij
4 Derikvand, Z; Riazi, M, *J. Molec. Liquids*, **B224**, 1311-8, 2016.
5 Mechraoui, A; Riedl, B; Rodrigue, D, *J. Cellular Plast.*, **47**, 2, 115-32, 2011.
6 http://www.mediacy.com/imageproplus
7 http://www.gatan.com/products/tem-analysis/gatan-microscopy-suite-software
8 Yu, J; Chen, F; Fan, P; Sun, L; Zhogn M; Yang, J; *Mater. Today Commun.*, **9**, 1-6, 2016.
9 Khan, I; Adrian, D; Costeux, S, *Chem. Eng. Sci.*, **138**, 634-45, 2015.
10 ISO 4590:2016 Rigid cellular plastics - Determination of the volume percentage of open cells and of closed cells.

11 https://www.anton-paar.com/ca-en/products/details/ultrafoam-and-pentafoam/
12 Bao, D; Liao, X; He, G; Huang, E; Yang, Q; Li, G, *J. Cellular Plast.*, **51**, 4, 349-72, 2015.
13 Wang, X; Liu, W; Li, H; Du, Z; Zhang, C, *J. Cellular Plast.*, **52**, 1, 37-56, 2016.
14 Huo, S-P; Nie, M-C; Kong, Z-W; Wu, G-M; Chen, J, *J. Appl. Polym. Sci.*, **125**, 152-7, 2012.
15 Solorzano, E; Antunes, M; Saiz-Arroyo, C; Rodriguez-Perez, M A; Velasco, J I; de Saja, J A, *J. Appl. Polym. Sci.*, **125**, 1059-67, 2012.
16 http://www.novadep.com/other-products/
17 Han, T; Wang, X; Xiong, Y; Li, J; Guo, S; Chen, G, *Compos. Part A: Appl. Sci. Manuf.*, **78**, 27-34, 2015.
18 Mechraoui, A; Riedl, B; Rodrigue, D, *J. Cellular Plast.*, **47**, 2, 115-32, 2011.
19 Ma, Z; Zhang, G; Shi, X; Yang, Q; Li, J; Liu, Y; Fan, X, *J. Appl. Polym. Sci.*, **132**, 42634, 2015.
20 Arjmand, M; Mahmoodi, M; Park, S; Sundararaj, U, *J. Cellular Plast.*, **50**, 6, 551-62, 2014.
21 Gedler, G; Antunes, M; Borca-Tasciuc, T; Velasco, J I; Ozisik, R, *Eur. Polym. J.*, **75**, 190-9, 2016.
22 Ma, H; Gong, P; Zhai, S; Li, G, *Chem. Eng. Sci.*, **207**, 892-902, 2019.
23 Ding, W D; Jahani, D; Chang, E; Alemdar, A; Park, C B; Sain, M, *Compos. Part A: Appl. Sci. Manuf.*, **83**, 130-9, 2016.

14

HEALTH AND SAFETY AND ENVIRONMENTAL IMPACT OF FOAMING PROCESSES

Blowing agents are relatively safe in use but have created environmental problems because of their ozone depletion potential and, most importantly, their global warming potential. The discussion below will include the following topics related to the health, safety, and environmental impacts of blowing agents:
- flammability and reactivity
- transportation restrictions
- general toxicity
- cancer risk
- mutagenic properties
- atmospheric lifetime
- ozone depletion potential
- global warming potential

Autoignition temperature is one of the characteristics of material flammability. Several currently used blowing agents have low autoignition temperature, low flash point, and a self-accelerating decomposition temperature. Table 14.1 presents data for the most flammable among blowing agents.

Table 14.1. Flammability data

Blowing agent	Autoignition, °C	Flash point, °C	Self-accelerating decomposition, °C
4-toluenesulfonyl hydrazide	90	60	60
4,4'-oxybis(benzenesulfonyl hydrazide)	150	115	90
2,2'-azodi(isobutyronitrile)	64	96.6	
cyclopentane	361	-37	
isopentane	420	-49	
N,N-dinitrosopentamethyl tetramine			76
neopentane	450	-51	
pentane	260		

The above substances should be handled with care, especially during transportation. Air transportation is forbidden in the case of 2,2'-azodi(isobutyronitrile) and azodicarbonamide.

Blowing agents do not belong to very toxic substances. Their LC50 values are usually reasonably high (>2000 or more). Only a few (toluenesulfonyl hydrazide – 283, N,N-dinitroso pentamethylene tetramide – >240 and p-toluenesylfonyl semicarbazide – 980) have lower LD50 values. None of the common commercial blowing agents is considered carcinogenic. They are also not reported to be mutagenic.

Carbon dioxide is the most used physical blowing agent due to its low toxicity, high stability, and low cost.[1]

The major hurdles in the application of the blowing agents are related to their environmental effects. Below we tabulate atmospheric lifetimes, global warming potentials, and ozone depletion potentials of commercial blowing agents that are a matter of concern.

Table 14.2. Environmental data

Blowing agent	Atmospheric lifetime, years	Global warming potential, 100 y ITH	Ozone depletion potential (CFC-11=1)
HCFC-141b	9.3	630	0.11
HCFC-142a	17.9	2310	0.07
HCFC-142b	17.9	2400	0.076
HFC-134a	13	1430	0
HFC-227ea	34.2	3360	0
HFC-245fa	7.6	950	0
HFC-365mfc	8.4	960	0
HFC-1233zd	26 days	5	0
Cyclopentane	a few days	7	0
Carbon dioxide	200	1	0
Methane	2	28-36	0
Nitrous oxide	115	265-298	0.017

HCFCs were used as temporary replacements for CFCs based on Montreal Protocol and subsequent amendments to Vienna Convention. European Union accelerated phase-out of HCFC, leading to a 75% reduction in 2004 and a complete phase-out in 2010.[2] The reason for phase-out was their high ozone depletion potential. Hydrofluorocarbons have zero ozone depletion potential, but some have high global warming potentials.[3] For this reason, volatile hydrocarbons, such as cyclopentane, and certain unsaturated fluorinated compounds, such as 1-chloro-3,3,3-trifluoropropene (1233zd), have been introduced due to their zero ozone depletion potential and very low global warming potential.[4] The last three rows show data for the most frequent pollutants, which affect atmospheric changes leading to global warming. Nitrous oxide is the most influential factor in the global warming increase. It comes from natural sources in addition to anthropogenic sources. Considering the amount of nitrous oxide emitted and its long atmospheric lifetime, it is the single most important ozone-depleting emission.[5] It can be safely concluded that the use of the present generation of blowing agents has negligible influence on global warming. Much stronger influence comes from natural sources and agricultural activities.

REFERENCES

1. Coste, G; Negrell, C; Caillol, S, *Eur. Polym. J.,* **140**, 110029, 2020.
2. Derwent, R G; Simmonds, P G; Greally, B R; O'Doherty, S; McCulloch, A; Manning, A; Reimann, S; Folini, D; Vollmer, M K, *Atmos. Environ.*, **41**, 4, 757-67, 2007.
3. Vo, C V; Fox, R T, *J. Cellular Plast.*, **49**, 5, 423-38, 2013.
4. Long, Y; Sun, F; Liu, C; Xie, X, *RCS Adv.*, **6**, 23726, 2016.
5. Ravishankara, A R; Daniel, J S; Portmann, R W, *Science*, **326**, 5949, 123-5, 2009.

INDEX

Numerics

1,1,1,2-tetrafluoroethane 189
1,1,1,3-tetrafluoropropene 189
1,1,1,4,4,4-hexafluoro-2-butene 150
1,1,1-trifluoropropene 189
1,1-difluoroethane 189
1,3,3,3-tetrafluoropropene 192
1,4-butane diol 103
1-chloro-3,3,3-trifluoropropene 251, 257
2,2'-azodi(isobutyronitrile) 270
2,4,4,4-tetrafluorobutene-1 193
2D X-ray image 194
2-ethyl hexanol 87
2-propanol 189
3D
  printing 154
  structure 221
  x-ray image 194
3-fluoropropene 189
4,4'-oxybis(benzenesulfonylhydrazide) 143, 147, 200

A

absorbed carbon dioxide concentration 158
absorption efficiency 254
accelerated cure 213
accelerator 1, 2, 209
  function 210
acetic acid 34, 210, 246
acetone 49, 234
acid
  curing catalyst 213
  scavenger 147
acidic reagent 34
acoustic
  absorbing foam 207
  absorption application 109
  damping 162
  foam 97
  insulation 149, 158, 162
acrylic 240
  process aid 200
acrylonitrile-butadiene-acrylate 134
acrylonitrile-butadiene-styrene 242-243
activated carbon 190

activation
  energy 124, 150
  temperature 111
activator 2, 32, 209
active carbon 43, 223
additives 209
adsorption 214
aerogel 129
  foam 190
aerospace 134
agglomerate 45
agglomerated state 167
aggregation 212
air
  bubble 41
  conditioning 207
  flow 237
  pycnometer 85
  temperature profile 69
  transportation 270
aircraft
  construction 141
  interior 168
airplanes 169
alcohol 37
aluminum
  flake 201
  oxide 45
amine-amide adduct 140
amine-functionalized iron oxide 197
aminosilane 149
ammonia 30, 211
ammonium
  chloride 32
  hydrogen carbonate 197
  polyphosphate 150
amorphous
  phase 251
  region 87
  thermoplastic 67
amphiphile 82
analytical methods 261
angular extrusion head 118
anisotropic
  mechanical strength 233

morphology 233
  nanocellular structure 224
anisotropy 113
  index 125
  ratio 224
annealing 37
anode material 190
antioxidant 176
antistatic agent 176
anti-wear effect 221
apparent
  core density 84
  overall density 84
Arburg 121
argon 243
Arrhenius equation 30
artificial leather 196, 210
aspect ratio 111
asphalt-based microporous organic polymer 159
asymmetric density profile 113
atactic polypropylene 211
atmospheric lifetime 269-270
attenuated total reflectance 263
attraction force 244
autoignition 269
  temperature 269
automotive 134, 149
  actuators 250
  industry 187, 253
  seat 196
auxiliary agent basic properties 3
average
  cell
    diameter 143, 152, 197
    size 58, 126, 145, 167, 175-176, 185, 205, 207
  pore diameter 149
aviation industry 168
azo compound 124
azobisformamide 190
azodicarbonamide 2, 29-30, 45-46, 51-52, 56, 61-63, 76, 95, 99-100, 103, 111-113, 118, 120, 125, 129-130, 135, 143, 145-147, 152-153, 162, 171-172, 176-177, 179, 184-185, 189-190, 199-202, 204, 207, 209-211, 215, 217-218, 220, 223, 225, 237, 239, 242-243, 245-246, 248, 251, 255, 270
  concentration 69
  decomposition 69, 203, 225

B

backbone chain 235
backpressure 39, 63, 186, 217
bagasse 253
baking
  powder 34
  soda 34
barium
  salt 209
  /zinc carboxylate 210
  -zinc salt 209
bark powder 217
barrier agent 43, 208
batch
  depressurization 90, 158
  foaming 140
    process 38
benzenesulfonyl hydrazide 207
benzoyl peroxide 212
benzyl butyl phthalate 200
biaxial
  rotational molding 69
  stretching 78, 217, 240, 252
biaxially stretched film 251
bimodal 158
  cell
    distribution 115
    morphology 89, 91, 190
    size distribution 40, 99
    structure 90-92, 244
  cellular foam 93
  distribution 190
  foam 87, 91, 93, 115
  morphology 87, 94, 98
  polycarbonate foam 114
  structure 89, 93, 109
binary system 48
binder layer 195
biobased 152
  phenol formaldehyde foam 150
biocompatibility 134
biodegradable 152
  resin 211
biofoam 129
biomass resources 208
biomedical applications 208, 234
bionanocomposite 42, 208
biopolyurethane 232

Index

birch bark 150
bis(2-ethylhexyl)-1,4-benzenedicarboxylate 203
bismaleimide 135
  resin 246
bismuth-based catalyst 210
bistetrazole diammonium 147
blend 218
blend
  formulation 145
  morphology 42, 235
  phase morphology 87
blistering 71
blow molding 152
blowing
  accelerator 209
  agent 1, 31, 33-34, 36, 43, 45, 48, 51-52, 55, 57, 109, 111, 115, 134, 150, 155-156, 158, 166, 176, 190, 213, 215, 218, 227, 243, 269
    activation temperature 111
    amount 30, 32, 51, 103, 246
    basic properties 3
    concentration 75, 171, 185
    content 42, 51-52
    decomposition 103
      temperature 120, 129
    dispersion 45
    enhancer 193
    evaporation 40
    formation 192
    injection location 116, 171
    mechanism of action 29
    reservoir 126, 170
    residence conditions 37
    selection 70
    sorption 59
  process 32
  ratio 31-32, 153
  time 102
blown
  film extrusion 111
  foam 33
bone
  cell adhesion 155
  defect 182
  regeneration 155, 182
  scaffold 138, 232
  tissue engineering 138, 220
borate crosslinking 221

borax 197
boron nitride 130, 147
bound water 43
Boyle's Law 85
branch 79
branched chain 264
branching modification 153
breathing mold 62
  technology 122
brittle
  failure 190
  transition 190
brittleness 152
bromobutyl rubber 136
bubble 30
  coalescence 57, 76-77, 82, 226
  collapse 76, 78
  density 39, 57, 76
  diameter 175
  dynamics 75
  expansion speed 249
  faith 77-78
  formation 77, 118, 124, 225
  growth 29, 37, 52, 58, 75, 78, 92
  internal pressure 77
  lifetime 68
  nucleation 57, 71, 76, 109, 115, 121, 190, 217, 262
    frequency 31
  nucleus 175
  radius 76
  ripening 76
  size 39, 249
    distribution 71, 225
  sphericity 254
  stability 78
  surface 78
  unrestricted growth 75
  wall 75-76
    geometry 175
building
  materials 169
  wall 256
bulk
  density method 84
  porosity 71, 126
bumper 237
buss kneader 152
butane 40, 48

C

cable
  coating 199
  separator 142
cadmium
  salt 209
  -zinc salt 209
calcium
  carbonate 116, 150, 189, 199, 204, 217-218, 223
    nanoparticle 189
  silica aluminate 45
  silicate 45
calendering 111, 152
cancer
  research 139
  risk 269
capillary
  forces 102
  tube 164
carbon
  black 99, 125, 186, 189, 248
  dioxide 2, 30, 34-35, 37, 48, 54, 87, 115, 118, 143, 155, 167, 169, 175, 179, 182, 185, 192, 243-244, 247, 251, 255, 262
    adsorption 140
    concentration 57
      threshold 55
    desorption diffusivity 235
    diffusion 55, 97
    diffusivity 37
    equilibrium concentration 57
    foaming time 91, 190
    induced crystallization 205
    gas emission 172
    mass-transfer rate 235
    nucleation rate 91, 190
      pressure 92
      reservoir 65, 187
      saturation 36
        pressure 115
      solubility 66, 91-92, 171, 182
        depression 87
    sorption 126
    uptake 60
  fiber 162, 227, 240
  foam 106, 177-178
    particle 213
    microballoon 135
  monolith 214
  monoxide 29-30, 209
  nanotube 87, 154, 181-182, 236, 240, 255
  oxide 29
  particle 220
carbonaceous foam 208
carbonization 178, 190, 240
carbonyl iron 143
carboxyl group 208
carboxymethyl chitosan 138
carcinogenic 270
cardiac rhythm management device 154
carrot nanofiber 232
cars 169
Casico-based foam 113
catalyst 2, 34, 180, 192, 196, 210, 246
  support 214
  system 34
cavity opening 62, 122
cell
  average size 83
  characteristics 204
  coalescence 52, 56, 66, 78, 81, 93, 119, 124, 152, 174, 179, 186, 226
  coarsening 31
  collapse 231
  density 30, 37, 39, 46-47, 49, 52, 54, 56, 58, 65-66, 68, 70-71, 87, 94-95, 102, 114-116, 119, 121, 140, 153, 171, 175-176, 184-185, 187, 189, 197, 204-205, 212, 216-217, 220, 222-223, 235, 261- 262, 265
    measurement 83
  diameter 40, 46, 49, 66, 89, 97, 176, 186, 192, 217, 223, 234, 261
  dimension 140
  distribution 70, 153, 217
  expansion 41, 46
  face 194
  formation 69, 119, 240
  growth 31, 42, 47, 56, 93, 114, 194-195, 218
    rate 58, 262
    /stabilization 195
  homogeneity 95
  morphology 31, 38, 46, 64, 92, 95, 185, 212, 216
  nucleation 31, 37, 47-48, 63, 67, 78, 90, 92-93, 116, 124, 176, 182, 186, 189, 195, 222
    density 66, 114, 205
    power 67

Index

rate 124
nucleus 47
opening 174, 227
　agent 196
orientation 96
ripening 77
rupture 67
shape 216
size 2, 31, 37, 39, 42, 46, 48, 52, 56-58, 65, 70, 87, 98, 101-102, 108, 114, 116, 119, 125, 140, 153, 155, 157, 171, 182, 186, 189-190, 202, 205, 207, 213, 217, 219-220, 227, 235-236, 247-248, 251, 262
　distribution 56, 71, 82, 98-100, 150, 152-153, 175, 197, 222, 232, 261, 262
　measurement 83, 101
　　method 83
　stability 82
stabilization 47
stabilizer 82, 222
structure 43, 56, 94, 99, 111, 115-116, 137, 152, 153, 166, 175, 182, 189, 226-227
uniformity 222
wall 49, 102, 169, 174, 185, 201, 218, 234, 248, 251
　rebounding 237
　rupturing 253
　thickness 87, 100-102, 137, 167, 251
　thickness
　　determination 84
　　distribution 101
　　variation 102
growth 47
cellular
　foam 47, 242
　material 84
　morphology 152, 170, 179, 201, 235
　PVC 199, 222
　structure 34, 56, 62, 87, 104, 120, 124, 143, 172, 178-179, 205, 220
　　degeneration ratio 62
　void 222
　volume 85
　wall 153
　window size 178
cellulose 118, 185, 200, 216
　acetate 42, 52, 117, 137, 223
　　butyrate 76, 82, 118, 137, 152
　ester 242
　fiber 70, 153, 223
　foam 221
cellulose
　nanocrystal 153
　nanofiber 197
cementitious foam 227
cenospheres 243
chain
　branching 235
　extender 34, 103, 107, 153, 157, 176, 179, 193
　extending agent 211
　length 148
　mobility 155
　scission 208
char 221
　residue 221
charcoal 43
charge storage ability 212
chemical
　blowing 35
　　agent 1, 31, 78, 121, 145, 199
　　content 61
　decomposition mechanism 29
　microfoam formation 56
China clay 232, 237
chitosan 129, 138, 220, 237
　hydrogel 232
chlorinated polyethylene 113, 251
　rubber 201
chlorofluorocarbon 210
chloroparaffin 162
chlorosulfonated polyethylene 177
circularity 261
cis-1,3,3,3-tetrafluoroprop-1-ene 137
citric acid 116, 139, 152, 157, 223
clamp pressure 55
clamping
　force 123, 167
　pressure 54-55, 71
classical nucleation theory 78
clay 112
　dispersion 46
　exfoliation 111
　　degree 112
　layers 112
closed-cell 42, 70, 87, 98, 102, 119, 152-153, 158, 186, 217, 247, 263
　content 103, 220
　foam 104, 112, 118

fraction 103
  structure 37, 48, 108, 121, 150
closed pores 128
coalescence 31, 47, 56, 68, 75, 103, 205, 224, 227
  mechanism 77
coarse cell 52
coarsening 31
coating 166
co-blowing agent 37, 49, 100
coextrusion 166, 212
collapse 103
  stress 238
colonization 155
color 52
comb-like cell orientation 113
combustion properties 150
compact layer 104
compatibility 64
compatibilization 104, 207
compatibilizer 95
composite 112, 140, 216
  electrode 190
compressed carbon dioxide 92
compression 190
  direction 72
  modulus 237
  molding 112, 114, 143, 153, 172, 179, 185, 212, 251
  performance 98
  pressure 184
  ratio 140
  resistance 108
  set 2, 211, 231
  strength 150, 163, 192, 211, 221, 232
compressive
  modulus 163, 217, 232
  properties 91, 189
  strain 231
  strength 98, 135, 140, 185, 216, 232
  stress 232, 234
  yield strength 233
concentration 52
  gradient 90
concrete 2, 257
condensation 220
conduction loss 205
conductive
  network 247

pathway 111
conductivity 167, 240
conduit 204
confocal micrograph 199
conjugated
  diene 212
  polymer chain 212
construction 172
  industry 134
continuous
  extrusion 38
    foaming 41
  phase 156
controlling factor 93
cooling 35, 92
  rate 124
  stage 66
coral reef-like 159
core
  layer 104
  region 105, 252
  thickness 171
  -shell structure 156
corrosion risk 150
cosmetic product 120
cost 152
counter-pressure 62
covalent bonding 211
crack
  direction 100
  growth 100
  propagation 159, 184
  toughening mechanism 93
crater 199
craze
  growth 93
  path 93
cream time 192-193, 216, 251
creep behavior 99
critical
  distance 106-107, 111
  energy barrier 217
  impact energy 194
  point 43
  pressure 37
  radius 78
  size nuclei 91
crosslink 79, 249
  density 207, 210, 213

crosslinked
  network 178
  polyethylene foam 172
crosslinker 34, 212, 231
crosslinking 192, 209-210, 227
  agent 32, 115, 190, 197, 211, 213
  degree 171
  monomer 82
  promoter 212
crosslinks 34
cryogenic scanning electron microscopy 41
crystal
  barrier 39, 61, 235
  interface 87
  nucleating agent 222
  nucleation 267
  perfection 176
  size 63
crystalline
  network 251
  phase 263
  region 38, 65, 87
  structure 39, 65, 267
crystallinity 36, 43, 56, 66, 82, 114, 155, 179, 216-217, 235-236, 251, 263, 267
  degree 235
crystallizable diluent 214
crystallization 92, 118, 124, 175-176, 181, 186, 223, 250, 267
  behavior 63
  half-time 216
  kinetics 66, 180, 235
  peak 91
  point 42, 65
  rate 2, 63, 82, 122, 152, 235
  temperature 90-91, 109, 176, 179, 216
cubic sodium chloride 223
cure 120
  accelerator 211
  characteristic 214
  mechanism 32
  rate 211
  time 120
curing 34, 87, 120
  agent 212-213
  coagent 112
  process kinetics 213
  rate 140
  reaction 32

  parameters 213
  shrinkage 213
  system 1
  temperature 213
curvature deformation 145
cyanoacrylate 139
cycle time 125
cyclic olefin copolymer 220
cycloisopentane 257
cyclopentane 40, 158, 270

D

damping motion 253
dark field-scanning transmission electron microscopy 183
dashboards 187
data communication cable 142
decomposition 1, 30
  catalyst 45, 125
  maximum temperature 225
  product 209
  rate 46, 67, 221, 227
  reaction 29
  temperature 1, 31, 32, 70, 113, 130, 145, 153, 166, 175, 209, 213
    range 221
decompression 115
defoaming 130
deformability 47
deformation 53
  history 237
  recovery 237
deformed cell 72
degassing pressure 116, 179
degradation temperature 120
delay time 55, 63, 123
demethylation 150
dense crazing 93
density 51, 54, 70, 127, 140, 145, 150, 157, 164, 167, 176, 179, 181, 187, 189, 233, 257
  change 51
  gradient 113, 240
  reduction 62, 82, 122, 152, 238
depolymerization 150
depressurization 48, 71, 76, 87, 89, 114, 244, 262
  degree 94
  process 92
  rate 47, 65-66, 87-88, 92, 115, 182, 197
  time 155

desorption 39, 61
  behavior 60
  coefficient 56
  diffusion coefficient 56
  diffusivity 61
  kinetics 61
  rate 56
  time 55, 60-61, 126
detergent 1
diaphragm pump 117
diatomaceous silica 45
diazirine 140
dibasic calcium phosphate 45
dicumyl peroxide 112, 146, 153, 211, 255
die 36-37
  exit 35, 56, 67, 223
  pressure 56
  temperature 56-57, 115-116, 119, 205, 208
dielectric
  constant 135, 140, 204, 240, 246
  permittivity 240
  properties 135, 140, 246
differential scanning calorimetry 221
diffraction peak 267
diffusion 35, 61
  coefficient 41, 55, 61, 189
  distance 41
  rate 33, 249
  time 59
diffusivity 49, 60, 66, 189, 243
dihydrooxadiazinone 167
diisononyl
  cyclohexane-1,2-dicarboxylate 203
  phthalate 203
diluent 214
dimensional stability 178, 192, 212, 251
dimethyl
  ether 34, 48, 58
  sulfone crystal 214
dimethylcyclohexylamine 210
diphenylamine 209
directional growth 182
disentanglement 111
disodium malate 244
dispersed phase nodules 41
dispersibility 167, 219
dispersing aid 45
dispersion 46, 140
  state 143

dissipation
  factor 140, 190, 216
  mechanism 205
dissolution 35
  temperature 182
dissolved gas 65, 249
  concentration 78
distribution morphology 204
ditolyl disulfide 32
diurethane dimethacrylate 190
divinylbenzene 82, 212
D-limonene 226
door frame 172
dopamine-modified graphene nanoplatelet 204
dosing equipment 119
double
  bond 146
  -layer extrusion coating 118
downpipe 116
driving
  dynamics 212
  force 93, 182
drug delivery 139
  system 166
dry ice 2, 116
dryblending 125
DSC 236, 263
durability 231, 237

E

eddy current loss 205
effective heat of combustion 150
elastic
  energy storage 249
  modulus 190, 194, 238
elastic recovery 232
elasticity 34, 36, 82, 164, 175, 190, 238, 248
electric conductivity 106
electrical
  cable 70
  conduction 106
  conductivity 111, 114, 140, 154, 205, 240
electrical resistivity 248
electromagnetic
  interference shielding 140
    shielding effectiveness 178
  percolation threshold 216
  radiation 240
  reflection 216
  waves 124

Index

electron
  beam radiation 201
  hopping 111
  transfer properties 212
electronic
  mobility 212
  package 176
  packaging 226
electronics 134, 166
electrostatic interaction 211
ellipsoidal cell 98
elongated cell 113
elongation 235, 239
elongation flow 42
EMI shielding 164, 240
  effectiveness 167, 240
emulsion polymerization 212
endothermic 1, 51
  blowing agent 32, 67, 254
energy
  absorption 72
  barrier 66, 90
  consumption 222
  harvesting 166
  state 251
  -management efficiency 220
Engel 121
entanglement 267
  density 244
  network 223
enthalpy of crystallization 263
environmental
  perspective 42
  problem 269
environmentally friendly process 39
epoxy 140, 179, 241, 243, 246, 250
  foam 140, 213, 217
  matrix 140, 213
  resin 210
equilibrium concentration 54, 60-61, 66
equipment adaptation 54
ethanol 49, 163, 169, 211
ethyl
  acetate 49
  hydroxyethyl cellulose 148
  lactate 54, 226
  vinyl acetate 251
ethylenediamine 223

ethylene
  -propylene diene monomer 32, 37, 64, 210, 213, 218, 227, 240, 242-243
  -vinyl acetate 30, 113, 120, 145-146, 207, 209, 211, 218, 220, 231-232, 242, 248, 251, 255-256
evaporating liquid 35
exfoliated 145
  layered-silicate 46
exfoliated montmorillonite 145
exfoliating additive 215
exfoliation mechanism 112
exfoliation process 112, 215
exothermal effect during curing 213
exothermic 1, 30, 46, 51, 192-193
  chemical reaction 69
  peak 32, 143, 213, 243
  process 225
  reaction 30
exothermicity reducer 140
expandable microspheres 156
expanded
  graphite 185
  microsphere 33
  polystyrene 140, 190
  foam 197
  sample 98
expansion 41, 55, 69, 78, 120, 122, 222
expansion
  kinetics 264
  rate 34, 140, 200
  ratio 37, 49, 52, 56, 58, 70-71, 113, 115-116, 137, 152, 169, 171, 179, 185, 187, 199, 208, 242-243, 251-252
  volume 153
extensional
  flow 36, 70, 185
  force 76
  stress 203, 225, 253
  -induced nucleation 90
  viscosity 203, 225, 248
external
  force 159
  pressure 103
  stimulus 250
extruded foam 76
extruder 35-37, 69
  barrel temperature 118
  die 69

length 117
port 59
extrusion 46, 89, 103, 112, 115, 117, 148, 153, 175, 182, 197, 199, 210, 215, 218, 220-221 224, 243
  barrel 116
  die 212
  direction 224
  foaming 36, 56, 67, 83, 84, 91, 115, 176, 189, 190, 220, 223, 242
    behavior 118
    process 89
  molding 147
  pressure 68
  process 70

F

fabric layer 195
fabrication process 168, 204
failure initiation 194
failure pattern 194
fast relaxation 175
feeding zone 116
ferrocyanide 190
ferromagnetism 167
fiber 63, 216, 253
  content 31
  length 217
  orientation 217, 240
  inter-connectivity 240
Fick's law 60
Fickian diffusion equation 60
field emission scanning electron microscope 83
filler 120, 125, 217
film density 55
filtration 166, 214
fine cell 122
fire
  fighting foaming composition 138
  protection 141
  retardant 221
    material 162
  retarding properties 149, 221
flake
  aluminum powder 200
  graphite powder 200
flame
  resistance 164
  retardancy 164, 169
  retardant 197, 221

    system 150
  -retardant
    foam 146
    resin 177
flammability 269
flammable material 35
flash point 269
flax fiber 30
flexible
  foam composite 224
  polyolefin foam 222
  modulus 176, 243
flexural strength 172, 176
floor covering 111
flooring 201, 242, 255
floral foam 150
flow
  boundary condition 254
  direction 217
fluorinated
  compound 138
  ethylene-propylene 147
fluorinated gas 49
fluorinated(meth)acrylate 189
fluorocarbon 57, 211
fluoropolymer 70
fluoropropene 189
fluororubber 37
foam
  bimodal 92
    nanocellular 92
  bubble 179
  cell
    air pressure 237
    morphology 40
    promoter 222
    size 119, 210
  density 2, 42, 66, 88, 92, 138, 167, 187, 205, 218, 233, 236, 240
    measurement 84
  expansion 75, 152, 233
  extrusion 42, 56, 116
  homogeneity 223
  injection molding 108, 116, 122, 124, 223
  microstructure 116, 171
  morphology 47, 56, 66, 87, 128, 187, 197
  nucleation process 62
  porosity 40
  quality 210

Index

resilience 231
rise direction 233
setting time 220
stabilization 75
stabilizer 196, 222
stiffness 125
structure 37, 47, 64, 66, 143, 214
temperature 192
thermal insulating properties 189
thickness 241
uniformity 119
volume expansion ratio 85
foamability 53, 140
foamed
  article 147
  core 105
  packaging material 148
foaming 34, 82, 175, 203
  agent 1, 123, 147, 156, 209, 249
    UV initiated 140
  behavior 43, 223
  composition 82
  efficiency measures 83
  extrusion process 36
  formulation 209
  gas 210
    dissolution 67
  injection molding 39
  mechanism 29
  parameter 51, 90, 113
  pressure 90, 92, 236
  process 43, 49, 54, 57, 87, 195, 210
  system 31
  temperature 39, 43, 47, 49, 54, 66, 67, 90, 92, 103, 116, 120, 126, 155, 171, 182-183, 204, 207- 208, 213, 236, 262
    window 169, 175, 186
  time 31, 126, 220
  window 70, 185
food
  packaging 153
  -contact application 31
footwear 146, 231, 238
formamide 223
Fourier transform infrared 263
fracture behavior 237
free
  energy barrier 115
  expansion 120

foaming 119-120, 177
  plasticizer 199
  radical 213
  volume 48, 61, 166
freeze drying 129
frequency range 254
fresh produce 129
freshwater invertebrates 150
froth spray insulation foam 120
fuel cells 129
fumed silica 45
furniture industry 193
fused deposition modeling method 166
fusion 31, 64, 199, 200
  factor 200
  time 200

G

gas
  adsorption 263
  bubble 104
    entrapment 208
  composition 30, 58
  concentration 39, 59-61
  content 57, 64
  counter-pressure 39, 54, 64, 122, 192, 254
  depletion 78
  desorption rate 61
  diffusion 31, 61, 93
  diffusivity 60, 78
  dissolution 47
  escape 57
  evolution 70
  expansion 70
  formation rat 227
  injection system 115
  input 58, 67, 115
  loss 57
  phase 254
  pressure 1, 30
  production 199
  pycnometry 263
  residence time 37, 116, 171
  retention 30
  saturation 93
    pressure 60
    temperature 66
  solubility 48, 68, 82
  sorption 59, 61
    model 60

supersaturation 76
uptake 60
yield 46
gaseous
  product 29
  reaction product 1
gas-laden pellet 39, 128
  shelf life 128
gas-rich region 78
gas-saturated melt 69
Gatan Microscopy Suite 262
Gaussian-like relationship 66
gel time 102, 192, 213
gelation 31, 64, 199-200, 203, 225
geometrical packing advantage 87
Gibbs free energy 224
glass
  fiber 185, 216
    orientation 40, 185, 216
    reinforced polyurethane foam 194
  microsphere 33, 140, 213
  transition temperature 32, 47, 54, 57, 66, 92, 114, 126, 129, 137, 140, 143, 153, 156, 158, 166, 176-177, 192-193, 200, 205, 213-214, 216, 243, 244, 251, 263
glassy carbon electrode 190
global warming 117, 119, 137, 223
  potential 35, 269-270
gloss 52
glycerol 43, 193, 219
glycidyl(meth)acrylate 189
glycol monostearate 222
gradient
  cell structure 167
  foamed material 240
grafting 222
graphene 37, 87, 106, 185, 215, 234, 240, 246, 265-266
  foam 204
    skeleton 204
  hollow sphere 146
  nanoplatelet 106-107, 186
  oxide 106, 197, 217, 220-221, 234, 240, 241
    nanoribbon 164
graphite 189, 221
  powder 190
greenhouse warming potential 189
grow time 211
growing bubble 78, 217

H

halloysite 42, 153, 217, 219, 234
  nanotube 140, 208
hard segment crystal 67
hardblock 34
hardening 1
hardness 39
HCFO-1233zd 238
heat 1
  conduction 116
  deflection temperature 170
  evolved 225
  flow transport mechanism 204
  generation 192
  of fusion 263
  propagation pathway 108
  release rate 177
  stabilizer 200
  transfer 47, 172, 250
    efficiency 108
heating 76
  system design 250
helium 243
  ion
    microscopy 107
    transmission 107
hemostatic activity 138
Henry's gas law constant 49, 60, 243
heterogeneous
  cell nucleation 31, 63
  effect 222
  nucleating
    agent 43
    point 264
  nucleation 39, 116, 182, 186, 218, 223-224
    effect 87
    point 114
    theory 90
heterophasic polypropylene 187
hexafluorobutyl acrylate 189
hexafluoropropylene/tetrafluoroethylene copolymer 147
hexagon closed cell 108
hexane 150
HFC-365mfc 223, 234, 237
HFO-1234ze 223, 238, 242, 257
hierarchical porous part 166
high
  impact polystyrene 40, 52

precision syringe pump 128
pressure
  foam injection molding 77
  vessel 126
holding
  stage 93
  time 114, 120
hollow
  filler 140
  microballoon 140
  microsphere 163, 169
  sphere 213
  structure 164
homogeneous
  cell nucleation 124
  cellular structure 67
  nucleation 47, 87
    process 68
    theory 65
homogenization 2
honeycomb
  polygonal cell 185
  structure 108
hopper 116
hot
  melt pressing 182
  press 126, 167
humidity environment 231
hybrid
  aerogel 197
  foam 149
hydrated magnesium silicate 45
hydrazide 130
hydrazine 130
hydrazodicarbinamide 29
hydrocarbon 40
  oil 163
hydrochlorofluorocarbon 210
hydrofluorocarbon 40-41, 48, 52, 189, 270
hydrogel 138
  foam 138
hydrogen
  bond 43
  bonded urea aggregation 103
  bonding 2, 194, 211, 240
  gas 140
  peroxide 208
hydrolysis 116, 231
hydrophile-lipophile balance 227

hydrophobic
  character 163
  polymer 187
hydrophobicity 164, 208
hydrotalcite 146
hydrothermal liquefaction 150
hydroxyalkylamide 140
hydroxyalkylurea 140
hydroxyapatite 125, 138, 217, 220, 232
hydroxyl
  component 43
  group 34
hydroxypropyl methylcellulose 118, 148
hysteresis loss 193, 219

I

ideal gas law equation 61
image
  analysis 83, 262
  capture card 262
  processing 264
ImageJ 261
Image-Pro Plus 262
imaging technique 134
immiscible polymer blend 42
impact
  direction 98
  duration 194
  modifier 176, 226
  strength 98, 125, 158, 159, 184, 226, 245
  toughness 235
impeller mixing 152
implant 134, 154, 182
  material 182
implantable device 154
indoor space 195
industrial scale 2
inert gas atmosphere 149
infrared
  attenuator 189
  opacifier dispersion 201
  radiation 220
initial growth 75
injection
  compression molding 113
  flow rate 122, 253
  location 36
  mold 161, 186
  molded part 104

molding 43, 62-63, 145, 153, 158, 171, 179, 185, 186, 192, 216, 220, 235, 243
  machine 121
  pressure 123
  speed 39, 63, 123, 217
injection-molded
  foam 98
  part 52
injector 121
ink adhesion 222
in-mold pressure 63
innerliner 136
in-plane thermal conductivity 204
instrument panel 187
insulating
  application 189
  billet 189
  effect 125
  foam 237, 250
insulation 172
  foam 242
  of pipes 149
  panel 64
  properties 204
insulator-conductor transition window 216
inter 167
interaction 49
  energy 169
inter-bead bonding strength 167
intercalation 175, 215
interconnected
  3D framework 172
  cells 98
  hollow cell 197
  pores 156
  porosity 125
  porous network 212
interconnecting pore 106, 108
interfacial
  adhesion 164
  adsorption 218
  bond 159
  bonding 253
  interaction 194
  mechanism 226
  polarization
    loss 205
  phenomenon 248
  shell 80

  tension 49
interior trim component 187
internal
  blistering 55
  mold pressure 62
  pressure 62
    after foaming 61-62
interpenetrating network 88
interphase region 90
interstitial voids 135
ion microscopy 107
ionic crosslinking 211
ionizing radiation 82, 212
IQ Foam® technology 121
isobutane 34, 40, 48, 120
isobutene 150
isocyanate 2
  conversion rate 263
  group 34
  -based polyimide foam 177
  -functionalized silica nanoparticle 43
isopentane 41, 120, 150
  boiling point 152
  droplet 41
isopropanol 163
isothermal temperature 176
isothermal treatment 176
Izod impact strength notched 176

J

joint integrity 111

K

Kelvin model 101
Kelvin's model 102
kicker 1, 199, 209
knife scratching 172
Krauss Maffei 121

L

1,1,1,4,4,4-hexafluorobut-2-ene 192
labeling 166
Laguerre model 101
lambda value 251
Langmuir
  capacity constant 61
  mode adsorption 60
large bubble 75
laser
  processing 128
  system 128

Index

lauryl sulfate. 138
layered particle 112
lead
  salt 209
  stearate 32
  -zinc salt 209
lightweight flexible foam 54
lignin 95, 193
  nanoparticle 150
  reactivity 150
limiting oxygen index 162, 177, 197
  value 221
linear
  molecular chain structure 124
  phthalate 199
liner 111
liquid
  blowing agent 119
  carbon dioxide 126
  nitrogen 265
  nucleating agent 234, 237
lithium-ion batteries 129, 190
long chain branch 248
low-energy repetitive impact 194
lubrication zone 254

M

machining 134
macropore 263
macroporous carbon monolith 214
macroscopic deformation 250
macrovoids 190
magnesium
  hydroxide 166, 253
  oxide 45
magnetic
  elastomer 143
  field 143
  properties 205
magnetism 250
malonic acid 103
  salt 107
marine 149
mass transfer coefficient 40
masterbatch 45
matrix 75
matrix
  crazing 93
  rheology 213
maximum decomposition rate 225

mean cell diameter 95
mechanical
  foaming 196
  strength 185
  stress 212
medical
  applications 172
  instruments 134
  products 152
melamine
  cyanurate 190
  foam 221
    skeleton 221
  sponge 149
  phosphate 43, 99, 221
  resin 149
melt
  blending 174, 179
    process 216
  compounding 112
  flow index 92, 125, 184, 234
  foamability 53
  front 124
  grafting reaction 187
  rheological properties 36
  strength 36, 47, 56-57, 63-64, 70, 75, 78-79, 91, 102, 104, 116, 122, 152, 175, 179, 185, 187, 206, 210, 216, 223, 226-227, 231, 236, 253
  temperature 122, 253
  viscosity 56, 67, 82, 119, 152-153, 175, 179, 204, 217-218, 248-249
  -compounding parameters 215
melting
  enthalpy 263
  peak 195, 235
  point 38, 48, 70, 149, 152, 240
    depression 66
  strength 43
  temperature 43, 66, 153, 176, 244
memory foam 249
mercury porosimetry 263
mesopore 263
mesoporous
  silica particle 39, 222
  structure 222
metering system 117
methyl formate 40, 227
methylmethacrylate 88

methylsiloxane resin 211
microbubble area 88
microcellular
　asymmetric structure 171
　foam 83-84, 93, 140, 158, 181-182
　　definition 115
　foaming 49, 77, 169, 223, 243
　　behavior 166
　injection molding 92, 124, 127, 175, 216
　molding 87
　part 254
　plastic foaming 190
　scaffold 115
　structure 92, 205
　transition layer 55
　urethane foam 83-84
microelectronics 204
microfoam 81, 265
microfoaming 254
micropore 263
microporous foam 166
microscale cell 93, 167
microsphere 33, 213
microwave 149
　absorber 125
　energy 125, 189
　heating 124
　irradiation 220
　radiation 125
　shielding performance 114
　transmitting 205
　-assisted foaming 167
midsole 125, 146, 231, 238
migration 176
Milacron 121
MIM process 63
mixing 119-120, 145
　rule 48
modification 214
modified graphene nanosheet 204
modular MuCell upgrade 121
moisture 45, 250
moisture permeation 177
mold 192
　breathing 55
　cavity 39, 54, 123
　opening 55, 63, 97, 99, 105, 122, 235, 252-253, 257
　　technique 98

pressure 62
　surface 254
　temperature 39, 63-64, 171, 186, 217
　-filling stage 124
molding
　cycle 125
　machine 121
molecular
　lubricant 61
　movement 244
　weight 87, 92, 152-153, 189, 199-200, 203, 225
momentum equation 75
montmorillonite 111, 145, 175, 215, 218
　loading 195
morphing structure 250
morphological
　change 75, 163
　parameter 82
　structure 265
　study 41
　transformations 37
morphology 82, 105, 155, 161, 247
morphometric characterization 194
morphometry evolution 194
motion-energy absorption 163
MuCell 121, 175, 257
　injection molding line 122
　　technology 121
　process 242
　technology 158, 245
multilayer
　interface distance 182
　structure 166
multiwalled carbon nanotube 81, 111, 164, 174, 189, 193, 202, 216, 240, 247, 249, 265-266
　functionalized 140
mutagenic 270
　properties 269

N

N,N'-dicyclohexyl-2,6-naphthalenedicarboxamide 175
N,N'-dinitroso pentamethylene tetramine 31, 143, 153, 270
nano-calcium carbonate 186, 217, 223
nanocarrier 139
nanocellular 71
　foam 71, 182
　foaming 55

structure 90, 158
nanocellulose 152, 211
nanoclay 187, 190, 200, 215, 235
   content 102
nanocomposite 215, 265
   foam 153, 232
nanocrystalline cellulose 95, 217, 220, 243
nanofiller 87, 249
nanofoam 54, 83, 85, 104, 118, 126
nanographite 95
nanoparticle 265
   dispersion 217
nanoribbon 164
nanoscale
   carbon nanofiber 114
   cell 93, 167
   pores 245
nanosilica 99, 189, 217, 222
   nucleation 99
   size 189, 219, 222
nanotube 216
   agglomeration 182
natural
   fiber 207
   polymer 118
   rubber 108, 120, 207, 210, 216, 218, 232, 237, 253, 256
NBR 210, 255
negative
   compression process 55
   gradient temperature profile 116, 152
Newtonian viscosity 76
nitrogen 29, 37, 56, 71, 108, 111, 116, 122, 136, 143, 145, 149, 179, 187, 209, 216, 218, 223, 235, 242-243, 267
   fraction 58
   gas injection 59
noise reduction coefficient 253
non-syntactic foam 220
norbornene segment 220
notched impact strength 158
nucleated bubbles 52, 63, 77
nucleating
   agent 2, 42-43, 52, 57, 116, 122, 137, 158, 162, 167, 175-176, 179, 186-187, 192, 204, 208, 221- 222, 224, 226, 232
   effect 216, 224
nucleation 37, 42, 65-66, 158, 182, 187, 189
   agent 91, 116, 212, 223

   density 205
   efficiency 46
   energy barrier 39, 57, 222
   homogeneous/heterogeneous 65
   mechanism 90
   points 68
   potential 189
   process 254
   rate 39, 57, 71, 167, 219, 223
   site 41, 61, 65
   supersaturation degree 57
   theory 65, 71, 115

O

octanoic acid 231
oil
   adsorption 164
   cleanup 109
   palm fiber 253
   spill 163
      cleanup 220
   /water separation 164
opacifier 201
open
   cell 87, 98, 102, 104, 108, 119, 122, 136, 140, 149, 150, 158, 163, 169, 174, 179, 182, 185, 187, 207, 217, 237-238, 240, 252, 263
      content 85, 98, 220, 253
      foam 42, 99, 108. 149, 180
      fraction 57
      morphology 109
      structure 43
   porosity 253
   structure 109
operating saturation temperature 66
operational window 32, 143, 213, 243
optical expandometer 264
organic
   montmorillonite nanoscale dispersion 194
   peroxide 152, 180
organoclay 112, 175, 218, 232, 237
   nanofiller 217
organogel 129
organotin stabilizer 200
orientation 111, 216
orthopedic implant 134
oscillating disk rheometer 120
oxidation
   degree 208
   thermal decomposition 221

oxygen consumption 150
ozone depletion potential 35, 41, 269, 270

P

p,p'-oxybis(benzene sulfonyl hydrazide) 32, 45, 243
packaging
  application 153, 155
  material 129
paint adhesion 222
paper mill sludge 197
paraffin 56
  wax 221
part thickness 125
particle
  diameter 175
  size 46, 125, 143, 261
peak
  heat release rate 164
  strain 194
  stress 194
penetration time 251
pentaerythritol 150, 175
pentamethyldiethyltriamine 210
pentane 56, 137, 138, 149-150, 226
percolation threshold 167, 189, 240, 247
perfluoroalkane 223, 234, 237
perfluoropolymer 147
performance
  criteria 3
  degradation 194
permselective medium 166
peroxide 146, 212
  crosslinking 212
    process 212
perturbed chain-statistical associating fluid theory 48
petroleum ether 150, 213, 227
pH 150, 220, 250
phase
  change material 195
  distribution 129
  morphology 109
  separation 38, 47, 65, 116, 167, 179, 214
  transition 1, 43
phenolation 150
phenol-formaldehyde
  resin 150, 237
  resole 150

phenolic
  foam 213, 237, 240
  resin 149
phosphoric acid ester 149
phthalimide 227
phyllosilicate surface 149
physical
  blowing agent 1, 39, 41, 76, 121
  foaming 54
phytic acid 212
Pickering emulsion 190
piezoresistive sensitivity 164
pinhole 42, 252
pipe 169
plasticization 41
  effect 37, 49, 91
plasticized starch 219
plasticizer 43, 45, 64, 190, 200, 203, 210, 214, 225
  compatibility 203, 210, 225
  molecular weight 203, 225
plasticizing
  action 58
  agent 140
  barrel 121
plastisol 31, 45, 203, 210, 242
  viscosity 64, 199
plugins 261
pneumatic tire 136
Poisson's ratio 190
polar function 189
polarization phenomena 248
polarizing optical microscope 264
polishing pad 72
pollutants 270
poly(3-hydroxybutyrate-co-3-hydroxyvalerate) 46, 67, 76, 82, 116, 118, 152, 223
poly(amic acid) 129
poly(biphenyl ether sulfone) 167
poly(butylene succinate) 64-65, 70, 113, 153-154, 212, 217, 219
poly(butylene terephthalate) 240
poly(ε-caprolactone) 61, 155, 237, 250
poly(ethylene 2,6 naphthalate) 126
poly(ethylene 2,6-naphthalate) 170
poly(ethylene terephthalate) 58, 65, 68, 99, 115, 175, 222, 242, 244, 248, 250, 257
  crystallization 222
  foam 100

Index

microcellular foam 176
poly(ethylene-co-octene) 48, 77, 81, 174, 216, 253
poly(3-hydroxybutyrate-co-3-hydroxyvalerate) 117
poly(lactic acid) 36, 47-48, 55-56, 61, 66, 79, 113, 116, 120-121, 153, 179-181, 216-217, 219, 223, 226, 235, 239, 245, 251, 256, 263, 265, 267
   foam 70, 122
   spherulite 264
poly(lactic-co-glycolic) acid 48, 115
poly(phenylene sulfide) 61, 235, 266-267
   phase 267
     crystalline ability 235
   /PES blend 39, 61
poly(p-phenylene oxide) 130
poly(styrene-co-acrylonitrile) 243
poly(ε-caprolactone) 180, 226, 256
   composite foam 125
polyacrylonitrile 156, 214
polyamide 104, 111, 157
   dendrimer 223
   -6 103, 172, 221
polyaryletherketone 134
polybutylmethacrylate 48
polycarbonate 43, 57, 61, 66, 90, 108, 114-115, 122, 158-159, 180, 223, 240, 242-243, 245, 252, 256-257, 263, 265-266
   -SAFT prediction 48
   /ABS blend 161
polychloroprene 162
polycrystallinity 67
polydimethylsiloxane 48, 65, 163, 187
   foam 164
polydispersity 43
polyester 243
polyetheretherketone 166
   /PEI 82
polyetherimide 37, 54, 60, 71, 126, 167, 193, 219, 237
polyetherketone 166
polyetherol 208, 211
polyethersulfone 100, 126, 166, 169-170, 235, 266-267
   /PEN blend 170
polyethylene 48, 51, 69, 103-104, 112, 130, 145, 171, 185, 199, 209, 212, 217, 223, 237, 239, 255
   chlorinated 251, 255
   high-density 30, 48, 63, 103, 111, 130, 171-172, 248
   low-density 40, 53, 57-58, 102, 130, 149, 171, 248
   /HDPE blend 96
   /PP blend 96
   linear low-density 53, 125, 172, 253
   linear medium density 125, 172
   ultra-high molecular weight 66, 114, 171
   -octylene 171
   -vinyl acetate 145
polyethyleneglycol 48
polyfunctional amine 211
polygonal cell 108
polyhedral
   cell 98
   oligomeric silsesquioxane 153, 223
polyhedron 102
polyhydroxyalkanoate 180
polyimide 129, 166, 177
   foam 177
polyisocyanurate
   reactive mixture 62
   rigid foam 64
polymer
   backbone 212
   blend 235
   composition 119
   crystalline zone 166
   crystallinity 179
   crystallization 223-224
   foamability 186
   matrix 75, 210, 215, 249
   melt 75, 119, 223, 254
   phase 235
   softening temperature 65
   structure 82
   swelling 49
polymeric
   gloves 162
   matrix 208
   nanofoam 119
polymerization cavity 62
polymethydroxysiloxane 140, 246
polymethylmethacrylate 48, 67, 87-88, 92, 115, 126, 156, 182, 190, 204, 216, 219, 244, 247, 255
   bimodal foam 126

foam 98, 114
polyol 2, 34
polyolefin 51, 124, 218, 223
   elastomer 231
polyoxymethylene 184, 245, 255
polyphenylsulfone 49
polypropylene 38, 40, 42, 47, 49, 51, 56, 63-65, 68, 70, 82, 88, 91, 98, 111-113, 117, 121, 125, 130, 185-187, 199, 211, 212, 215-216, 218, 220, 226, 236-237, 239-240, 242-243, 245, 254-255, 265
   foam 93, 104, 108, 121, 124, 185
      density 84
   foaming behavior 38
   glycol 43
   isotactic 57, 91, 186, 217, 223, 245, 255
      crystals 91
      foaming 62
   linear 38, 186
   maleic anhydride grafted 157
   nanoporous foam 92
   phase 88
   /NaCl composite 93
   /polystyrene
      blend 89
      interface 89
polypyrrole conductive network 212
polystyrene 37, 39, 42, 46, 48, 53, 56, 87-88, 98, 100, 111-112, 115, 130, 189-190, 204, 215-219, 222, 226, 232, 243, 247, 255, 257, 262, 265-266
   beads 124, 189
   brush 222
   cell 90
   expanded 226
   foam 41, 221
   foaming 49
   phase 88, 90
   porous microsphere 130
   /PE blend 95-96
   /SBS 218
polysulfone 49
   insulation 169
polytetrafluoroethylene 38, 70, 79-80, 95, 117, 122, 154, 175, 179, 185, 226
   fibril 175-176, 222-223
   powder 99
polyurethane 1, 34, 43, 54, 62, 98, 103, 106, 108, 119, 125, 146, 156, 192-193, 209-211, 214, 216, 218-221, 223, 227, 231, 234, 237, 240, 243-244, 249-251, 253, 257, 263
   foam 82, 107, 114, 120, 208, 211, 221, 233
      blowing 82
      density 84
      layer 195
      rigidity 34
   foaming 39
   formulation 2
   rigid foam 41
   system 210
polyvinylalcohol 43, 197, 221, 223
poly(vinyl chloride-co-vinyl acetate) 203, 210
polyvinylchloride 1, 46, 51, 64, 99-100, 118, 199, 201-202, 209, 217, 222, 225, 239, 242, 255-256
   foam 49, 212
   foaming 199
   plastisol 46, 199, 210, 225
   /NBR 210
   -VA resin 225
polyvinylidenechloride 156
polyvinylidenefluoride 56, 119, 204, 236, 239
   melt 56
pore 54, 56, 177, 222
   deformation 36
   elongation 211
   growth 36
   morphology 56
   size 82, 197, 200, 233, 247
      distribution 37, 155, 167, 200
   structure 36, 154, 214
   volume 200
porogen agent 125
porosity 37, 47, 114, 116, 138, 149, 155, 167, 172, 182, 194, 205, 208, 234, 242, 246
   distribution 146
porous
   carbon 190
   material 211
   morphology 212
positive temperature effect 248
potassium
   oleate 207
   silica aluminate 45
   silicate 45
precision
   mold opening 158
   opening 63

part 114
pre-foaming 56
premature nucleation 36
pressure 49, 114, 119, 148, 267
  difference 90
  drop 35, 37, 55, 62-63, 66-67, 69, 119, 122-123, 167, 223
    rate 39
  fluctuation 78
  loss rate 37, 167
  profile 69
  -temperature regime 62
printing 166
processing
  aid 176
  conditions 92, 127
  range 170
  temperature 112, 126, 244
  window 58, 64
product appearance 123
production yield 130
propane 34
propellant 149
propylene carbonate 234
pull and foam
  method 122
  process 186
pulp foam 221
purification 166
pyritohedron 153
pyromellitic dianhydride 175

Q

quick response 164

R

radiation crosslinking 78
radiative thermal conductivity 200
rapid
  cooling rate 124
  depressurization 99, 114
    foaming 115
rate constant 30
reaction
  heat 143, 213, 243
  kinetics 108, 264
  order 30
  rate 264
    constant 264
  temperature 143, 213, 243

time 263-264
reactive
  blowing 35
  extrusion 175
  processing 107
reactivity 46, 269
rebound resilience 138, 171
recovery
  rate 237
  ratio 237
redoxase immobilization 214
refined crystal 64, 124
reflection loss 186
reflectivity 175
refrigeration 207
refrigerator 256
reinforce 217
reinforcement 174, 234
relative
  density 248
  permittivity 246
relaxation retarder 249
released gas 210
reliability 164
renewable
  resources 118, 148, 172
  sugarcane ethanol 172
reservoirs 41
residence time 59, 67, 69, 116, 148, 171
resilience 231
retarder 227
rheological
  percolation 176
  properties 210, 227
rheology 248
  modifier 211
rigid
  cellular plastic 83
  closed-cell foam 201
  polyurethane 115
    foam 40, 192-193, 195, 221, 227
rise time 193, 216
rotational molding 103, 125, 172, 209
  equipment 125
roughness depth 124
rubber 1, 209
  elasticity 249
rubbery state 249
rupture 205

S

sample thickness 61, 194
Sanchez-Lacombe equation of state 48-49
satellite bubble 77
saturation 39, 60
   pressure 39, 47, 57, 60-61, 64-67, 102, 114, 153, 182-183, 187, 207, 235
   temperature 60, 64, 66-67, 90-91, 114, 126, 153, 182-183, 235
   time 47, 60, 65-66, 90, 236
   vapor pressure 62
scaffold 155, 197, 220
   architecture 88
scanning electron microscopy 95-97, 262
   image 76, 79
   micrograph 159
   photograph 75
scrap EVA foam 146
screw
   feeder 116
   revolution speed 67-68, 115
   speed 223
sea-island
   micro-structure 130
   morphology 42
   structure 53
sealing composition 140
secondary
   cell nucleation 186
   operation 212
segregated structure 168
self-accelerating decomposition temperature 269
self-healing
   composite system 250
   mechanism 250
sensors 129
separation 166
sepiolite 224
settling 45
shape
   fixity 249
   memory 216, 238
      effect 193, 250
      foam 2, 33, 43, 106, 193, 250
      performance 249
      recovery 219, 249
sharp hole structure 178
shaving gel 120

shear 70, 253
   force 112
   mixing 67
   modulus 101, 250
   rate 249
   thinning 67, 217
   viscosity 36
sheet 114
   foaming 67
shielding effectiveness 205
shish-kebab 245
shock wave 140, 213
shoe
   foam material 146
   sole 145, 211, 240
shrinkage 42, 178, 222, 251
shrinking 209
silane 212
silanized starch 208
silica 87
   aerogel layer 177
   aluminate 45
   dispersion 219
   gel 45
   loading 189
   nanoparticles 249
   particle 190
silicon
   oil 126
   oxycarbide foam 211
silicone 227
   foam 231
   rubber foam 164
siloxane-based
   amphiphile 82
   surfactant 163
single
   phase liquid 121
   -screw
      extruder 40, 167, 182, 219
      extrusion 152
   -step process 182
   -walled carbon nanotube 213
sintering 167
size distribution 89
skin 55, 63, 121, 123, 186, 210, 265
   layer 55, 105, 146, 195, 196, 252
   removal 134
   thickness 52, 87, 104, 186, 265

Index

sliding angle 172
slip
  condition 254
  layer 254
small bubble diffusion 52
smart textiles 250
smoke
  growth rate 162
  specific optical density 177
smooth surface 200
sodium
  bicarbonate 32, 34, 51, 67, 111, 116, 124, 130, 138, 143, 152, 157, 197, 200, 207, 210, 220, 223, 243, 246
    decomposition rate 140
    +citric acid 171, 185, 187, 199
  chloride 223
  citrate 218
  hydrogen carbonate 32, 34
  hypochlorite solution 217
  montmorillonite 207, 218, 232, 237
  silica aluminate 45
  silicate 45
soft foam 209
softening phenomenon 163
software 261
soil gripping properties 111
solid
  blowing agent 29, 120, 209
  phase conduction 220
  state nanofoaming 126
  -state foaming 66, 126
solidification 42, 63
solubility 40, 42, 48, 56, 118, 223
  parameter 37
  pressure 56
solubilized gas 59
sonication 197
sorbitol 219
sorption 66
  pressure 87
  time 59
sound
  absorption 108, 253
    coefficient 108, 207, 253
    efficiency 216, 253
    material 253
    peak 253
    performance 98

    profile 254
  energy 108
  insulation 221
  wave
    absorption 252
    transmission loss 108, 158, 252
spacecraft 149, 250
  protection 140, 213
  shielding 140
specific compressive strength 234
spherical
  bubble 75
  cell 106
    size 184
spherulite 267
  size 2, 235
spherulitic
  morphology 264
  orientation 214
spongy sole 211
sporting goods 172
spray foam 34, 119
sprayable PU foam 210
stabilizing system 1
stable
  cell growth 63
  gas content 166
stannous octoate 210
starch 43, 95, 116, 118, 125, 155, 187, 193, 211, 217, 250
  foam expansion 208
static
  charge 45
  electricity 119
  mixer 115-116
statistical fluid theory 48
steam chest molding 99, 186
stearic acid 30, 145-146
  amide 222
stick condition 254
stiffness 169, 173
strain
  energy 219
    storage 193, 249
  hardening 36, 81, 174-175, 217, 222, 248
  -induced crystallinity 251
streaks 124
stress
  concentrator 93

growth 53
relaxation 249
softening 237
structural
  defect 204
  foam 105, 179
structure collapse 75
strut 102
  joint 194
  morphology 106
  thickness 106
  volume fraction 102
styrene-acrylonitrile 48, 182
styrene-butadiene polymer 210, 221, 255
styrene-butadiene-styrene 189, 217, 226, 232, 255
styrene-ethylene-butylene-styrene 95-96, 190
styrene-methylmethacrylate 48
styrofoam 189
  production line 118
sublimate 29
sucrose 163, 240
sulfonated aromatic acid 209
sulfosuccinate 162
sulfur 213, 227
Sulzer mixture unit 115
superabrasive resin 156
superabsorbent foam 138
supercapacitors 129
supercooling 117
supercritical
  carbon dioxide 37, 39, 46-49, 54-57, 61, 64-66, 77, 87-91, 95, 114-116, 121, 126, 130, 140, 153, 155, 167, 169-170, 174, 179, 182, 186-187, 189, 190, 193, 207, 211, 216-217, 219, 222, 226, 237, 245, 263
    concentration 204
    extrusion foaming 176
    foaming 166, 182
  drying 129
  fluid 48, 121
    -laden pellet injection molding foaming technology 39, 128
  injection molding foaming technology 127
  liquid 2
  nitrogen 37, 175, 185, 187, 216
superhydrophobic
  film 190
  polyethylene foam 172

superhydrophobicity 172
supersaturation 39, 65, 68, 78, 115
supersaturation degree 124
suppressed cell ripening 78
supramolecular nanostructure 223
surface
  appearance 56
  area 45, 186, 218
  coverage 226
  electric resistance 176
  energy 79, 80, 82, 179, 226
  free energy 218
  layer 104, 121
  quality 124
  roughness 164, 254
  tension 1, 30, 68-69, 76, 78, 80, 82, 163, 217, 223, 227, 254
  treatment 219, 222
    layer 195
  void 227
surfactant 1, 34, 82, 106, 112, 138, 227
surrounding gas 30
swelling 203, 225
  extent 61
  heat 203
  process 225
swirled pattern 124
switching behavior 248
syndiotactic
  foam 135
  polystyrene 190
synergistic
  effect 48, 221
  lubrication 221
  mixture 54
syntactic foam 135, 140, 169, 213, 243

T

talc 2, 42, 52, 56-57, 130, 137, 147, 167, 179, 187, 212, 217, 223, 226, 235, 242
  concentration 223
tandem extrusion 38
  system 117
tear strength 255
temperature 37, 46, 70, 148, 167, 237
  gradient 104, 112, 219
  profile 64
  window 211
template 214

Index

tensile
  modulus 112, 184, 255
  strength 112, 176, 182, 197, 235, 256
  stress 78
  testing 93
tension 190
termomechanical cycle 249
tertiary amine 210
tetrabutyl titanate 180
tetrahydrofuran 169
tetrakaidecahedron 102
tetramethylthiuram disulfide 146, 210
textured surface 111
thermal
  conduction 108
thermal
  conductive pathway 204
  conductivity 37, 64, 119, 150, 153, 158-159, 162, 169, 185-186, 189, 192-193, 201, 204, 207, 218, 220, 252, 256
    anisotropy 113
  decomposition 143
    temperature 153, 212
  degradation 116
  diffusivity 64
  insulation 87, 91, 108, 149, 153, 162, 177, 185, 207, 220-222
    application 38
    foam 99
    layer 169
    material 134
    performance 134, 205
  management material 204
  radiation 220
  regulation 195
  stability 111, 164, 221
thermally-induced phase separation 155
thermodynamic
  condition 78
  instability 35, 47, 67, 69
  state 78
thermo-expandable microsphere 164
thermoforming 128, 152
thermogravimetric analysis 221
thermomechanical cycle 250
thermoplastic
  cellulose ester 101
  polyolefin 56, 71
  polyurethane 49, 87, 194

cap 212
  foaming behavior 67
  shell 156
  starch 42, 242
thermosetting polyimide foam 178
thickness 145
  direction 253
  ratio 253
thin-walled product 124
thymol 61
tile system 201
time 60
  lag 87
  lapse acquisition 262
tin octoate 231
tire noise reduction 136
tissue engineering 182
titanium
  dioxide 45, 209
  hydride 134
toluene
  diisocyanate 43
p-toluene sulfonyl
  hydrazide 45, 136, 270
  semicarbazide 32, 45, 204, 270
tool wall 124
topological network 194
tortuosity 248
total
  heat release 150
  smoke release 164
  solubility 49
toughening
  additive 226
  agent 226, 249
  mechanism 159
  theories 93
toughness 92-93, 184, 186
toxicity 269
trains 169
transesterification 180
transition layer 55, 71, 104
transmission electron microscopy 265-266
transparency 175
transportation 270
  restrictions 269
Trexel 121
tribasic calcium phosphate 45
tribological properties 221

triethanolamine 209
triethylamine 211
triethylenediamine 210
trimethylolpropane trimethacrylate 112, 153, 212
tunnel
  conductivity 107
  mechanism 106
  speed 64
twin-screw
  extruder 40, 120-121, 212, 216
  extrusion 152
twisted pair of conductors 167
two-roll mill 120, 152
two-step
  depressurization 88, 93
  foaming 114, 182
    process 183
  pressurization 115

U

ultra-high saturation pressure 114
ultrasonic atomizer 139
ultrasonication 172
  bath 125
uniaxial compressive load 72
uniform
  cell structure 92
  distribution 2
unimodal foam 90, 114, 158
unit cell shape 182
unmelted crystal 66
untextured edge 111
upholstery industry 149
urea 32, 240
  activator 153
uretonimine linkage 109
UV pulsed laser 129

V

vacuum
  drying 129-130
  vent 40
vapor-grown carbon fiber 114
variotherm mold 62
vehicle
  seat pad 196
  tire 212
ventilation 40
vertical foam growth direction 113

vibration damping 162
Vicat softening point 169
vinyl
  ester 243
  flooring 199
  foam extrusion 200
  plastisol 45
vinylsilane 212
vinyltriethoxysilane 189, 219, 222
viscoelastic
  normal stress 76
  properties 179
  response 75
viscoelasticity 69, 124
viscosity 30, 36, 76, 179, 208, 213, 217, 224, 225-226, 249
  build-up 107
  increase 199
  reduction 175
visualization 118
  component 223
  mold 187
  system 118
vitrification 250
void 71, 72
  concentration 72
  fraction 54, 56, 70, 113, 119, 167, 185, 205, 216, 220
  fraction determination 85
  structure 107
  volume 72
    fraction 72
volatile capture 166
volume
  expansion 68, 192
    ratio 37, 47, 58, 65, 153, 175, 219
  fraction 140
  increase 55
  resistivity 247
Voronoi cell 102
vulcanization 32, 64, 120, 143, 153, 212-213, 243
  temperature 112
  tunnel 64

W

wall
  roughness measurement 107
  thickness 100, 124, 213
  thickness variation 107

wallcovering 210
water 2, 34, 42, 49, 91, 100, 116, 118, 143, 148, 189, 193, 197, 208, 210-211, 219-220, 223
  absorption 43, 208
  adsorption 93
  carrier 42, 109, 187
  condensation 177
  contact angle 172
  content 43
  droplet 190
  flow impact 172
  foaming time 91
  mixing state 87
  solubility 187
  uptake 138
  /salt solution 158
  -binding capacity 138
wavelength of visible light 67
Weaire-Phelan model 101
wear surface porosity 146
weatherability 201
Weber number 254
wheat gluten 129
  biofoam 138
white corundum 146
wide-angle x-ray diffraction 267
wire and cable 172
  coating 130
  jacketing 130
wood
  dust 145
  flour 125, 145, 217, 231
  treatment 193
wound dressing 138

X

x-ray
  analysis 267
  diffraction 267
  microtomography 95

Y

yellowing 222
yield
  point 72
  strength 72
Young's modulus 101-102, 158

Z

zeolite 200
zero pressure intercept 61

zinc
  borate 162
  oxide 2, 111-113, 125-126, 146, 150, 162, 179, 209-210, 212, 215
    /zinc stearate 210
  stearate 112, 146, 153, 212
zirconium hydride 134